SUPERIONIC CONDUCTORS

PHYSICS OF SOLIDS AND LIQUIDS

SUPERIONIC CONDUCTORS
Edited by Gerald D. Mahan and Walter L. Roth • 1976

SUPERIONIC CONDUCTORS

Edited by
Gerald D. Mahan
Indiana University
Bloomington, Indiana

and

Walter L. Roth
General Electric Research and Development Center
Schenectady, New York

PLENUM PRESS · NEW YORK AND LONDON

Library of Congress Cataloging in Publication Data

Conference on Superionic Conductors—Chemistry, Physics, and Applications, General
  Electric Research and Development Center, 1976.
  Superionic conductors.

  (Physics of solids and liquids)
    1. Ions—Migration and velocity—Congresses. 2. Superionic conductors—Congresses.
  I. Mahan, Gerald D. II. Roth, Walter L. III. Title.
  QD561.C64 1976   537.6'23   76-28538
  ISBN-13: 978-1-4615-8791-0   e-ISBN-13: 978-1-4615-8789-7
  DOI: 10.1007/978-1-4615-8789-7

Proceedings of a Conference on Superionic Conductors: Chemistry, Physics,
and Applications held at the General Electric Research and Development
Center, Schenectady, New York, May 10-12, 1976

© 1976 Plenum Press, New York
Softcover reprint of the hardcover 1st edition 1976
A Division of Plenum Publishing Corporation
227 West 17th Street, New York, N.Y. 10011

All rights reserved

Preface

A hundred and eighty five chemists, physicists, and engineers met in Schenectady, New York, for the three days May 10-12, 1976, to discuss the subject of Superionic Conductors. This International Conference was held at the Research and Development Center of the General Electric Company.

The subject of the Conference was fast ion transport in solids. These materials have potential application in new types of batteries, fuel cells, and sensors. Some like beta alumina are under active development in novel new systems. Their study has also become a popular area of scientific investigation. One objective of the Conference was to provide a forum for interdisciplinary communication between chemists, physicists, and engineers. The Conference was an attempt to bring these groups together, in order to listen to each others problems and progress.

We began organizing the Conference in the spring of 1975. It was suggested to General Electric managers Drs. Craig S. Tedmon, Jr. and Roland W. Schmitt. They provided immediate and enthusiastic support. They also provided the advice, staff, and backup which were necessary at all points in the planning and duration of the Conference. We were also pleased that they could participate in the Conference: Dr. Tedmon welcomed the participants and officially opened the Conference, and Dr. Schmitt gave the after banquet address. We thank them. Additional and invaluable help, and advice, were also provided by Drs. D. Chatterji, J. B. Bush, G. W. Ludwig, and J. B. Comly.

We were joined on the program committee by Drs. B. A. Huberman (Xerox), R. A. Huggins (Stanford), J. T. Kummer (Ford), M. O'Keeffe (Arizona State), and D. B. McWhan (Bell). This committee worked hard, not only in selecting the invited and contributed papers, but also

in arranging the schedule. The chapters in this book are
the talks given by the invited speakers. Each speaker was
asked to survey his area, and the composite picture should
provide an up-to-date view of this rapidly evolving field.
Papers on applications and needs were scheduled early in
the Conference in order to place the entire field in
perspective, and to stimulate awareness of the significance
of each investigator's work to the whole. These were
followed by papers on theory, the physical properties of
single crystals, and the properties of polycrystalline
ceramic materials used in practical systems. The program
committee decided to represent the contributed papers by
just their abstract. These results are invariably published
in complete form elsewhere, and it was felt unnecessary to
republish them here.

The contributed papers were divided into two classes.
Some were presented on the program as ten minute talks with
five minutes of discussion. Due to lack of time, since
there were no parallel sessions, others were presented in
poster sessions: results were pinned on posters which were
on view in a seminar room adjacent to the main Conference
lounge. The division between these two classes was made
chronologically--those abstracts submitted before the
official January 5, 1976 deadline were nearly all accomo-
dated into the main program. Abstracts submitted late
were put into the poster sessions. Before the meeting,
some authors of poster papers complained that they were
receiving second class treatment. But during the meeting
it became apparent that the poster papers were a more
successful format for presenting the contributed papers.
They could be viewed at leisure, and discussed in depth.
After the meeting, many felt that next time, if we hold
another meeting, all contributed papers should be presented
in the poster sessions. Then more leisure time could be
scheduled into the program to encourage their viewing and
discussion. They also have the advantage that one can be
flexible in accepting later papers for poster sessions
since no scheduling is involved. We were still accepting
papers ten days before the Conference, when the program
booklet finally went to press. All contributed papers have
been treated equally in preparing the abstracts for publication. We also permitted authors to revise them at the
meeting, and these contributed abstracts are often quite
different than those in the program booklet.

PREFACE

The trick to organizing a successful conference is to start a bandwagon effect, whereby everybody thinks it is going to be a success and wants to participate. Special thanks must go to Dr. J. R. Birk, who was the first person to join the bandwagon. His early and generous support, from the Electric Power Research Institute, was crucial in the early days of proposing the Conference. It gave concrete backing to the Conference proposal. Soon afterwards we received additional support from the Air Force Office of Scientific Research and the National Science Foundation. Cosponsorship and valuable publicity were provided by the American Physical Society and the Electrochemical Society.

A special thanks must also go to those who did the work. Foremost we thank Elizabeth Allen, who came out of retirement to help us with the local details. She is the conference and publication expert at the General Electric Corporate Research and Development Center. She was invaluable in organizing buses, motels, meals, and printing. Her efforts in overseeing the preparation and printing of the Conference Program were crucial: thanks to her they were available on time. She also ran the registration and information table, organized the travel reimbursing, and helped everybody. We also thank the secretaries who did the endless typing: Mrs. Karen Gibson at Indiana University, and Mrs. Mary Sammler and Stephenie Smaldone at General Electric. We also thank Professor Harold Story at SUNY-Albany, for helping arrange the Conference banquet at his beautiful campus.

Schenectady, New York　　　　　　　　　G. D. Mahan
May, 1976　　　　　　　　　　　　　　　W. L. Roth

Contents

* Articles marked with an asterisk appear in the form of abstracts only.

B: THEORY

C: SILVER CONDUCTORS

E: OTHER SUPERIONIC CONDUCTORS

Electrochemical Systems

ENERGY STORAGE, BATTERIES, AND SOLID ELECTROLYTES:

PROSPECTS AND PROBLEMS

James R. Birk

Electric Power Research Institute

Palo Alto, California 94303

INTRODUCTION

Historically, storing energy has generally been accomplished by containment of raw fuel. This has been satisfactory for the transportation sector since the fuel,
petroleum, is portable and readily converted into the desired result -- motion. In the electric utility industry,
the necessity to supply energy on demand (which fluctuates
seasonally, weekly, and daily) has been successfully accomplished by using different classes of generating equip-
ment; namely, capital intensive base-loaded units using
nuclear or coal fuel, older and smaller intermediate cycling plants fired by coal or petroleum, and relatively in-
expensive peaking units fired by oil or gas. Recently,
however, the high cost and limited reserves of petroleum are
forcing a reconsideration of these approaches. With energy
storage it would be possible to mitigate these concerns by
indirectly using coal or nuclear energy to supply peaking
and motive power which are the major uses of petroleum. In
the longer term when petroleum-derived energy must be replaced by that derived from coal, fusion, nuclear, and/or
solar, new chemical fuels must be produced and/or electrical energy must be stored for applications where direct use
of these new energy sources is not possible. From the
foregoing discussion it is apparent that the objectives for
storage energy are:

   1. To meet discrepancies that can exist between the
      supply and demand for energy and to meet the
      fluctuating demands for power (see Figure 1).

1

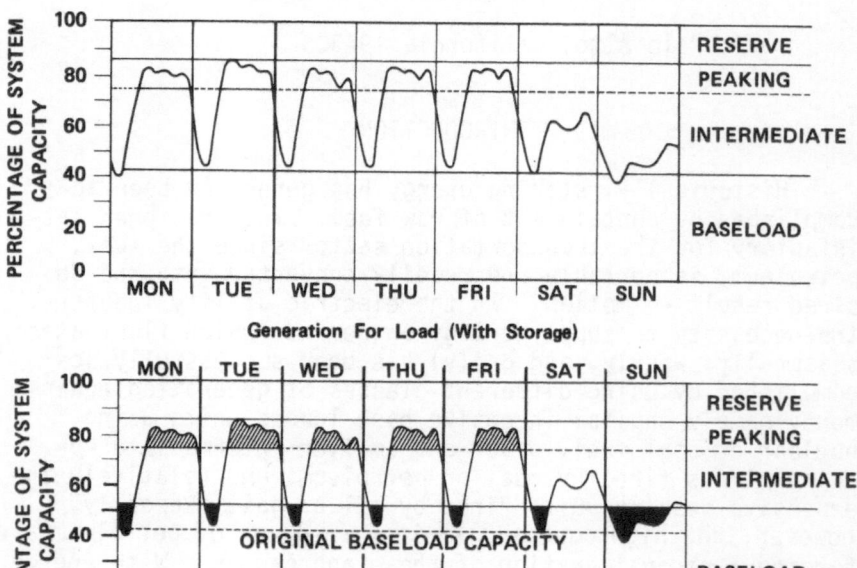

Figure 1

2. To provide for a more versatile source of energy than can be achieved with the conversion of raw fuels.

3. To change the relative use of the various raw fuels.

To accomplish these objectives several energy storage schemes have been proposed as noted in Table I.

Table I

Energy Storage Schemes

Approach	Technology
Electrochemical	Batteries
Chemical	Hydrogen, Methanol, etc.
Thermal	Steam or Hot Oil
Potential Energy	Compressed Air or Pumped Hydro
Electrical	Superconducting Magnets
Kinetic Energy	Flywheels

BATTERY ENERGY STORAGE

Of the approaches to energy storage, electrochemical storage with batteries is enjoying the greatest interest. For example, a survey of the nation's utility planners has shown that batteries have the most "favorable outlook (of) the various energy storage technologies" and, therefore, warrant the highest R&D priority. In addition, the amount of utility and government funding directed toward the development of batteries (over $10 million in 1976) is larger than the support of all other energy storage technologies combined. On top of this, the U.S. House of Representatives has passed a $250 million bill for electric vehicle development and demonstration. The bill is said to have at least an 50-50 chance of being passed into law.

The major reason for this interest in battery energy storage is that batteries appear to be promising for both vehicular and utility application. For utility application the advantages of batteries, compared to most alternative storage schemes, are shown below:

1. Functional advantages resulting from electrical versatility and rapid response: Load-leveling, peaking, spinning reserve, regulation, and load-following.

2. Physical advantages resulting from the modular construction of batteries: Few siting restrictions, dispersed siting (T&D savings and enhanced system reliability), short lead time, and installation on an as-needed basis.

3. Environmental advantages in that there is no anticipated water or air pollution.

The interest in vehicle batteries results, in large part, from the overall improved use of raw energy (as shown in Figure 2) and the limited environmental impact.

The prospects for batteries depends, of course, on success of the R&D community. As noted below in Table II, battery development is faced with a variety of technical problems -- most of which are materials related. Although considerable progress has been achieved over the past couple years, technical success is by no means assured. However, chances are good that technical success can be achieved

Table II

Technical Problems Encountered in
Battery Development

- Degradation of Practical Positive-Electrode Current Collectors
- Technically Acceptable and Economically Practical Separators
- Inhomogeneous Current Distribution Which can Cause Materials Degradation
- Effects of Impurities on Life
- Technically and Economically Acceptable Seals and Current Feed-Thrus
- Sufficiently Uniform Cell Characteristics to Permit Practical and Safe Operation of Batteries

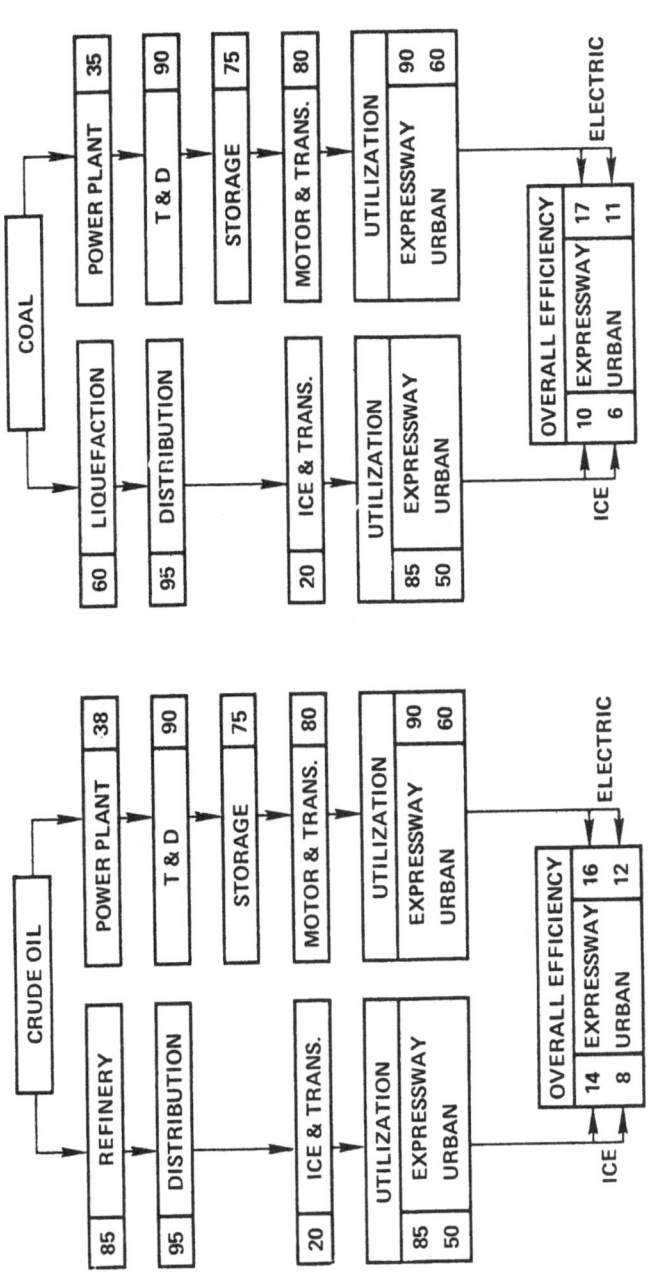

COMPARISON OF ENERGY CONVERSION EFFICIENCIES:
INTERNAL COMBUSTION ENGINE VERSUS ELECTRIC VEHICLES

Figure 2

since financial support for battery R&D appears stable and
sufficient.

Most difficult to project is our ability to achieve
the economic goals for utility or vehicular application.
The cost of batteries are materials sensitive (e.g. half
the price of lead-acid batteries are materials costs) and
material costs are extremely difficult to project since
tolerances and purities have yet to be specified, future
costs are unpredictable, and materials requirements to
achieve long life are unknown. Even if techno-economic
success can be achieved, large financial commitments are
necessary for the manufacturing capital required to achieve
the necessary low costs obtained with high-volume production.
However, due to the need for the U.S. to rely on alternatives
to petroleum, this latter impediment should only delay not
stop the availability of commercial battery systems.

What does the market look like for the new battery
systems? the personal electric vehicle does not appear a
likely development in this century. Techno-economic success of battery development will not eliminate the relatively low performance, slow "refueling", and high first
costs of the electric vehicle (said to be at least 50%
greater than their ICE counterpart). On the other hand,
utilities are buying and using energy storage today (in
the form of pumped hydro). In addition, many urban electric vehicle applications (particularly fleet, commercial,
and delivery) can take advantage of the lower cost over
life of electric vehicles without having to be concerned
with the sacrifice in performance. Fuel availability and
economics suggest that these two markets -- amounting to
several billion dollars by 1985 -- will be a reality for
the right battery systems.

The High-Temperature Sodium Batteries

The sodium-solid electrolyte battery systems are one
of the most scientifically intriguing and widely studied
class of battery systems ever discovered. One major reason
for this interest is the potential of these systems to
achieve the desired low costs for use in vehicular and
utility application. Table III outlines the characteristics
which are desired to achieve low cost and compares how well

Table III

Approaches To
Achieve Low Cost

Characteristic	Pb/PbO2	Na/S
High Utilization of Active Materials	25%	60-80%
High Current Densities to Lower Area Related Costs	15 ma/cm^2	100 ma/cm^2
Low Specific Cost of Active Materials ($/kWh)	10	0.49
High Energy Density to Lower Volume-Related Costs	8-15 WH/lb 0.8 WH/in^3	70-100 WH/lb 1.5 WH/in^3
Low Specific Cost of Materials for		
Container	Molded Plastic ($10/kWh)	Coated Al or SS
Separator	Microporous Rubber ($3/kWh)	--
Electrolyte	Sulfuric Acid	Beta Alumina
Current Collector	Lead ($10/kWh)	Carbon ($0.44/kWh)
Price	$65/kWh	?

Table IV

Preferred Option for the High-Temperature Sodium Batteries

Specification or Component	Organization			
	Dow	ESB	Ford	GE
Positive Electrode	S	$NaAlCl_4$ $SbCl_3$	S	S
Electrolyte (Tubular)	Borate Glass	MgO-βAl_2O_3	β''-Al_2O_3 (Li_2O)	βAl_2O_3
Current Collector	Carbon-Coated Aluminum	Powdered Carbon	Graphite Felt	Carbon and Graphite Mattes
Cell Case	Aluminum	Stainless Steel	Coated Stainless	Coated Aluminum
Seals	Glass	Polymeric	Braze and Glass[1]	Glass and Mechanical
Cell Design (and size-kWh)	Multitube (?)	Multitube (15)[2]	Single Tube (0.4)	Multitube (2.5)
Operating Temperature (^{O}C)	300	200	300-350	300

(1) Shrink alpha to beta seals are being investigated.

(2) "Inside-Out" configuration - positive mix is inside the tube.

the lead-acid and sodium-sulfur battery meet these characteristics. While the comparison is rather simplistic, the
obvious conclusion, concerning the excellent potential for
sodium-sulfur batteris to have a low cost, is difficult to
refute. This potential low cost is unsurpassed by any of
the other advanced battery systems. In addition, the use
of liquid active materials should provide the reproducible
morphology required to achieve long life.

Although the sodium-sulfur battery is the most well
known of the high-temperature sodium systems, two other
modifications are being developed. Dow is using glass capillaries instead of beta alumina as the electrolyte and
ESB has, instead of sulfur, a mixture of sodium chloroaluminate and antimony trichloride as the active positive
electrode material. Table IV compares the characteristics
among these systems and within the sodium-sulfur system.

The sodium-antimony trichloride system sacrifices current density to achieve the many materials benefits --
particularly with respect to seals -- that are possible with
lower operating temperatures. From the electrolyte stand-
point, the operating conditions are strikingly different
conditions than those used for the sodium-sulfur system
(see Table V). These differences in potential, current
density, temperature, and electrolyte environment can greatly
impact the electrolyte; however, far too little fundamental
information is available to establish the relative benefits
and drawbacks (with respect to the electrolyte viability)
of the two alternatives.

Table V

Operating Conditions Affecting the Electrolyte of
Sodium Sulfur and Sodium Antimony Trichloride Batteries

Condition	Na/S	Na/SbCl$_3$
Temperature (^{o}C)	300-350	200
Current Density (mA/cm^2) (10-hour rate)	100	25
Potential (V)	2.0	2.9
Positive Electrode	S/Na$_2$S$_3$	SbCl$_3$/NaAlCl$_4$

The Dow sodium-sulfur battery appears to be an engineering delight. Design modifications are easy to implement and, already, the laboratory cells are produced by semiautomated methods. However, while major strides have been made in cycle life capability, the high temperature coefficient of resistivity of the borate glass electrolyte may severely limit life for deep-cycle, large engineering cells. What can happen is that small temperature deviations within the cell can cause major changes in current density among the individual electrolyte capillaries. This effect, in turn, tends to further amplify the temperature deviations. The net result can be overutilization of the electrolyte and formation of solid sodium polysulfides -- potentially causing the electrolyte tubes to break.

One must view the outlook for sodium/beta-alumina batteries with cautious optimism. Worldwide, these systems are receiving more R&D support than any other battery. In addition, the potential for long life and low cost is excellent. However, the cell life is still almost an order of magnitude away from the goals for utility application and problems of scale up have yet to be addressed.

While the potential outlook is great for the high-temperature sodium batteries, R&D of these systems has been plagued by every one of the problems indicated previously in Table II. Most notably are the seals problems, electrolyte degradation, and container corrosion. The synergism associated with these problems often results in poor cell life (100-300 AH/cm^2) even though several developers have achieved outstanding electrolyte life (2000 AH/cm^2). Cell failure is commonly caused by electrolyte failure, thus problems of seals degradation and container corrosion can manifest themselves as failure of the electrolyte. For this reason the life of sodium-sulfur systems is always stated in total charge passed per unit area as opposed to the more conventional use of time or cycle life.

BETA-ALUMINA ELECTROLYTES

The key to the sodium-sulfur battery and a key to its success or failure is the beta-alumina electrolyte. For an electrolyte to be adequate some rather obvious specifications are important to achieve as noted in Table VI.

Table VI

Objectives For Solid Electrolytes

Parameter	Specification
Fracture Strength	30-40K PSI
Density	>98%
Resistivity (at 300°C)	5 ohm-cm (vehicular use) 15-20 ohm-cm (utility use)
Cost	$3.30/ft^2

The technical specifications listed above can be achieved and it is believed that the cost goal is realistic. However, electrolytes having these specifications are no assurance of long cell life as discussed previously. Thus, it's apparent we lack the information as to what makes an electrolyte "good" and/or as to what makes it fail. Thus, with respect to the sodium-sulfur battery, most questions are electrolyte related and deal with its use and interaction with its environment.

One of the greatest areas of uncertainty with beta alumina is the effect on cell performance of certain physical and chemical changes in the electrolyte with time. There appears to be some relation between these changes and cell life because life of sodium-sulfur cells is still best depicted by total charge passed through the electrolyte. Let's take a look at some of these changes: (1) discoloration of the electrolyte (Figure 3), (2) loss of sodium from the positive-electrode side of the electrolyte (Table VII), (3) change in structural characteristics (Table VII), and (4) chemical interactions at the surface (Figure 4). With respect to this latter point is the question of impurity penetration, surface films, and pitting. Another question is the effect of magnesia on the life of magnesia doped beta alumina. These questions and most others are all related to time-dependent characteristics and changes.

Clearly, basic research is needed and encouraged. However, if the basic research is conducted without the awareness of operating problems, progress will be limited.

Figure 3

TIME-DEPENDENT CHANGES OF BETA-ALUMINA ELECTROLYTE

Tube fragments from cell displayed with the dark (inside)
surface up and in cross-sectional view
showing the three color bands

from General Electric Company

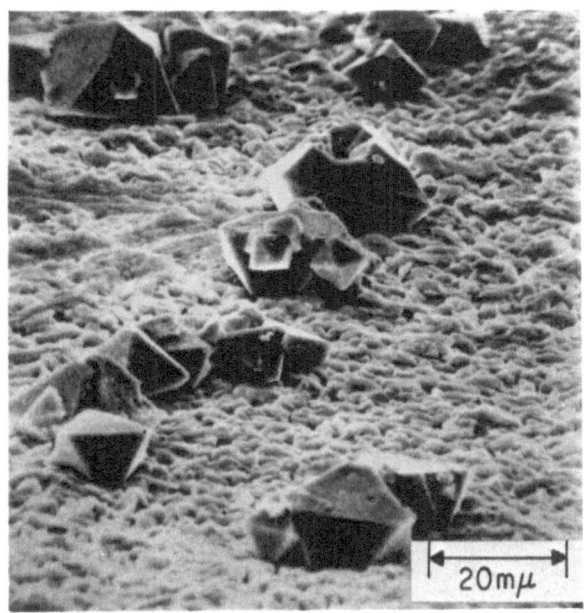

Figure 3, continued

Electron micrograph showing small octahedral crystals
deposited on the outside surface of
the beta-alumina tube from cell

from General Electric Company

Table VII

Chemical and Structural Changes Upon Cycling*

Examined Articles	Virgin State	After 1300 Hrs. C & D. at 100 mA/cm^2	Remarks
Lattice Parameter a_o b_o	5.61Å 22.38Å	5.61Å 22.35Å	Scarcely Changed
Line Broadening (210) (217)	0.306^o 0.338^o	0.331^o 0.356^o	Spread
Na Concentration Through the Section of β-Al$_2$O$_3$ (In Arbitrary Scale)	1.00~1.05	Na Side 0.98~1.13 S Side 0.71~0.78	Decreased at S Side

* From Toyota Motor Company

HIGH TEMPERATURE SOLID ELECTROLYTE FUEL CELLS

T. L. Markin, R. J. Bones, R. M. Dell

U.K. Atomic Energy Authority

Harwell, Oxfordshire, U.K.

The High Temperature Fuel Cell (HTFC) is unique among fuel cells in being an all-solid-state device which operates at temperatures of 800-1100°C. Conventional, low temperature fuel cells possess a liquid electrolyte (an acid or alkali) which gives rise to problems of corrosion (acidic electrolyte) or carbonation (alkaline electrolyte). In addition, low temperature operation favours electrode polarisation and poisoning by impurities. Finally, the difficulty of maintaining a stable 3-phase contact (liquid-solid-gas) creates design problems at the electrodes. For these reasons researchers turned their attention to the solid electrolyte fuel cell.

The HTFC is based upon the use of stabilised zirconia as an oxygen ion conducting electrolyte at temperatures above 800°C and may be represented as
Fuel gas, anode|ZrO_2 (stabilised)|cathode, air.
Oxygen from the air is ionised at the cathode to form $O^=$ ions which diffuse through the oxide electrolyte and then react with the fuel gas at the anode to liberate electrons and form water and/or carbon dioxide. The use of high temperatures eliminates the electrode activation and polarisation problems, while the absence of a liquid electrolyte avoids the three-phase contact problem and the corrosion problems experienced with acid electrolytes. Instead, one is faced with a series of materials science and technology problems which stem from the use of high temperatures.

During the late 1960's the HTFC concept was investigated by research teams in several different countries.

Although the U.K. programme lasted for four years (1967-71) only a single paper has been published (1). In the present paper we seek to review briefly other published information on the HTFC and to present some results and conclusions from the U.K. research programme. Such a review is regarded as timely in the light of the present widespread interest in hydrogen as an energy vector and storage medium for the next century (2) and as the ultimate fuel of the post-fossil fuel age(3). If this view of the future gains acceptance, then the perfection of an efficient fuel cell for hydrogen utilisation will assume increasing significance.

2.   CHOICE OF ELECTROLYTE

A satisfactory oxide electrolyte must meet three criteria:
(i)   its ionic conductivity should be a maximum and its electronic conductivity negligible.
(ii) the ionic conductivity should be stable at cell operating temperature.
(iii) the permeability of the oxide to gas must be very low and remain so during the entire life of the cell, including many cooling and heating cycles.

The first two criteria may be met by a suitable choice of electrolyte composition, while the third criterion relates more to the fabrication method used to prepare a fully densified oxide.

There are three simple fluorite-type oxides which, when doped with suitable divalent or trivalent cations, might be considered as candidates for the electrolyte, viz: stabilised ZrO_2, CeO_2 and ThO_2. Doped CeO_2 is known to be reduced to CeO_{2-n} in the fuel gas atmosphere, thereby introducing an electronic component to the conductivity, while ThO_2 solid solutions have lower ionic conductivities than ZrO_2 based oxides at the cathode. Attention has therefore focussed on stabilised zirconia and it has been shown that the effect of a divalent or trivalent cation on its ionic conductivity increases as the ionic radius of the dopant ion decreases. Thus, the conductivity of doped ZrO_2 increases in the order La<Ca<Y<Yb<Sc. At 950°C ZrO_2 - 13% CaO has a resistivity of $\sim$ 30 ohm-cm, ZrO_2 - 8% Y_2O_3 12 ohm-cm and ZrO_2 - 6% Sc_2O_3 6 ohm-cm(4). Unfortunately Sc_2O_3 is too expensive for general use and, as ZrO_2 - CaO has too high

a resistivity, the U.K. programme concentrated exclusively on ZrO_2 - 8 mole % Y_2O_3. In Germany the Brown Boveri Co. showed that by replacing half of the Y_2O_3 by Yb_2O_3, to give $(ZrO_2)_{0.92}$ $(Y_2O_3)_{0.04}$ $(Yb_2O_3)0.04$, the resistivity at 1000° was reduced from 8 ohm-cm to 6 ohm-cm[5].

A considerable amount of work on new oxide ionic conductors has failed to produce a material which can match doped ZrO_2, although CeO_2 - Gd_2O_3 [6] and some of the perovskites (eg. $LaAlO_3$ doped with Ca and Ba)[7] may warrant further investigation.

3.   ELECTROLYTE AGEING

A phenomenon which is now well recognised with stabilised zirconia is the steady increase in resistivity which occurs on annealing at ∿ 1000°C. This ageing process was first investigated by Carter and Roth for ZrO_2 - CaO and it was concluded that defect ordering processes must be taking place [8]. Since HTFC's operate in this temperature range and are required to have long lives with stable performance, the observed ageing is a cause for concern and it was necessary to investigate the magnitude of the effect for ZrO_2-8m/o Y_2O_3. Measurements were made in air and in hydrogen at 1000°C. (Figure 1). The maximum rate of ageing occurred during the first few days and was more pronounced in hydrogen than in air. Thereafter the rate was similar in both gases. The resistance, which almost doubled on annealing in hydrogen, returned to its original value after firing at T> 1400°C, but this is too high a temperature for practical use in a fuel cell.

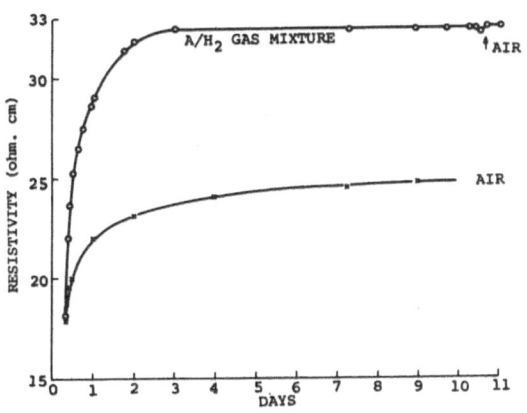

Fig.1.

Ageing of ZrO_2 - 8m/o Y_2O_3 in air and A/H_2 gas mixture at 1000°C

Recently, two separate investigations of the ordering processes which take place in ZrO_2-CaO have been carried out
by Allpress and Rossell at Melbourne[9] and by Hudson and
Moseley at Harwell[10] using electron diffraction techniques.
As well as the normal Bragg scattering, both sets of workers
observed a diffuse scattering which arose from microdomains
orientated with equal probability along the 12 equivalent
<110> fluorite directions of the parent crystal. Addition-
ally, Hudson and Moseley found symmetry-forbidden "superlattice" reflections with mixed even/odd indices. Both
effects are clearly shown in the <11$\bar{2}$> fluorite diffraction
pattern of ZrO_2-15 m/o CaO annealed for 40 days at 1000°C
(Fig.2). Analysis has shown that the "superlattice" spots
stem from orientated intergrowths of monoclinic zirconia
within the cubic parent crystal, while the diffuse scattering
arises from microdomains of an ordered, calcia-rich structure,
probably $CaZr_4O_9$, also nucleating in the cubic fluorite
crystal. When a specimen was heated at 1600°C for 24h and
quenched, the superlattice reflections disappeared as the
crystal adopted the stabilised fluorite structure, but the
diffuse scattering remained unchanged. The experiments
show that the "ageing" of ZrO_2-CaO at 1000° is associated
with the coherent intergrowth of monoclinic ZrO_2 nuclei.
Rossell has observed also diffuse scattering in ZrO_2-Y_2O_3
and attributed it to microdomains of $Zr_3Y_4O_{12}$.[11]

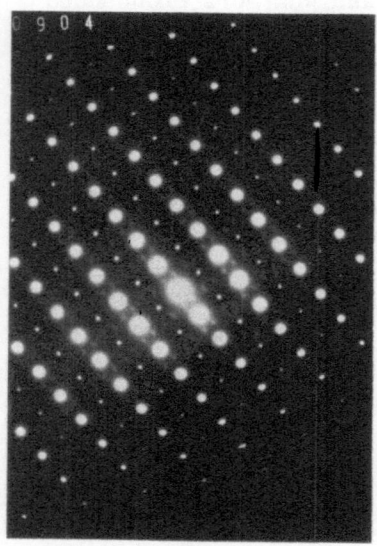

Fig.2

<11$\bar{2}$>fluorite diffraction
pattern from ZrO_2-15% CaO
showing both diffuse
scattering and reflections
forbidden by the symmetry
of the fluorite structure.

Russian workers have studied phase relations and electrical conductivity in the system ZrO_2-Sc_2O_3 [12]. For low contents of Sc_2O_3 (6-10 mole %), where the conductivity at 800-1000°C is a maximum, the phase diagram is complex. Above 800° a mixture of stabilised fluorite phase and tetragonal ZrO_2 is reported to exist. On cooling, tetragonal ZrO_2 reverts to monoclinic ZrO_2 at 750-800° and on further cooling below 600° the fluorite phase changes to rhombohedral $Sc_2Zr_7O_{17}$. This Russian work throws some doubt on the use of Sc_2O_3 to stabilise zirconia as temperature cycling is likely to result in phase changes and consequent cracking of the electrolyte.

4.   ELECTROLYTE FABRICATION

The electrolyte fabrication problem is that of developing a technology for manufacturing reproducibly and cheaply thin non-porous membranes of the specified electrolyte composition. The first two questions which come to mind are:

-   what is the optimum configuration of the electrolyte and its linear dimensions?
-   how thick should the membrane be?

Although there are no theoretical constraints on the electrolyte configuration, in practice only simple tubes and plates have been considered. Planar electrolyte, in the form of discs, poses difficult gas sealing problems and most effort has been directed towards tubular electrolyte, with hydrogen passing down the interior of the tube and oxygen around the outside.

The length of the unit cells is determined by considerations of cathode resistance (vd. Section 5.3). The electrolyte thickness is also a **key** parameter and this may be deduced as follows:

The hydrogen/oxygen fuel cell operating in a H_2/H_2O atmosphere has a theoretical voltage of around 1.1V. For commercial applications we require a current density of $\sim$ 1A.cm^{-2}. The internal resistance of the cell is made up in part of the electrolyte resistance and in part of the electrode resistances. If we aim for a voltage drop across the electrolyte of 0.1V, then its resistance is required to

be 0.1 ohm.cm^2. Assuming the resistivity of the electrolyte
at cell operating temperature to be 10 ohm.cm, we calculate
the required membrane thickness to be 100μm. This is, of
course, only an order of magnitude calculation for the electrolyte thickness and will vary somewhat according to the
resistivity of the electrolyte and the commercial targets
set for the fuel cell. We may say that electrolyte thickness in the range 100-500μm will be of interest in the
development phase, with the true commercial target nearer
the low end of this range.

    Although it is possible to fabricate free standing
(ie. unsupported) ceramic electrolyte tubes down to at least
200μm wall thickness, and these are sufficiently robust to
be used in the laboratory, it is certain that for commercial
applications electrolyte tubes of 100-200μm wall thickness
would need to be mechanically supported. In principle this
support could be either a relatively thick, porous electrode
(anode or cathode), or an inert porous support, such as a
ceramic tube on the outer surface of which a thin porous
electrode is deposited followed by the electrolyte layer.
The fabrication problem can therefore be seen in two stages:

    (1)   The fabrication of non-porous, free-standing
          electrolyte tubes of wall thickness 200-500μm
          for laboratory studies.
    (2)   The fabrication of non-porous, supported electro-
          lyte layers of thickness 100-200μm for commercial
          applications.

The first of these stages is considerably simpler than the
second as the sintering process which is necessary to form
a non-porous electrolyte does not involve interactive effects
with other materials.

4.1   Freestanding Electrolyte Tubes

    A number of techniques have been successfully employed
to prepare non-porous tubes of stabilised zirconia for
laboratory study. Westinghouse have made short lengths of
thin walled tube by conventional ceramic sintering and
machining techniques.[13] Brown Boveri prepared relatively
thick-walled tubes (700-800μm) of the composition ZrO_2, 4m/o
Y_2O_3 4m/o Yb_2O_3 by isostatic pressing and sintering at 1600°
for 24h. [5, 14]

Compagnie Generale d'Electricité (France) have patented
the preparation of thin-walled ZrO_2-Y_2O_3 tubes by an electrophoretic method.[15] Particles of ZrO_2-Y_2O_3 of 5μm diameter
are slurried in nitro-methane and electrophoretically deposited
(400V, 30 seconds) on to polished stainless steel tubes.
The oxide is then isostatically pressed, the steel mandrel
removed and the freestanding ZrO_2-Y_2O_3 tube fired for 2h at
1800°C. Tubes of wall thickness 100-200μm have been
fabricated.

At Harwell, most attention was directed to plasma spraying as a technique for fabricating free-standing electrolyte
tubes. For this to succeed it is important to use a care-
fully prepared feed powder which is free-flowing and of
controlled particle size. ZrO_2-8m/o Y_2O_3 prepared by the
sol-gel process [16] was found to be ideal for this purpose
and was available in a range of particle sizes. Fig.(3) is
a photomicrograph showing microspheres in the size range
10-50μm. Unsupported tubes of wall thickness 200-1000μm were
prepared by plasma spraying on to a salt-covered mandrel,
which was then removed and the ceramic tube sintered in air.
Sintering temperatures of at least 1800° were found necessary
to produce electrolyte which was gas-tight to air and hydrogen.
When tubes sintered at lower temperatures were studied in
test H_2/air fuel cells it was observed that the open circuit
voltage measurement declined with increasing tube permeability.
Non-porous tubes of ZrO_2. 8m/o Y_2O_3 prepared by plasma
spraying and sintering at 1800-1900° were employed in all
the Harwell fuel cell studies.

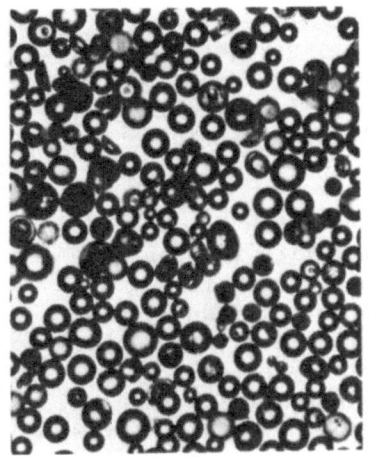

Fig.3

ZrO_2- CaO prepared by
Sol-Gel method.

4.2   Supported Electrolyte

In addition to the fabrication techniques already
mentioned, there are further possibilities for producing
thin, supported layers. These include chemical vapour
deposition, electron beam evaporation and radio frequency
sputtering on to substrate surfaces. These were all investigated briefly at Harwell.

Chemical vapour deposition was attempted by gas phase
hydrolysis of mixed $ZrCl_4$ + YCl_3 vapours in steam and also
by the thermal decomposition of mixed zirconium and yttrium
alkoxides on heated anode substrates. In each case either
highly porous layers were formed or there was poor adhesion
of the electrolyte to the substrate. Electron beam evaporation gave thin (10-60μm), adherent films on nickel which
could be temperature cycled without spalling, but when
porous nickel was employed the films were invariably porous.
R.F. sputtering was found to be too slow, with films deposited
at a rate of only 1μm.h^{-1}.

Following the success of plasma-spraying in producing
free-standing electrolyte tubes, this technique was also
investigated for fabricating thinner electrolyte layers
supported on anode substrates. Cobalt and nickel substrates
in the form of porous discs and tubes were plasma-sprayed
with electrolyte layers 150μm thick. These layers were
adherent and could be temperature cycled, but were highly
porous and did not sinter at 1100°C, the maximum temperature
acceptable for the anode support. Similarly, when porous
Al_2O_3 or TiO_2 tubes were employed as substrates, cobalt
anodes being painted on to their external surfaces, it was
found that sintering temperatures were limited to 1300°C,
above which the Co layer diffused into the electrolyte.
Again, the electrolyte layer was still porous at this
temperature.

Although a great deal of development work has been done
on the fabrication of ceramic substrates with the desired
porosity (~30%) and pore structure, this is not really the
central problem. The key issue is how to lay down upon such
a substrate tube a thin metal anode followed by an electrolyte layer and then to sinter the electrolyte layer to a
non-porous state without modifying the porosity in the
substrate and anode. One possible approach to this, investigated at Westinghouse, is to reduce the electrolyte sintering

temperature by means of reaction sintering.[17] In this work, slurries of ZrO_2 and $CaZrO_3$ in butyl acetate were applied to Co anodes on porous substrate tubes. After drying, the composite was fired at 1400°C when reaction sintering led to non-porous electrolyte films which, although only 60μm thick, gave full theoretical open circuit voltage when tested in a H_2/O_2 cell. There were, however, problems with poor reproducibility and cracking of the layers.

The supported electrolyte problem is seen as one of the key HTFC problems remaining to be solved before a commercial fuel cell can be developed.

5.   ELECTRODE MATERIALS

5.1   Anode

The anode is exposed to a H_2/H_2O atmosphere at ∿ 1000°C and the composition of the atmosphere becomes progressively more oxidising towards the exhaust end of the cell stack. Of the base metals, thermodynamic considerations show that only nickel and cobalt will withstand the fuel gas environment and remain non-oxidised.

Several methods of preparing the anode/electrolyte bond were investigated at Harwell. The most successful was found to be the painting or screen printing of Co_2O_3+ ZrO_2/CaO onto the electrolyte or porous substrate and firing in air at 1320°C. The Co_2O_3 was then reduced to cobalt in hydrogen and the anode was found to adhere well to the ceramic surface. Over 60 single fuel cells were operated with cobalt anodes prepared in this way.

5.2   Cathode

The choice of cathode material is the second of the major unsolved problems of the HTFC. A material is required which is a good electronic conductor, is porous to oxygen and is stable at 800–1000°C. The noble metals fulfill most of the requirements but are too expensive. At first sight silver appears to be a possibility and has indeed been used in experiments. Calculations based on oxygen diffusion rates in silver showed that for a 1μm thick layer of metal the limiting current would be 0.85A.cm^{-2} at 850°C and 0.39A.cm^{-2}

at 725°C.   At the same time, silver volatility measurements
in air made at Harwell showed that 1µm of Ag would be lost
in 6 months at 800°C.   This eliminates silver as a non-
porous cathode material.   There remains the possibility of
using a porous silver cathode, stabilised against creep
and sintering by addition of ZrO_2, at a lower temperature
where volatility is less important.   This might be possible
with a new electrolyte operating at 750°C.

The most promising alternative to noble metal cathodes
appears to be the well-known electronically conducting oxides.
An experimental study of some 35 oxides identified from the
literature led to 5 with bulk resistivities of 1-2 x 10^{-3}
ohm.cm at 950°C.   These were In_2O_3 - 3m/o SnO_2, $PrCoO_3$,
$SrRuO_3$, CdO - 3m/o ZnO and $(La_{0.6}Sr_{0.4})CoO_3$.   Of these,
$SrRuO_3$ was eliminated on grounds of cost and CdO-ZnO because
of volatility of CdO above 800°C. $PrCoO_3$ was found to have
a very high expansion coefficient and spalled from the
electrolyte on cooling, while the perovskite structure of
$(LaSr)CoO_3$ was destroyed in contact with the electrolyte.
This left In_2O_3 - 3m/o SnO_2 as the only satisfactory oxide
cathode.

Several methods of applying this cathode to the electrolyte were studied at Harwell.   Plasma spraying and chemical
vapour deposition gave adherent, but non-porous cathodes.
Best results were obtained by painting or screen printing
the mixed oxide solid-solution in an organic solvent on to
the electrolyte surface, followed by drying and sintering.
The resulting porous In_2O_3 - SnO_2 cathode had a bulk resistivity of 2.2 x 10^{-2} ohm.cm at 970°C.   Power tests carried
out in experimental cells (vd. Section 6) showed that the
power density improved as the particle size of the cathode
oxide decreased.   One of the problems with using an indium
based oxide as cathode is that indium is available only in
limited quantities as a byproduct of zinc extraction.   This
could be a limiting factor for large scale commercial applications of the HTFC.

Other laboratories have favoured different cathodes.
Brown Boveri ran fuel cells successfully using $LaCoO_3$ and
$LiNiO_3$ cathodes.[14]   W Baukal (Battelle, Frankfurt) suggested that $LaCrO_3$ may be a satisfactory cathode material[18]
but this is not in accord with published resistivity data
for this oxide.

5.3   Current Collection Problem

The maximum acceptable length of a unit fuel cell is determined by the resistance of the electrodes, in particular the cathode. If the electrodes serve also as the current collectors, then all the current passes along the length of the electrodes. The voltage drop along the electrodes is then given by

$$V_d = \frac{Q \rho_a L^2}{2\delta_a} + \frac{Q \rho_c L^2}{2\delta_c}$$

where Q is the current density $(A.cm^{-2})$, ρ_a and ρ_c the resistivity of anode and cathode (ohm.cm), δ_a and δ_c are the anode and cathode thicknesses (cm), and L(cm) is the length of the unit cell. For a porous Ni anode (ρ_a = 1.4 x $10^{-4} \Omega$ cm at 1000°C) of thickness 200μm, the first term is 35 x 10^{-4} Q L^2. For a porous In_2O_3 - SnO_2 cathode (ρ_c = 7 x 10^{-3} Ω cm at 1000°C) of thickness 500μm, the second term is 7 x 10^{-2} QL^2. The resistance of the cathode is clearly dominant. If the cell is to operate at a current density of 1A. cm^{-2} and the voltage drop associated with the electrodes is restricted to 0.1V, then .07 L^2 + .0035 L^2 = 0.1 ie. L = 1.17cm.

This calculation shows that the use of conducting oxides as cathode limits the length of unit cells to 1-2cm if the voltage drop in the cathode is not to be unacceptably large. One way to overcome this limitation would be to separate the electrode and current collection functions of the cathode and to use an independent current collector. This might take the form of a metal grid or bus bar system attached to the cathode at frequent points along its length. The entire ceramic tube would then represent a single fuel cell giving, say, 50-100A at 0.7V. Adjacent tubes would be series connected. Relatively massive current collectors would be required to keep the resistance losses small. For cells operating at 1000°C, the only possible metallic current collectors would be oxidation resistant alloys (eg. nichrome, nimonic etc.). Even then the resistance of the protective oxide film at the contact points with the cathode is likely to be high. This could possibly be overcome by interposing a precious metal contact strip metallurgically bonded to the base metal current collector.

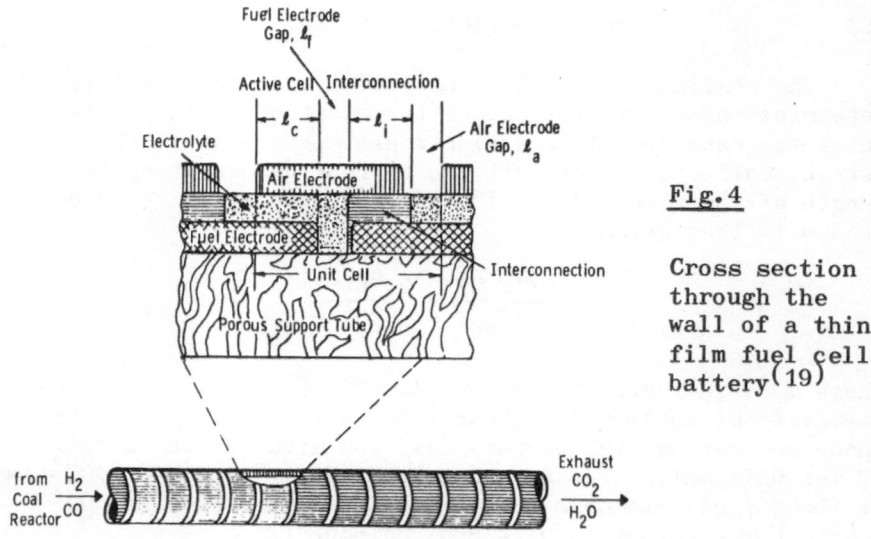

Fig.4

Cross section
through the
wall of a thin
film fuel cell
battery[19]

The second general solution to the current collector
problem is the so-called "Cell Overlap Concept". This
concept, which has been investigated in several laboratories,
involves restricting individual cells to ∼ 1cm. in length
and series-connecting these cells to form a cell stack.
When using free-standing electrolyte, short lengths of shaped
tube were employed with rebated ends which fitted into
adjacent cells so that the anode of one cell over-lapped the
cathode of the next. With supported electrolyte the connection between cells is easier: using appropriate masking
and screen printing techniques it is possible to form any
number of short series - connected cells along the length
of a single ceramic substrate tube as shown in Fig.4[19]
Each tube with its series connected cells now produces a
small current at relatively high voltage.

This overlap concept requires the use of an interconnector material to join the air electrode of one cell to
the fuel electrode of the next cell. This interconnector
must be an electronic conductor and must be chemically stable
at 1000°C in air and in hydrogen. Precious metals might
be considered as interconnectors but are unacceptable on
cost grounds. Electronically conducting oxides seem the
most likely materials and these have been investigated in
several laboratories. Westinghouse workers[19] showed that
cobalt chromite doped with 2% manganese is a suitable

interconnector. At 1000°C this compound has a stable resistivity of 6ohm.cm in air and 50 ohm.cm in H_2/H_2O mixtures having an oxygen partial pressure of 10^{-15}atm. These resistivities are very high and the material will be suitable only if thin gastight layers can be fabricated. Furthermore, there is a considerable thermal mismatch between cobalt chromite and the anode and substrate materials.

Possibly better inter-connector materials, identified by Brown Boveri,[20,21] are cobalt-doped $LaNiO_3$ and gadolinium-doped $LaCrO_3$. The former has a resistivity of <0.1ohm.cm at 1000°C at both the anode and cathode oxygen potentials. $LaCrO_3$ doped with Gd has a resistivity of 0.1ohm.cm at the cathode oxygen potential rising to 0.8ohm.cm at the anode oxygen potential. Both of these seem suitable candidates for the inter-connector material provided that there are no compatability problems and the oxides show long term stability.

At Harwell a wide variety of ternary oxides were investigated as possible inter-connector materials. These included $NiCr_2O_4$, $CoCr_2O_4$, $MgCr_2O_4$, $MnCo_2O_4$, $TiCo_2O_4$, $MgIn_2O_4$, $TiZn_2O_4$, $Sr_{1-x}La_xTiO_3$, $UO_2-Y_2O_3$ etc; none of these were found to be satisfactory. Eventually, it was found that TiO_2 doped with certain transition metals was suitable. The resistivities of these doped TiO_2 specimens at 1000°C in air and in fuel gas were as follows.

Resistivity at 1000°(ohm.cm)

Doped oxide	Air	Fuel Gas
TiO_2 + 5m/o Nb_2O_5	0.7	.03
TiO_2 + 5m/o Ta_2O_5	3.5	.033
TiO_2 + 5m/o WO_3	8.0	.14

X-ray analysis confirmed that these oxides retained the rutile structure after extended annealing at 1000°C. TiO_2 doped with Nb_2O_5 has a clear advantage over Ta_2O_5 and WO_3 doping as regards its resistivity in air. The resistivity was found not to be critically dependent upon the precise mole % of dopant, and the expansion coefficient is intermediate between that of In_2O_3 - SnO_2 and ZrO_2 - Y_2O_3. Hydrogen has a diffusion coefficient of $\sim 10^{-7}$ cm^2/sec in TiO$_2$ but its solubility is small.

TiO$_2$ - 5m/o Nb$_2$O$_5$ was chosen for further development work at Harwell. The mixed oxide was prepared in spherical particle form by sol-gel techniques and plasma-sprayed on to the cobalt anodes. The In$_2$O$_3$ - SnO$_2$ cathodes were then painted on to the interconnector and electrolyte regions. Thin films (100μm) of TiO$_2$-5m/o Nb$_2$O$_5$ sprayed on to a non-porous substrate were shown to be completely non-porous towards hydrogen.

6.   LABORATORY CELL PERFORMANCE

Most of the laboratories active in the development of HTFC's succeeded in assembling laboratory unit cells or cell stacks which had a creditable performance considering that they were early attempts at solving the many problems. As early as 1969 workers at General Electric reported respectable power densities of 0.30W/cm^2 at 1000°C for simple cells employing PrCoO$_3$ cathodes.[22] These cells operated for more that 6 months with only slight loss of power but, unfortunately, they could not be temperature cycled as the large thermal expansion coefficient of PrCoO$_3$ led to spalling of the cathode on cooling.

Westinghouse employed In$_2$O$_3$ - SnO$_2$ cathodes and, for single cells at 1000°C, obtained current densities of 0.77A/cm^2 (equivalent to 0.38W/cm^2 at maximum power) over 1100h operation.[23] These authors also series-connected cells by anode-cathode overlap using Mn-doped cobalt chromite as interconnector, but only poor power densities were observed (<0.07 W/cm^2). More recently, Westinghouse has undertaken a detailed analysis of the best geometry for the HTFC based upon these materials.[24] It is concluded that for maximum power output the lengths of unit cells are extremely short eg. only 2mm each with 1mm gap between unit cells on 12mm diameter support tubes. To fabricate cell stacks to these dimensions on a porous substrate tube would require very precise processing techniques. From banks of such cells Westinghouse hoped to achieve efficiencies (thermal $\longrightarrow$ D.C. electrical) of 80% and an overall power system efficiency of >60%.

CGE has concentrated its attention on silver cathodes and attempted to reduce the volatility by alloying. Porous silver cathodes were deposited on thin-walled (100-200μm) ZrO_2-Y_2O_3 electrolyte tubes formed by electrophoresis and cell stacks were made up using silver as interconnector.[25] Good power densities were reported from this system.

Considerable research has also been carried out at Brown Boveri using relatively thick electrolyte (0.8-1.0mm) and a variety of oxide cathodes with nickel anodes.[26] Single cells at 1000°C gave maximum power of 0.35W/cm^2, but the power density obtained from series-connected cells was only 0.15W/cm^2 after 10 days.

At Harwell a number of unit cells were operated at 960°C and 1060°C based upon cobalt anodes and In_2O_3 - SnO_2 cathodes deposited on ZrO_2-8m/o Y_2O_3 electrolyte tubes (free standing) of thickness 430μm. Observed maximum power densities ranged from 0.15 to 0.40W/cm^2, the values at 1060° being 0.03-0.05W/cm^2 higher than at 960°C. The power density was found to increase significantly with decreasing particle size of the cathode, the maximum being achieved with In_2O_3 - SnO_2 particles of 2-5μm diameter. It is of interest that these unit cells did not decline in power with time, as expected because of electrolyte ageing. Either ageing did not take place under fuel cell operating conditions, or it was off-set by improving electrode behaviour. Substantially higher values for maximum power might be expected for thin(50-100μm) supported electrolyte layers.

A series-connected stack of free-standing electrolyte cells was life tested for > 4 months at 960°C with periodic cycling to room temperature. As these cells were not optimised for power the observed power density was low (0.06W/cm^2) but remained stable throughout the test. Examination of the electrolyte/electrode interface at the end of the experiment showed no deterioration. It is anticipated that the power output could be improved by optimising the geometry of the series connection.

Westinghouse has reported assembling 20 cell batteries of this type which gave a maximum power output of 6.7 ± 0.8W with complete combustion of hydrogen at a current density of 435 mA/cm^2.[13] The open circuit voltage was 19-20V. Twenty of these 20 cell batteries were used to construct a 100W fuel cell power supply.

7.   ENGINEERING CONSIDERATIONS

So far we have discussed only unit cells and simple series-connected cell stacks, hand-made for laboratory studies. Next we consider the engineering problems associated with designing and building fuel cell modules of realistic size-say 50kW - intended to burn natural gas or gasified coal. Here we encounter a new range of problems to be solved which are outlined briefly below.

7.1   Reforming of Fuel Gas

As the HTFC is unable to convert directly hydrocarbon fuels, methane must first be reformed with steam at 800-1000°C

$$CH_4 + H_2O \rightleftharpoons CO + 3H_2 \quad \Delta H = 226kJ. \ mole^{-1}$$

This highly endothermic process is carried out over a nickel based catalyst. Normally the reformed gases would be converted to hydrogen by a catalytic shift reaction at 350°C, followed by scrubbing of CO_2

$$CO + H_2O \rightleftharpoons CO_2 + H_2 \quad \Delta H = -33kJ. \ mole^{-1}$$

The incorporation of a shift reactor is not mandatory as the HTFC is able to consume CO directly, with only a small increase in polarisation losses over those measured in hydrogen.

A design and costing exercise carried out in the U.K. for a gas reformer of a size suitable for matching to a 500kW (e) HTFC gave costs which were unacceptably high and also a low fuel efficiency. The high costs for "conventional" gas plant led to a radical redesign of the entire system. Advantage was taken of the fact that the cell operates at the same temperature as a steam reformer to combine the two and incorporate both into a single unit.

7.2   The Integral Cell - Reformer Unit

For electricity substation use, a fuel cell plant of output ~ 500kW (e) is required. Engineering analysis led to the proposal that this should be of modular construction, with each fuel cell module of size 50kW(e). Design studies

were therefore carried out on a 50kW unit in which the reformer is integral with the fuel cell. The overall objective of the study was to produce an optimised module design
with maximum power density, high fuel efficiency and low
constructional costs.

The design which eventually emerged is shown pictorially
in Figure 5. It is based upon the thin-film supported -
electrolyte concept and each module contains seven vertical
substrate tubes, 210mm in diameter and 3m long. These are
placed as closely as possible with Incalloy reformer tubes
packed with catalyst on the inside of each cell element
tube. The design calculations assumed an average power
density of 0.45W cm^{-2}. and, on this basis, the overall dimensions of the 50kW module would be 5.5m high and 1.5m diameter
(ie. a power density of 5.1kWm^{-3}). The fuel inlet temperature was taken as 950°C and the outlet temperature as 1050°C.

Under most operating conditions the combined (cells +
reformer) will not be thermally self-sustaining unless heat
is supplied to the air and fuel intakes. This is accomplished by means of preheaters (figure 5). The tail gas from
the fuel cell module is mixed with spent air (both at high
temperature) and burnt in an external combustion chamber.

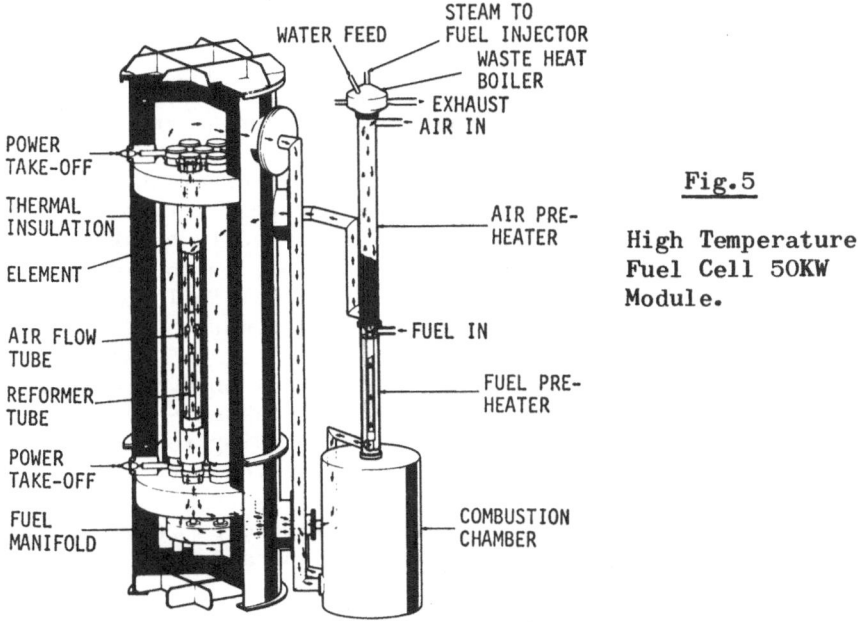

Fig.5

High Temperature
Fuel Cell 50KW
Module.

The combustion products are then passed through the fuel preheater, air preheater and waste heat boiler. These auxiliary units, which are necessary for a practical system, do add substantially to the overall cost.

7.3   Overall System Efficiency

The operating voltage of a fuel cell (V) is less than its electrochemically reversable voltage (E_o) because of its internal resistance (Rc) and electrode polarisation losses(Vp)

$$V = Eo - Vp - IRc$$

The electrical efficiency (ηe) of a fuel cell is given by

$$\eta e = \frac{V}{(Eo - Vp)}$$

In addition there are non-electrical losses, for example in the unconverted tail-gas and also thermal losses, and the overall system efficiency (ηo) is given by

$$\eta o = \frac{P}{Ho}$$

where P = electrical power output of battery
      Ho = thermal power supplied to battery
      (ie. calorific value of fuel supplied per second.)

Since V decreases as I increases, it is clear that a graph of useful power output (P = VI) against current density passes through a maximum. It can readily be shown that at maximum power output, ηe = 0.50; this is too low an efficiency to be acceptable for a practical fuel cell and so operation must be at less than peak power.

In the Harwell design calculations the mean d.c. output of the cell stacks has been assumed to be $0.45 Wcm^{-2}$ at 70% electrical efficiency (ηe = 0.70), corresponding to $0.53 Wcm^{-2}$ at peak power. The calculated overall efficiency (ηo) then ranges from ηo = 0.40 for 85% fuel gas utilisation to ηo = 0.32 for 65% fuel gas utilisation. These efficiencies relate to H_2/CO mixtures prepared by reforming methane and to d.c. out-

put of the battery. Inversion to produce a.c. leads to a
further 10% reduction in efficiency.

These values are considerably lower than those quoted
by Westinghouse workers for the fuel battery operating on
gasified coal.[27] They believed that an electrical efficiency ηe = 0.80 and an overall efficiency ηo = 0.60 were
reasonable goals. More work is required to resolve this
discrepancy, but our present view is that the Westinghouse
efficiency targets are unduly optimistic and that a practical
HTFC is more likely to yield an overall efficiency ηo of
0.3 - 0.4.

8.   CONCLUSIONS

Although HTFC's have been built and tested in the
laboratory, both as single cells and as short series -
connected stacks, there are still technical problems to
be solved. These include the manufacture of suitable porous
substrate tubes, the deposition of thin, non-porous electrolyte layers on a metal anode and the identification of a
better conducting cathode material stable in air at $1000^{\circ}C$.
There is also the problem of making terminal connections to
the cathode and developing the thin film overlap concept
as a routine mass production procedure. Integration of the
gas reformer into the fuel battery has not advanced beyond
the conceptual design stage and there are undoubtedly difficult engineering problems to be solved.

Although the overall efficiency of electricity production from fossil fuels seems unlikely to exceed that of
conventional steam turbines, the small unit size at which
it is attainable may be a substantial advantage. On this
scale conversion can take place near the point of use, leading to greater flexibility and minimal transmission losses.
Waste heat from combustion of the tailgas can be utilised
locally. Compared to the diesel generator, the lack of
moving parts in the HTFC is also an advantage. We conclude
that in certain applications the HTFC still has attractions,
but that at the present time these do not justify the development effort required to bring it to commercial realisation.
In the future, particularly if hydrogen plays a major role
in the energy scene, the situation could well be different.

ACKNOWLEDGEMENT

Acknowledgement is made to colleagues in the U.K.A.E.A. who contributed to this project, especially Dr L E J Roberts, Dr B J Wood, Mr M E Clarke and Mr R Gower.

Sponsorship of the programme by the U.K. Department of Industry, and by the British Gas Corporation is acknowledged also.

REFERENCES

1.   T L Markin "Power Sources" 4 (Ed.D H Collins) Oriel Press 1973, P.583

2.   "Hydrogen Energy" (Ed. T N Veziroglu). Plenum Press 1975.

3.   R M Dell and N J Bridger. Applied Energy 1, 279(1975)

4.   B C H Steele "Electrode Processes in Solid State Ionics" (Eds. M Kleitz and J Dupuy). D Reidel Publishing Co. 1976. P.367

5.   R Steiner. Energy Conversion 12, 31 (1972)

6.   T Kudo and H Obayashi J Electrochem. Soc.122, 142(1975)

7.   T Takahashi and H Iwahara Energy Conversion 11, 105(1971)

8.   R E Carter and W L Roth "EMF Measurements in High Temperature Systems" (Ed. C B Alcock 1968) P.125

9.   J G Allpress and H J Rossell. J. Solid State Chem. 15, 68 (1975)

10.   B Hudson and P T Moseley. Harwell Report AERE R.8236 (1976)

11.   H J Rossell, private communication.

12.   F M Spiridonov, L N Popova and R. Ya. Popil'skii J. Solid State Chem.2, 430 (1970)

13. D H Archer, L Elikan and R L Zahradnik. Hydrocarbon Fuel Technology, Academic Press 1965, P.51

14. W Fischer, H Kleinschmager, F J Rohr, R Steiner and H H Eysel. Chem.-Ing.-Techn. $\underline{43}$, 1227 (1971)

15. U.K. Patent Specification 1,261,317 (1972)

16. J Woodhead. Science of Ceramics, $\underline{4}$, 105 (1968)

17. N J Maskalick and C C Sun. J Electrochem.Soc.$\underline{118}$, 1386 (1971)

18. W Baukal "Power Sources" $\underline{4}$, (Ed. D H Collins) Oriel Press 1973, P.594

19. C C Sun, E W Hawk and E F Sverdrup, J Electrochem.$\underline{119}$, 1433 (1972)

20 H Kleinschmager and A Reich Z Naturfors. $\underline{27}$, 363 (1972)

21. H H Eysel, H Kleinschmager and F J Rohr. 4th Int.Symp. on Fuel Cells, Antwerp, October, 1972.

22. C S Tedmon, H S Spacil and S P Mitoff. J Electrochem. Soc.$\underline{116}$, 1170 (1969)

23. E F Sverdrup, D H Archer and A D Glaser. "Fuel Cell Systems II" Amer.Chem. Soc. Advances In Chem.Ser. $\underline{90}$, 301, (1969)

24. E F Sverdrup, C J Warde and R L Eback. Energy Conv. $\underline{13}$, 129 (1973)

25. U.K. Patent Specification 1,261,317 (1972)

26. R Steiner, F J Rohr and W Fischer. 4th Int. Symp. on Fuel Cells. Antwerp, October 1972.

27. E F Sverdrup, C J Warde and A D Glasser. Westinghouse Scientific Paper 71 - 9E6 - FCELA - P1 (1971)

APPLICATIONS OF BETA ALUMINA IN THE ENERGY FIELD

Neill Weber

Research Staff
Ford Motor Company
Dearborn, Michigan 48121

During the past decade, several devices based on beta alumina have been proposed for energy storage and conversion: a sodium-sulfur storage battery; a sodium-halogen primary battery; a solid state cell; and a sodium heat engine or thermoelectric converter.

The attractive properties of beta alumina solid electrolytes are high ionic conductivity, small electronic transference number, chemical passivity and capability of being formed into useful shapes.

In each application the electrolyte membrane is a polycrystalline sodium ion conductor. Large single crystals of beta alumina have been produced and are available [1] but they are uneconomic for most purposes. Sodium has the greatest mobility of all ions in the conducting planes of the beta alumina structures. Except for the silver derivative, no method is known for making dense polycrystalline membranes of ion substituted beta aluminas.

Beta alumina now refers generally to either of two closely related compounds with layer structures, β and β'', found in the binary system $Na_2O-Al_2O_3$ and within the boundaries of certain ternary systems. Ternary oxides of special interest are spinel formers with Al_2O_3 and include MgO, NiO, CoO, ZnO and Li_2O. Small amounts of these oxides have the effect of stabilizing the β'' phase to high temperature where rapid sintering is possible. Stabilized β'' has higher ionic conductivity than the unstabilized compound. Solid solution of MgO and Li_2O in the β phase also can improve the ionic

37

conductivity. The dopant cations do not act as current
carriers and presumably occupy aluminum sites in the spinel
blocks in both β and β'', charge compensation being supplied
by Na^+ ions in the conducting planes.

The improvement of ionic conductivity by doping may
occur simply from a higher density of sodium ions in the
conducting planes, indirectly from a change in the c-axis
spacing or from annihilation of aluminum vacancies in the
parent structures.

The variation of properties of technological interest
with composition have been extensively studied in the systems $Na_2O-Al_2O_3$[2,3], $Na_2O-Al_2O_3-MgO$ [3,4,5,6], $Na_2O-Al_2O_3-$
Li_2O [3,7] and $Na_2O-Al_2O_3-Li_2O-MgO$ [3,8]. A minimum in resistivity is found at compositions which sinter to fully
stabilized β''. The location of this phase in a proposed
equilibrium diagram for the magnesia ternary system at
1700°C is shown in Fig. 1. In this same system, minima in
resistivity also are found in the β alumina solid solution
field [5,6].

In these studies, a good correlation between resistivity
and phase composition is often found, but the influences of
grain size, impurity content and homogeneity on resistivity
also are shown to be important.

Sintering temperatures as low as 1480°C and as high as
1830°C have been used, depending on composition and formulation. A liquid phase in small amounts promotes sintering
and compositions which form low melting eutectics are used
for achieving densification below 1600°C. Typical values
of resistivity at 300°C for well prepared polycrystalline
membranes are: 3-6 Ωcm for β'' stabilized with Mgo,Li_2O or
both oxides; 10-16 Ωcm for MgO doped β; 15-20 Ωcm for binary β. These values are not highly sensitive to grain size.
At lower temperatures the effect of grain size on resistivity
is more pronounced [9]. The currently preferred beta alu-
mina compositions are binary β or MgO doped β, and β'' fully
stabilized with MgO, Li_2O or both oxides. Usually all but
trace quantities of other phases are avoided, especially
those with transitions below the sintering temperatures and
those subject to attack by sodium metal. The preference of
some research groups for materials with the β or two block
structure is based on their viewpoint that between β and β'',

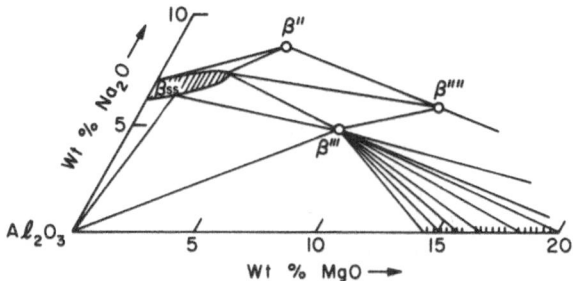

<u>Fig. 1.</u> Proposed equilibrium diagram for the ternary sys-
them Al_2O_3-$NaAlO_2$-$MgAl_2O_4$ at 1700°C. [10]

differences in properties other than conductivity are of
paramount concern for certain applications or processing
conditions. Differences in thermal expansion anisotropy
and ion exchange equilibria in molten salts, for example,
have been noted [10].

SODIUM-SULFUR CELL

Work on the application of this storage cell to batteries for load leveling and vehicle power now is entering
the development and demonstration phase in several countries.
Most of the applied research on beta alumina has been directed towards use in this sytem. As shown in Figure 2,
the beta alumina electrolyte, usually in the form of a
closed end thin walled tube, separates compartments of sodium and sulfur/sodium polysulfide. The cell is operated
between 300°-400°C to keep reactants and products molten.
The large fraction of reactant utilization realized in a
cell with liquid reactants and products is an attractive
feature of the Na-S cell. Achieving good utilization at
high rates is complicated by a region of phase separation
in the sulfur-polysulfide system. Liquid sodium/beta alumina cells using molten salt oxidants other than sulfur
have been proposed. One of these, based on a solution of
metal chlorides in molten sodium chloroaluminate which operates at 200°C, has been described [12].

In the operating temperature range of these cells,
chemical corrosion of beta alumina by sodium and most molten salts is not a problem. Polarization at the beta alumina interfaces is small, and membrane resistivities of

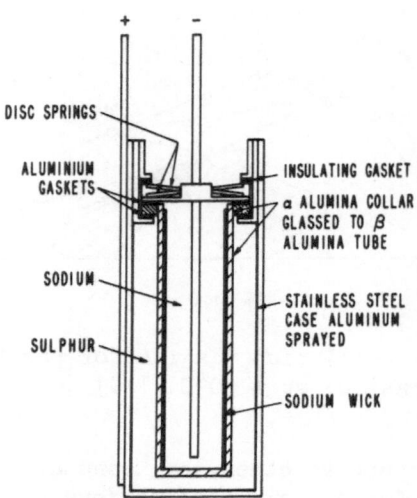

<u>Fig. 2</u>. Sodium-sulfur tube cell with combination glass and gasket seal assembly (after Tilley [11]). Porous positive electrode structure not shown.

0.5 Ωcm^2 consistent with high power operation are commonly achieved. In the early stages of Na-S cell development, a dismaying phenomenon was observed. Cells on fast charge-discharge cycles lost capacity and behaved in a non-faradaic manner. Dissection of failed cells revealed a network of dark cracks extending through the wall of the electrolyte membrane causing electrical shorts, loss of strength and often fracture. This behavior, observed independently in many laboratories, results from the propagation of sodium filled cracks originating at the interface where sodium ions are neutralized to form bulk metal during charge. There is evidence from short-term tests that there is a charging current density threshold below which this electrolytic degradation is imperceptible [13]. The threshold current is sensitive to electrolyte formulation and processing and does not always correlate well with the modulus of rupture of virgin material. Systematic studies of composition and process variables have led to the development of electrolytes which in sodium-sodium cells resist degradation at charge current densities up to 1-2 amp/cm^2, well above a useful range for use in Na-S cells [7]. The maximum current density during the charging cycle particularly in

the two phase region must never exceed the critical
value at any point. Two hypotheses have been advanced to
account for electrolytic degradation. The first of these
[14] proposes that the charging ion current is focused
along lines of electric field to the tips of liquid sodium
filled cracks pre-existing in the electrolyte surface.
These cracks then propagate by repeated fracture at the
crack tip from stress induced by the hydrodynamic pressure
required to sustain the efflux of sodium metal. According
to a second hypothesis [15], cracks propagate by selective
removal of electrolyte from crack tips by effluxing sodium.
The electrolyte dissolution depends on the interaction between capillarity and stress. Ion current focusing and
stress induced by sodium flow are also considered in this
mechanism. It has been suggested that electrolytic degradation is superimposed on a conditions of stress corrosion
of beta alumina in certain media. Stresses arising from
the presence of a second phase, from absorption of impurity
ions such as K^+, from seals, from Joule heating and from
other sources all have been suspected of contributing to
crack growth. In some instances [6,16,17] a correlation
between cell life or other property such as electrolyte
strength and a suspected contributor to stress was found.
Electronic conductivity in beta aluminas, however small,
also has been suspected of playing a role in electrolytic
degradation, but quantitative evidence is lacking [17].
Possible detrimental effects of local high current densities
[18] and electric field gradients at grain boundaries [19]
also are being investigated.

The current program of research on the sodium-sulfur
battery supported by the U.S. government includes a compre-
hensive study of electrolyte properties likely to affect durability. Within this program, procedures for preparing and
characterizing electrolytes with good homogeneity and controlled microstrcutres have been developed [17].   This
generation of beta alumina material has not yet been tested
in cells. Both β'' and β electrolyte tubes in individual
Na-S cells have survived as many as 6000 cycles and over
1400 Ah/cm^2 in life tests [19,20]. Incidence of electrolyte failure in cells, however, can be high when judging
from published reports [17,18]. Part of the necessary improvement is expected to come from processing refinements
and prooftesting, and part by minimizing stress in the
electrolyte, charge current density, impurity concentration

and other detrimental conditions in the cell as they become identified. The absence of electrolyte failure in
recent cell testing is noted in one report [20].

SODIUM AMALGAM-HALOGEN CELL

The development of a new class of high energy density
primary cells using beta alumina membranes has been completed recently [22]. Intended for application in heart
pacemakers, electronic watches, etc., near ambient temperatures, they exhibit long shelf and operating life. A beta
membrane separates a sodium-mercury amalgam anode from an
aqueous or nonaqueous catholyte. Suitable oxidants include bromine, iodine, water and air. A low power capability is a consequence of the relatively high resistivity
of beta alumina at low temperatures and polarization at the
electrolyte/catholyte interface. Practical specific energies as high as 660 Whr/Kg, and lifetimes of four years,
have been achieved.

SOLID-STATE CELL

A solid-state cell using beta alumina was described
some years ago [23]. One version of this cell has a beta
alumina separator sandwiched between two solid electrodes
fitted with metal leads. The electrodes were mixed conducting alkali ferrite-beta alumina solid solutions doped
with TiO_2. For voltages less than 0.6V the cell behaved as
a capacitor with a capacity of $180F/cm^3$. The electrode reactions were presumed to be:

$$Fe^{3+} + e + Na^+ \underset{\leftarrow}{\overset{\rightarrow}{}} Fe^{2+} \cdot Na^+$$

One other study of the solid electrode in a low temperature
cell was reported [5]. The specific energy density of these
cells are low compared to other electrochemical systems,
and as a capacitor the cell as originally described suf-
fered from a high dissipation factor even at elevated temperatures. Unless the frequency response at 60 Hz can be
improved, practical uses for this device appear to be limited.

SODIUM HEAT ENGINE

A beta alumina ceramic membrane is the heart of a thermoelectric converter or heat engine shown schematically in
Figure 3. The system can be regarded as a sodium vapor

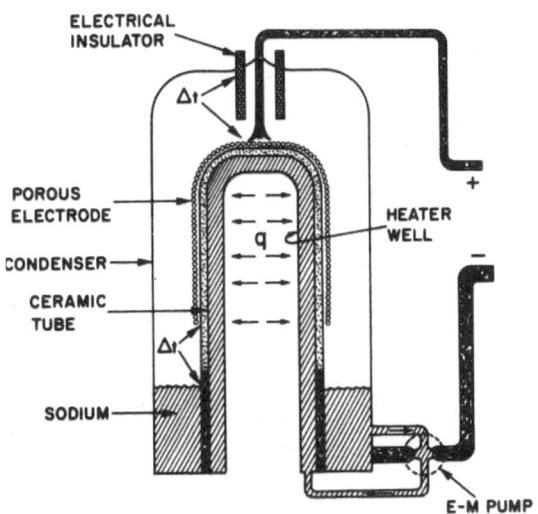

Fig. 3. Tube cell configuration of sodium heat engine

concentration cell where a pressure differential is maintained by two heat reservoirs. By the passage of ions
through the solid electrolyte and electrons through a load,
sodium (the working fluid) is expanded nearly isothermally
at some elevated temperature (500°-900°C). Ion neutralization occurs at a porous metal electrode on the low pressure side at the membrane. Neutral atoms evaporate and
pass through a vapor space to a condenser. Cooled liquid
sodium is then pumped electromagnetically back to the high
temperature region. Except for circulating liquid, there
are no moving parts and the work output is electrical only,
The thermal to electrical conversion efficiency of the
idealized (no thermal losses) cycle is close to the Carnot
limit for small loading and decreases with increasing load.
For certain estimates of radiative and conductive heat
losses, an elementary theory of operation [24,25] predicts
thermal to electrical conversion efficiencies between 20-30%
at power levels near 0.5 watts/cm^2, as indicated in Figure
4. Beta alumina is particularly suitable for this application since the most attractive features of the heat engine
derive from the properties of the work fluid.

The sodium heat engine application requires an electro-

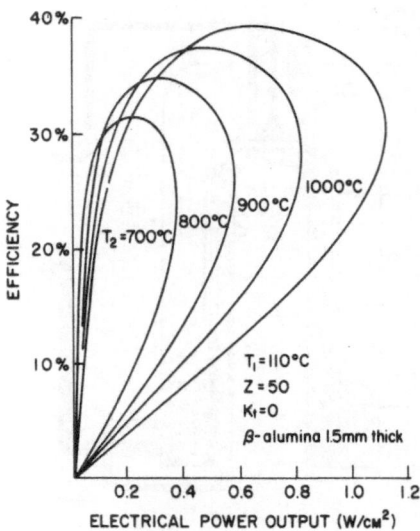

Fig. 4. Calculated power-efficiency-temperature curves for sodium heat engine with concentric cylinder geometry. The radiation resistance z is defined as the factor by which net blackbody radiation between two surfaces at T_2 and T_1 is diminished by reflection.

lyte with long-term durability in hot sodium. Electrolyte resistivity and electrode polarization are characteristically low at the high operating temperature. Because bulk sodium cannot form at the ion neutralization surface, this device should be immune to electrolytic degradation if either of the proposed mechanisms for this effect are correct. The possibility of stress corrosion, however, cannot be dismissed.

Because the conversion efficiency is practically independent of size, the sodium heat engine, in modular form, is a candidate device for local generation of power in a total energy system. Applications in solar-thermal-electric systems also are being explored. To date, only a few small recirculating cells have been tested. One of these operating near 800°C delivered 9 watts at 10% thermal efficiency. Operating life beyond 1200 hours has not yet been demonstrated.

AREAS FOR FUTURE WORK

If the commercial promise of beta alumina electrolytes is to be realized, methods for large-scale production of tubes will be required. There are three main steps in the currently practiced fabrication from powders: powder preparation, green body formation, and sintering. For each step several options have been evolved. Solutions to problems of producing beta alumina with adequate strength and conductivity at reasonable cost appear to be close at hand. The problems of durabilty of β and β'' electrolytes, particularly in the secondary cell applications, are not understood completely and may not be identical. The study of beta alumina durability in the heat engine should be revealing since electrolytic degradation ought to be absent, only one active component is present, and there is low potential for absorption of impurity ions in the electrolyte. At the high operating temperature, however, equilibration of the ceramic with sodium is fast and electronic conductivity, though not a practical liability, is readily observable. The heat engine configuration is suitable, in fact, for measuring electronic conductivity [24].

Information about certain fundamental properties of beta alumina is still not available. The equilibria, kinetics, and volume changes associated with sodium absorption in β and β'' at high temperatures (>500°C) are not known. The electronic conductivities of the sodium-beta single crystals at high temperatures have not been measured. Few studies of ion exchange equilibria between β'' and molten salts have been made. There is a challenge to devise techniques for further study of electrolytic degradation. Low temperatures storage cell applications with non-aqueous solvent systems deserve further attention. Participation in these areas by interested scientists has been frustrated, to some extent, by the difficulty of obtaining well prepared polycrystalline material. It is to be hoped that the production scale-up expected in the near future will relieve this situation.

References

[1] For example, Union Carbide Corp., Crystal Products Dept., 8888 Balboa Ave, San Diego, CA.
[2] J. Fally, C. Lasne, Y. Lazennec, Y. Le Cars, and P Margotin, J. Electrochem. Soc. 120, 1296 (1973).

[3] I. Wynn Jones and L. J. Miles, Proc. Brit. Ceram. Soc. 19, 161, 179 (1971).

[4] A. Imai and M. Harata, Jap. J. Appl. Phys. 11, 180 (1972).

[5] John H. Kennedy and Anthony F. Sammells, p. 563, Fast Ion Transport in Solids, W. van Gool, Editor, North-Holland Publishing Company (1973).

[6] W. Haar, W. Fischer, H. Kleinschmager and G. Weddigen, p.B-25 in Proceedings of the Symposium and Workshop on Advanced Battery Research and Design, March 22-24, 1976, ANL-76-8.

[7] S. Weiner et al., Annual Report for June 30, 1973-June 29, 1974, NSF Contract C805.

[8] M. D. Hames and J. H. Duncan, SAE Preprint 750375, 1975.

[9] R. W. Powers and S. P. Mitoff, J. Electrochem. Soc. 122, 226 (1975).

[10] J. T. Kummer, Progress in Solid State Chemistry, Vol. 7, J. O. McCaldin, Howard Reiss, Editors, Pergamon Press, New York (1972).

[11] A. R. Tilley, Op. Cit. ref [6] B231.

[12] Report for EPRI Contract RP109, Dec. 1974.

[13] G. J. Tennenhouse, R. C. Ku. R. H. Richman and T. J. Whalen, Am. Ceram. Soc. Bull. 54, 523 (1975)

[14] R. D. Armstrong, T. Dickinson, and J. Turner, Electrochim. Acta. 19 187 (1974).

[15] R. H. Richman and G. J. Tennenhouse, J. Amer. Ceram. Soc. 58, 63 (1975).

[16] Y. Lazennec, C. Lasne, P. Margotin, J. Fally, J. Electrochem. Soc. 122, 734 (1975).

[17] S. Weiner et al., Annual Report June 30, 1975-Dec. 30, 1975, NSF Contract C-805.

[18] L. C. DeJonghe and M. Y. Hsieh, Op. Cit. ref [6] p. B-13.

[19] Steven A. Weiner, Op. Cit. ref [6[p. B-219.

[20] A. J. Appleby and J.-P. Gabano, Op. Cit. ref [6] p. A-49.

[21] Interim Report for EPRI Contract 128-2 (1975).

[22] F. G. Will and S. P. Mitoff, J. Electrochem. Soc. 122, 457 (1975).

[23] K. O. Hever, J. Electrochem. Soc. 115, 830 (1968).

[24] N. Weber, Energy Conversion 14, 1 (1974).

[25] T. K. Hunt, N. Weber, T. Cole, Proc. 10th IECEC Conf. Addendum, 231 (1975).

PROPERTIES AND APPLICATIONS OF SOLID SOLUTION ELECTRODES

B.C.H. Steele

Department of Metallurgy and Materials Science

Imperial College, London, SW7, U.K.

INTRODUCTION

Selected materials can function as a host lattice for the incorporation of elements such as hydrogen, oxygen, copper, lithium, sodium which may be employed as the electroactive species in a variety of electrochemical devices. Providing the solubility of the relevant species is relatively large and that the rate of dissolution is fast then these materials can be used as SOLID SOLUTION ELECTRODES (S.S.E.) in appropriate electrochemical cells. The criteria for a non-stoichiometric S.S.E., $A_x MX_y$ incorporating electroactive species A have been discussed by Steele (1) and may be summarised as follows:

1. Good electronic conductor
2. Large range of homogeneity e.g. $0 < x < 1$
3. $\Delta \bar{G}_a$ ($\Delta \bar{H}_a$) relatively constant over wide range of composition
4. Rapid chemical diffusion ($\tilde{D}a$) of electroactive species within non stoichiometric electrode in both directions. As a shallow $\Delta \bar{G}$ - composition curve is specified (3rd criterion), then the magnitude of the relevant chemical and self diffusion coefficients will be similar ($\tilde{D}_a \sim D_a^* \sim 10^{-6}$ cm^2/s). The S.S.E. should therefore be a good ionic conductor.
5. Rapid transfer of electroactive species A across the S.S.E. / electrolyte interface.
6. Ability to form stable interfaces with relevant electrolyte phase.
7. For mass production constituents should be cheap, abundant, and easily fabricated into desired component.

EXAMPLES OF SOLID SOLUTION ELECTRODES

(a) Incorporation of Hydrogen

It is well known that large quantities of hydrogen can dissolve in certain metals and alloys and often the hydrogen diffusion coefficient is of the appropriate magnitude $(10^{-7} - 10^{-6}$ cm^2/s) for the system to be considered as a viable solid solution electrode. Many of the simple metal hydrides are too stable thermodynamically (2) to be used in electrochemical systems incorporating aqueous electrolytes. However using nickel or iron as an alloying element it is possible to produce ternary hydrides in which $\Delta \bar{G}_H$ has values suitable for use in aqueous systems. LaNi$_5$H$_x$ for example dissolves about 6.0 atoms of hydrogen per mole LaNi$_5$ at 25°C when the imposed hydrogen pressure is 2.5 atm (3), and N.M.R. measurements (4) confirm that the protons are very mobile. This material and other rare earth metal alloys are being investigated as hydrogen storage electrodes (5). There is also considerable interest in iron-titanium and nickel-titanium hydrides.

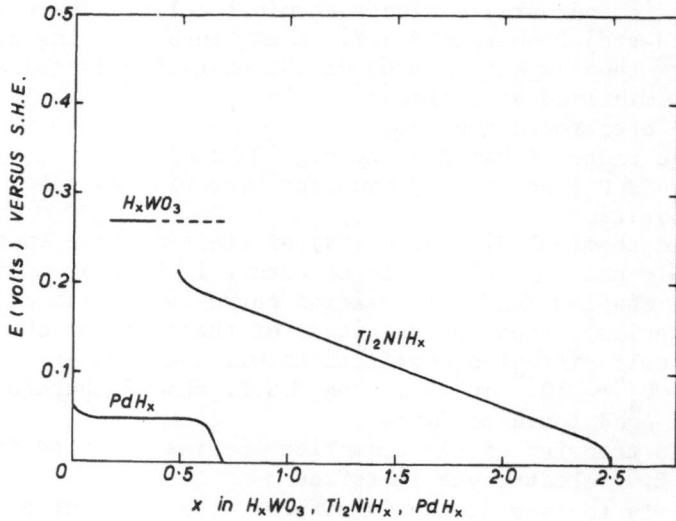

Fig. 1 Voltage composition curves for Ti$_2$NiHx, PdHx, and HxWO$_3$

The hydride, $Ti_2 Ni H_{2.5}$, for example, has been successfully employed as an electrode in aqueous systems (6). In
the presence of small quantities of the second phase Ti NiH
approximately two atoms of hydrogen per mole Ti_2Ni can be
rapidly and reversibly incorporated into the electrode.
The open circuit voltage (O.C.V.) recorded during a discharge cycle for this electrode as a function of composition
is depicted in Fig. 1, which also includes for comparison
the emf composition for palladium hydride which also ex -
hibits a high proton diffusivity (5×10^{-7} cm^2/s) at room
temperature (7)

Protons can also dissolve in non-stoichiometric compounds such as oxides but very little quantitative data are
available (8). It is known, for example, that protons can
be incorporated into MnO_{2-x} (9), and Gabano (10) has reported a value for D_{H_2} in MnO_{2-x} of 10^{-19} cm^2/s. Higher values
($10^{-10} - 10^{-11}$ cm^2/s) have been reported (11) for proton
diffusion in nickel hydroxide. More appropriate are the
transport data (12) for Hx WO_3 based on N.M.R. and electrochemical relaxation measurements which suggest D_H values
approaching 10^{-6} cm^2/s. The emf values derived from avail -
able thermodynamic data (13) for Hx WO_3 are therefore also
included in Fig. 1.

(b) Incorporation of Silver and Copper

The use of silver sulphide as a solid solution elec -
trode has been exploited for many years particularly by
Wagner and Rickert (14) to provide many elegant experiments
which illustrate the various measurement techniques possible
in solid state electrochemistry. The high temperature form
of Ag_2S can be deformed relatively easily by small stresses
and so excellent solid/solid interfaces (e.g. Ag_2I/Ag_2S or
Rb Ag_4 I_5/Ag_2S) can be prepared. This facility when combined with the remarkably high D_{Ag} values ($10^{-2} - 10^{-1}$ cm^2/s)
ensures that very high currents can be passed across this
solid/solid interface. However, although the transport
properties of Ag_2S are excellent it has a poor storage cap -
acity because of the limited range of stoichiometry ($Ag_{2.000}S$
- $Ag_{2.002}S$ at $573°K$). Similar comments apply to other
silver S.S.E. such as Ag_2Se and $Ag_x Se_{0.925} (PO_4)_{0.075}$ (15)

The copper chalcogenides can also exhibit high electronic and ionic conductivities. Moreover the ranges of
composition available for the incorporation of copper are

usually higher than the corresponding silver compounds and
copper has the additional advantage of being lighter and
cheaper.

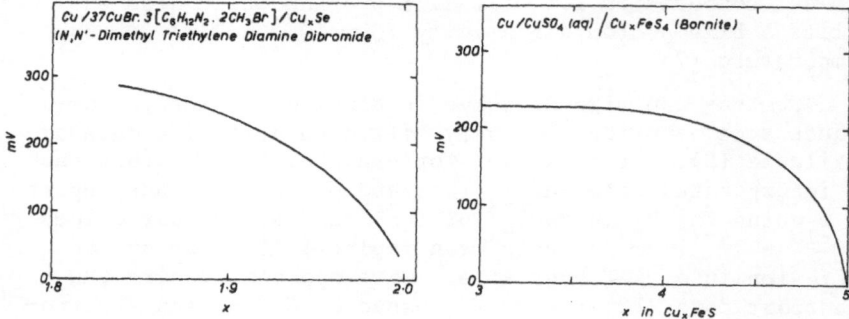

Fig. 2a Voltage composition   Fig 2b Voltage composition
         curve for Cu_xSe curve for Cu_xFeS_4

The thermodynamics of the $Cu_{1+x}S$ system have been studied
at ambient temperatures using aqueous electrolytes (16).
It also appears (17) that the high temperature disordered
form of chalcocite ($Cu_{1.99}S$) can be stabilized at lower
temperatures with increasing copper defect to produce a
value for the copper diffusion coefficient much greater
than that (10^{-10} cm^2/s) reported by Etienne (18). This
situation certainly prevails for $Cu_{1+x}Se$ which transform
from the β-phase (low temperature) to the α -phase at 110°C
for the stoichiometric compound Cu_2Se, and at 25°C for the
composition $Cu_{1.85}Se$. (19). Emf-composition data for this
compound are reported in Fig. 2   and the associated transport properties are discussed in section 4. Also included
in Fig. 2 are recent results obtained in the writer's
laboratory for the compound Cu_5FeS_4 (Bornite). Attention
is drawn to the wide range of stoichiometry exhibited by
this material in that almost 2 atoms of copper per mole can

be rapidly and reversibly incorporated at room temperature, and this process is associated with an emf (i.e. $\Delta \bar{G}_{Cu}$) value which is almost independant of composition.

(c) Incorporation of Alkali Ions

(i) Graphite and Fluorographite. It is well known that graphite can intercalate alkali metals and emf data (20) for the potassium graphite system are included in Fig. 3. The maximum potassium content is represented by the formula C_8K, and attention is drawn to the high activity of potassium in this phase. In contrast the intercalation compound C_8CrO_3, prepared by Armand (21) behaves as a very deep electron sink and consequently lithium atoms incorporated into this structure have a very low activity ($\sim$3.9V).

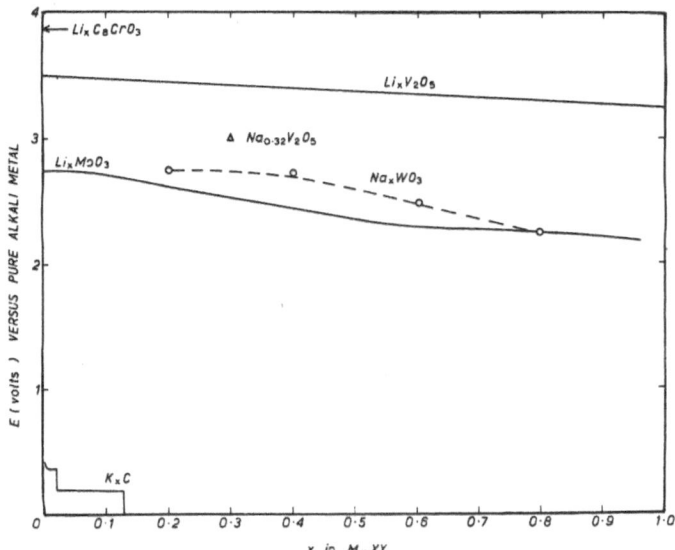

Fig. 3. Voltage composition curves for alkali metals incorporated into selected non-stoichiometric electrode materials.

The range of stoichiometry in this compound, $Li_x C_8CrO_3$ appears to be very large and the lithium diffusion coefficient approaches 10^{-6} cm^2/s. The principal problem with this type of compound is the additional incorporation of solvated species into the compound when the material is used as an electrode in electrochemical cells incorporating

organic electrolytes such as propylene carbonate. It should
also be noted that the fluorographite (CF) n electrode
probably behaves as a solid solution electrode during the
initial stage of its reaction with lithium. This suggestion (22) could explain why the voltage developed by
Li - CF_x cells is considerably less than that calculated
for the Li - LiF couple; and also why the discharge charac -
teristics are markedly influenced by the organic electrolyte
used in the cells (23).

(ii) Transition metal Chalcogenides. The metal di-
chalcogenides (MX_2) of Groups IV, V, VI adopt a layer
structure (CdI_2) in which slabs of XMX layers are held together by Van der Waals bonding. Recent work (24, 25, 26)
has confirmed that alkali metals can be rapidly and reversibly intercalated within the Van der Waals layer at ambient temperatures and emf composition results (26) for the
incorporation of sodium into single crystals of Ti_xS_2 are
depicted in Fig. 4. The emf results shown were obtained
after 3-4 cycles and were slightly dependant upon the Ti/S
ratio in the original crystals.

Lithium can also be rapidly and reversibly electro-
intercalated into TiS_2 (26). It is interesting to note

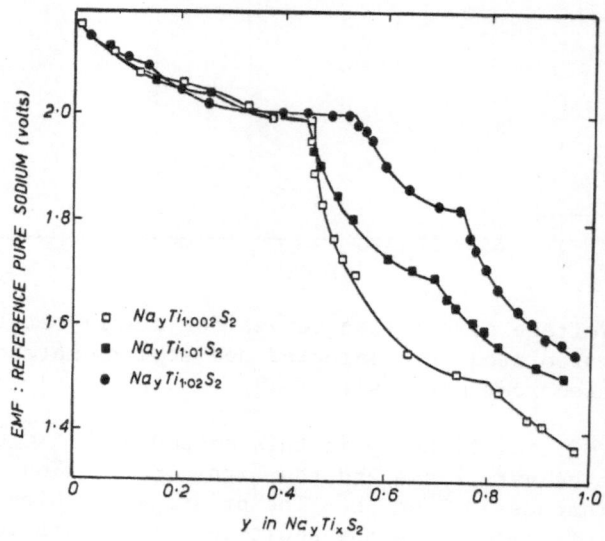

Fig. 4. Potential composition curves for Na - Ti_xS_2 system

that the voltage differential for the incorporation of 1 g
atom of lithium is only 0.5 V (2.4 - 1.9V) and thus the
theoretical specific energy density (500 WH/Kg) is even
more favourable than that calculated for the Na - TiS$_2$
system (400 WH/Kg). It is interesting to note that TiS$_2$ has
recently been successfully employed (27) as a reversible
solid solution cathode in the high temperature cell LiAl/KCl
- LiCl/TiS$_2$. Niobium diselenide has also been used (24) as
a solid solution electrode for lithium in similar cells
using organic electrolytes.

The transition metal trichalcogenides MX$_3$ can also intercalate lithium (24, 28). Upto three lithium atoms may
be incorporated but as discussed by Whittingham (29) this
process may be accompanied by structural reorganisation of
the host lattice in which chemical bonds are irreversibly
broken thus preventing the non-stoichiometric compound from
behaving as a secondary electrode. For example the formation
of Li$_3$ Nb Se$_3$ appears to be reversible whereas TiS$_3$ cannot
be used as a secondary electrode.

(iii) Electrodes Incorporating MoO$_3$, WO$_3$, and V$_2$O$_5$
The alkali metal tungsten and vanadium bronzes (e.g. Na$_x$WO$_3$
and Na$_x$V$_2$O$_5$) have structures which contain channels of
various diameters in which the alkali ions reside. The
channels in the tetragonal and hexagonal structures should
be large enough to permit translational motion of the alkali
ions and so these materials have been investigated as possible S.S.E. as they also exhibit high electronic conductivities. Thermodynamic and kinetic measurements made with cells
of the type:

          Na / Propylene carbonate (NaI) / Na$_x$WO$_3$
     or   Na / beta alumina   / Na$_x$WO$_3$

quickly revealed (30) that although the $\Delta \bar{G}_{Na}$ values were
very promising (see Fig. 3) the chemical diffusion coefficient were very low ($<$ 10^{-15} cm^2/s) which prevented their
application as S.S.E. It appears that transport along the
tunnels can be impeded by the presence of impurity ions, or
more likely crystallographic imperfections exist which
effectively block the tunnels. High resolution electron
micrographs (31) of other tetragonal bronze structure reveal complex distorted arrangements of the WO$_6$ oxtahedra in
contrast to the simple idealised structure originally proposed for these materials. It is relevant to note that
disappointing results have also been obtained for ionic

transport in other structures having relatively larger
tunnels such as hollandite ($K_{1.6}Mg_{0.8}Ti_{7.2}O_{16}$).

In contrast to these results it does appear possible
to inject lithium ions rapidly and reversibly into MoO_3,
and V_2O_5 at ambient temperatures to produce metastable
bronzes, e.g. $Li_x MoO_3$. These oxides have been employed as
cathodes in lithium organic batteries (32) and the conventional interpretation of the observed current voltage curves
is to assume a cathode reaction of the type:

$$MoO_3 + 2Li \longrightarrow MoO_2 + Li_2O$$

However, thermodynamic data indicates that the maximum
voltage to be expected for reactions of this type is about
2V. A more plausible explanation is to assume that lithium
is dissolving in the MoO_3 host lattice and the cell voltage
reflects the activity of lithium in this non-stoichiometric
phase. The transport of lithium in these compounds appears
to be sufficiently high for them to be regarded as S.S.E.'s
although there is little information about the ability of
these materials to tolerate repeated cycling. For $Li_x V_2O_5$
it does appear (29) that bonds are broken and structural
distortion occurs for lithium contents greater than 0.25
atom, and so secondary electrode behaviour is unlikely.

(iv) Beta Ferrite Electrodes. Potassium beta ferrite
($K_{1+x}Fe_{11}O_{17}$) is one of the isomorphous iron analogues of
beta alumina and exhibits electronic as well as ionic conductivity. Additional potassium ions can be incorporated
into the structure and charge compensated by the formation
of Fe^{2+}. Emf composition relationships have been obtained
(33) using beta alumina as the electrolyte phase and results
obtained at 523°K are shown in Fig. 5. Attention is drawn
to the rapid rise of voltage as the potassium content
approaches the stoichiometric value $K_{1.0}Fe_{11}O_{17}$ It does
appear, however that the limiting composition is $K_{1.09}Fe_{11}O_{17}$
and that excess potassium is always present in the layers
which is similar to the situation reported for beta alumina
(34).

(v) Incorporation of oxygen. The non-stoichiometric
fluorite phases, UO_{2+x}, CeO_{2-x} were among the first S.S.E.
systems to be investigated. Emf composition data was first
measured in 1962 (35) and the associated chemical diffusion
coefficients reported in 1970 (36) using solid state cells
of the following type at elevated temperatures (800 - 1000°C)

$$Cu, Cu_2O/Zr_{0.85}Ca_{0.15}O_{1.85}/UO_{2+x}$$

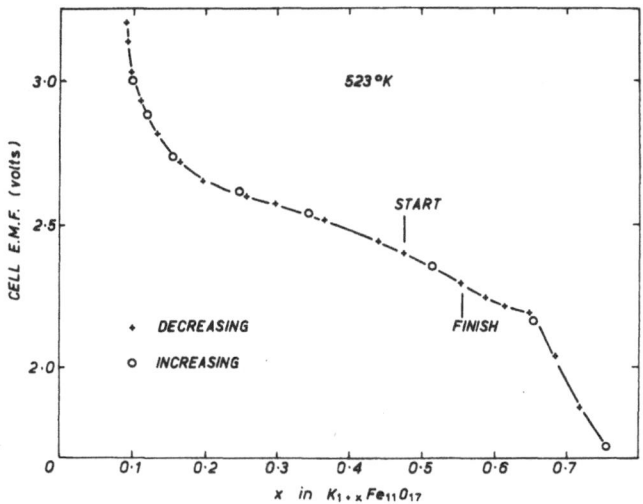

Fig. 5. Potential composition curves for $K_{1+x}Fe_{11}O_{17}$

More relevant to the problems of electrochemical storage
are the investigations of Kudo et al (37). Examination of
their paper indicates that the pervoskite solid solutions
$Nd_{1-x}Sr_xCoO_{3-g}$ can behave as S.S.E.'s in alkaline solutions
at $25^{\circ}C$, although the diffusion coefficients reported were
low, $10^{-11} - 10^{-14}$ cm^2/s. This type of behaviour, however,
has obvious implications for the proposed application (38)
of this material as an oxygen electrode in fuel cells.

3. INTERPRETATION OF PROPERTIES OF SOLID SOLUTION ELECTRODES

(a) Thermodynamics

The characteristics of the preceding examples of emf
composition curves are controlled by the partial molar free
energy of solution ($\Delta \bar{G}_a = \Delta \bar{H}_a - T\Delta \bar{S}_a$) of the relevant
solute. The type of behaviour exhibited shows marked di-
vergence from ideality ($\Delta \bar{G}_a = -T\Delta \bar{S}_a = -RT \ln X_a$), and in
fact the dominant term will be the partial molar enthalpy
of solution ($\Delta \bar{H}_a$). This situation arises because the entropy term ($\Delta \bar{S}_a$) is unlikely to exceed $50J/^{\circ}K$ mol., and so
the $T\Delta \bar{S}_a$ term at ambient temperatures (~ 15 kJ) will be
small compared to $\Delta \bar{G}_a$ values (~ 200 kJ) corresponding to
voltage in excess of 2 volts. It is appropriate to consider

therefore the energetics of the following reactions which
will influence the magnitude of $\bar{H}_a$.

$$A(s) \longrightarrow \left[A\right]_{Mx} \longrightarrow \left[A^{+} + e^{-}\right]_{Mx} \quad \Delta\bar{H}_a$$

$$A(s) \longrightarrow A^{+}(g) + e^{-}(g) \qquad \Delta\bar{H}_1$$

$$A^{+}(g) + e^{-}(g) \longrightarrow \left[A^{+} + e^{-}\right]_{Mx} \quad \Delta\bar{H}_2$$

i.e.

$$\Delta\bar{H}_a = \Delta\bar{H}_1 + \Delta\bar{H}_2$$

A detailed examination of the factors controlling $\Delta\bar{H}_2$ for
dilute solutes in liquid and solid metals often reveals
useful correlations. O'Keeffe (39), for example, has demon-
strated that $\Delta\bar{H}_2$ and thus $\Delta\bar{H}_a$ for the dissolution of hydrogen in transition metals and alloys can be correlated with
the density /number of valence electrons. There is at present insufficient data to attempt this type of analysis for
alkali metals in such compounds as V_2O_5, MoO_3, TiS_2, etc.
but general trends can be discerned in certain thermodynamic values.

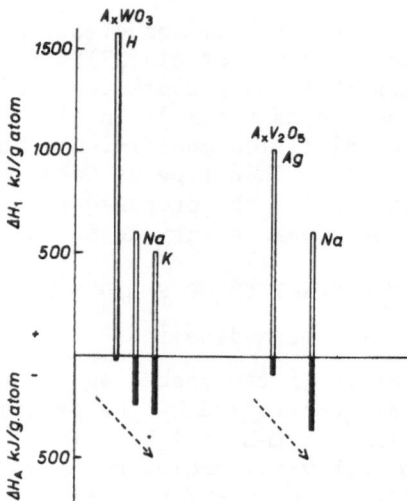

Fig. 6. Plot of ionisation energies (ΔH_1) and heats of
solution ($\Delta\bar{H}_a$) for selected bronzes.

For example from the information contained in Fig. 6 it can
be seen that the smaller the value for the ionisation energy
($\Delta\bar{H}_1$) the larger the magnitude of the partial molar heat

of solution ($\Delta \bar{H}_a$). Attention is also drawn to the fact
that the value of $\Delta \bar{H}_{Na}$ is more negative in $Na_xV_2O_5$ compared to Na_xWO_3.

Models to account for the thermodynamic of concentrated
solutions (i.e. highly defective non stoichiometric solids)
are much more speculative as it is necessary to account for
the interaction between the solute species and the formation
of defect complexes (i.e. $\Delta \bar{H}_a$ will usually vary with composition). A variety of models exist (40) and probably the
one devised for Atlas (41) for disorded fluorite systems
is at present the most useful. However it is interesting
to note that the $\Delta \bar{H}_a$ values measured (42) for the systems
Na_xWO_3, H_xWO_3, $Na_xV_2O_5$, $Ag_xV_2O_5$ are independant of composition which suggests that interactions between the inserted
ion in the oxide matrix are vanishingly small.

(b) Transport Properties

Solid solution electrodes can be incorporated in cells
with a variety of electrolytes to alloy potentiostatic,
galvanostatic, and other electrochemical relaxation techniques (7, 30, 43) to be employed to investigate how the chemical diffusion coefficient ($\tilde{D}$) varies as a function of composition. The chemical diffusion data can be supplemented
by measurements of ionic conductivity using blocking electrodes, and by determination of self diffusion coefficients
(D^*) using radio-active tracers and other techniques such
as N.M.R. The various diffusion coefficients should be
consistent with relationships of the type:

$$\tilde{D}_i = D_i^* \frac{d \ln a_i}{d \ln x_i} ,$$

and so a check on the measurements is often possible.

The results of a current pulse experiment (26) on the
three electrode cell,

$$\begin{array}{c} Li \ (ref) \\ Li \ (counter) \end{array} \Big| \ Propylene \ carbonate \ (NaI) \ \Big| \ Na_xTi_gS_2$$

are as shown in Fig. 7, and the $\tilde{D}_{Na}$ values derived from
similar experiments are summarised in Fig. 8 for two single
crystals having the composition $Ti_{1.002}S_2$ and $Ti_{1.02}S_2$
respectively. The excess titanium residing in the
normally vacant Van der Waals layer will occupy a fraction
of the possible octahedrally co ordinated sites available
for the intercalated sodium. However the small fraction of

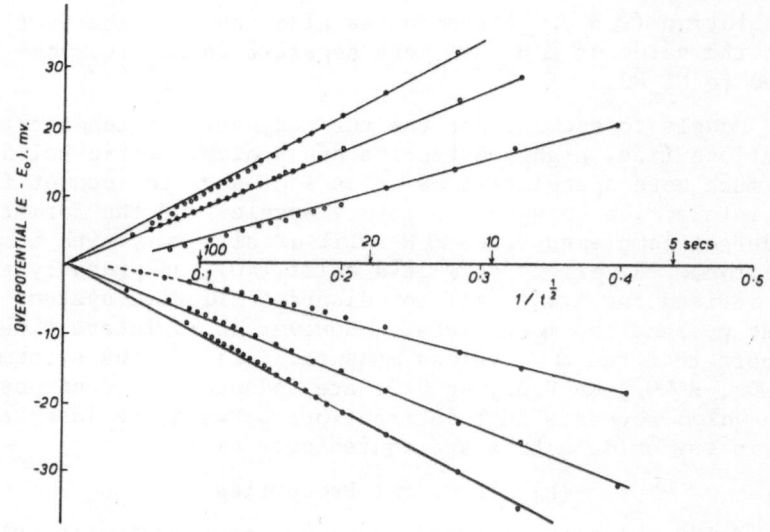

Fig. 7. Plot of overpotential (η) versus $1/t^{\frac{1}{2}}$ for pulse
experiments

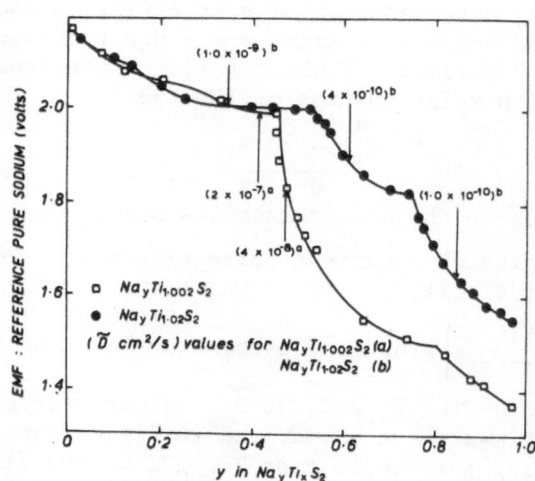

Fig. 8   Chemical diffusion coefficients for sodium in
$Na_7Ti_{1.002}S_2$ and $Na_7Ti_{1-02}S_2$

sites occupied cannot by itself explain the decrease in
sodium diffusion coefficients by more than two orders of
magnitude. Another important factor must be the development of strong co-valent bonding between the titanium and
the adjoining sulphur layers. The stronger bonding would
significantly increase the migration energy and thus lower
the diffusive flux. The variation in diffusion coefficient
with sodium content reflects the increasing occupancy of
the available sites and the structural changes associated
with the discontinuities in the emf composition curves.

Ionic conductivity measurements can also provide interesting information as this property can be measured as
a function of composition in S.S.E. materials. As the concentration of charge carriers and their consequent interaction is thus varied it should be possible to use the results to test various theoretical models devised to account
for fast ion conductivity such as the path probability approach of Sato and Kikuchi (44) and ionic polaron model of
Pardee and Mahan (45).

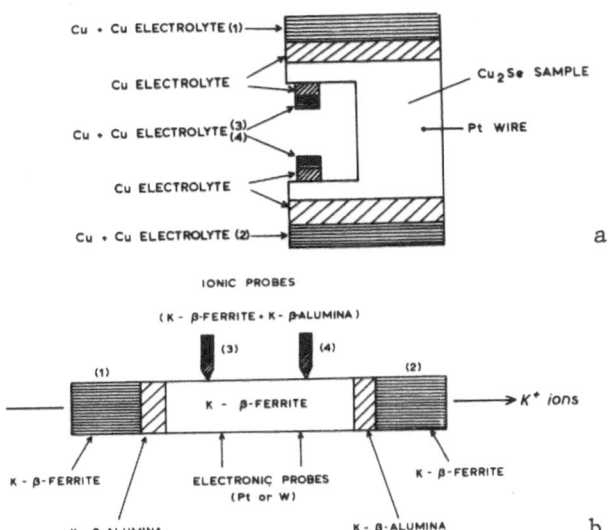

Fig. 9. Experimental arrangements for measuring ionic conductivities in S.S.E. (a) Investigation on Cu_xSe, (b) Investigations on $K_{1+x}Fe_{11}O_{17}$.

Typical experimental arrangements for measuring ionic conductivities in S.S.E.'s are shown in Fig. 9a and 9b and particular attention is drawn to the necessity of incorporating electrolytes as electronically blocking electrodes and using ionic probes to obtain values for the D.C. ionic conductivity. The concentration dependence of the ionic conductivity for $Cu_x Se$ (19) is depicted in Fig. 10 and for this material the conductivity appears to be relatively insentive to the concentration of charge carriers at $150^{\circ}C$ but exhibits a marked decrease in conductivity at $25^{\circ}C$ due to the appearance of the ordered β phase at the composition $Cu_{1.85}Se$.

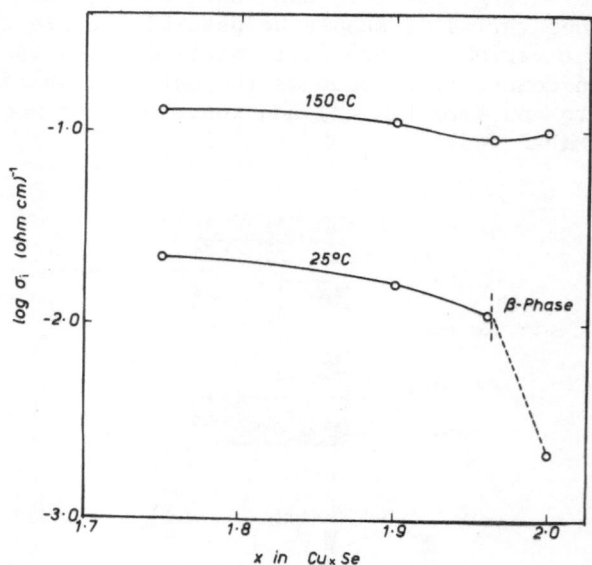

Fig. 10   Ionic conductivity of $Cu_x Se$ as a function of composition at $150^{\circ}C$ and $25^{\circ}C$.

The factors controlling fast ion transport in solid electrolytes are reviewed elsewhere in the Proceedings but it is interesting to speculate whether it might be easier to prepare S.S.E.'s exhibiting fast ion conduction because of the possibility of introducing covalent and metallic type bonding which might produce screened mobile ionic species.

(c) Interfacial Properties

The transfer of the electro-active species across the electrolyte/S.S.E. interface has so far received little attention. It is well known (7) that the electrode kinetics for the transfer of protons at the palladium/aqueous electrolyte interface can be rate-limiting unless special precautions are taken to activate the metal surface, and that the anodic reaction involving the removal of hydrogen from the palladium is slower than the reverse cathodic process. Transfer of lithium from molten salts into titanium disulphide is very rapid (27) whereas the transfer of alkali metals into the same S.S.E. from organic solvent electrolytes can be influenced by the type of solute employed.

Mention has already been made of the Ag_2S/AgI interface which involves contact between two 'soft' materials which can be readily plastically deformed. It is much more difficult to ensure good interfacial contact between two hard ceramic materials such as K-beta-alumina and K-beta-ferrite. In principle it should be possible to fabricate satisfactory interfaces by hot pressing techniques. However for these particular materials the situation is complicated due to a large difference in thermal expansion between K-beta alumina (7×10^{-6}) and K-beta-ferrite (13×10^{-6}) which produces cracks at the interface on cooling. There is no doubt that investigations in this area of interfacial kinetics are urgently required.

4. APPLICATIONS OF SOLID SOLUTION ELECTRODES

(a) Electrochemical Energy Storage Systems

Mention has already been made of the development of metal alloy solid solution electrodes for the incorporation of hydrogen (e.g. $NiTi_2H_x$) using aqeous electrolytes. Ambient temperature secondary batteries of the type:

$$Li \ / \ propylene \ carbonate \ / \ Li_x \ TiS_2$$

are being investigated, and high temperature batteries, incorporating molten salt electrolytes,

$$Li \ / \ molten \ salt \ / \ Li_x \ TiS_2,$$

also appear to satisfy many of the design criteria. Completely solid state batteries of the type $Ag/RbAg_4I_5/RbI_3$ $Li/LiI/PbI_2$ have been described in the literature (46,47) but these systems involve phase reactions at the electrodes which are often rate limiting, and very little work has been reported using solid electrolytes in conjuction with S.S.E. materials.

(b) Electrochemical Coulometers, Timers, Analogue Memories
Cells for these applications have been reviewed (48, 49) but again the special properties of S.S.E.'s have not been utilised except in the analogue memory device proposed by Takahashi (50), namely

$$Ag/RbAg_4I_5/S.S.E. \quad (Ag)/RbAg_4I_5/Ag$$

(c) Electrochromic display devices

Cells of the type:

$$Ag \ / \ RbAg_4I_5 \ / \ WO_3, \ SnO_2$$
$$Li \ / \ Li - B-Al_2O_3 \ / \ WO_3, \ ShO_2$$

can be used to reversibly inject ions (H^+, Ag^+, Li^+, etc) into thin films of WO_3, MoO_3 electrodes. The colour centres produced can be used as a display device (51, 52) requiring only a small energy input. The main problem at present appears to be the response time.

REFERENCES

1. B.C.H. Steele, p.103, 'Fast Ion Transport in Solids' Ed W. Van Gool (North Holland, Amsterdam, 1973).
2. G.G. Libowitz, 'The Solid State Chemistry of Binary Metal Hydrides' (Benjamin, New York, 1965).
3. J.H.N. Van Vucht, F.A. Kuijpers and H.C.A.M. Bruning Philips, Res. Rept. 25, 133, 1970.
4. T.K. Halstead, J. Solid State Chem. 11, 114, 1974.
5. H. Ewe, E.W. Justi and K. Stephan, Energy Conversion 13, 109, 1973.
6. M.A. Gutjer, H. Buchner, K.D. Beccu, and H. Saufferer, Power Sources 4, ed. D.H. Collins (Oriel Press, Newcastle, U.K., 1973).
7. H. Zuchner and N. Boes, Ber. Bun. Gesell. 76, 783, 1972
8. L. Glasser, Chem. Reviews 75, 21, 1975
9. J. McBreen, Electrochemica Acta, 20, 221, 1975.
10. J.P. Gabano, J. Seguret, and J.F. Laurent, J. Electrochem. Soc. 117, 147, 1970.
11. D.M. MacArthur, J. Electrochem. Soc. 117, 729, 1970.
12. P.G. Dickens, Dept. of Inorganic Chem. University of Oxford. Private communication.

13. P.G. Dickens, J.H. Moore, and D.J. Neild, J. Solid
    State Chem. 7, 241, 1973.
14. H. Rickert 'Einfuhrung in die Elektrochemie fester Staffe
    (Springer Verlag, Berlin, 1973)
15. T. Takahashi and O. Yamamoto, J. Electrochem. Soc. 119,
    1735, 1972.
16. H.J. Mathiess and H. Rickert, Z. Physik Chem. N.F. 79,
    315, 1972.
17. P. Kubaschewski, Ber Bunsen, Gesellschaft 77, 74, 1973.
18. A. Etienne, J. Electrochem. Soc. 117, 870, 1970.
19. T. Takahashi, O.Yamamoto, F. Maturyama, and Y. Noda,
    J. Solid State Chem. 16, 35, 1976.
20. S. Aronson, F.J. Salzano and D. Bellafiore, J. Chem.
    Phys. 49, 434, 1968
21. M.B. Armand p.665 'Fast Ion Transport in Solids' ed.
    W. Van Gool (North Holland, Amsterdam, 1973).
22. M.S. Whittingham, J. Electrochem. Soc. 122, 526, 1975.
23. R.J. Gunther, p.729, Power Sources 5, ed. D.H. Collins,
    (Academic press, London, 1975).
24. J. Broadhead and F.A. Trumbore, Power Sources 5, ed.
    D.H. Collins (Academic Press, London, 1975).
25. M.S. Whittingham and F.R. Gamble, Mat. Res. Bull, 10
    363, 1975.
26. D.A. Winn and B.C.H. Steele, Mat. Res. Bull 11, (5),
    1976.
27. D. Inman and Y.E.M. Mariker, Abs. 2, Symp. on Novel
    Electrode Mats, Sept. 25/26, Brighton, U.K.
28. R.R. Chianelli and M.E. Dines, Inorg. Chem. 14, 2417,
    1975
29. M.S. Whittingham, J. Electrochem. Soc. 123, 315, 1976
30. B.C.H. Steele, p.269, 'Mass Transport Phenomena in
    Ceramics' ed. A.R. Cooper and A.H. Heuer (Plenum Press,
    New York, 1975).
31. S. Iijima and J.G. Allpress, Acta, Cryst. A30, 22, 1974.
32. F.W. Dampier, J. Electrochem. Soc. 121, 656, 1974.
33. G.J. Dudley, B.C.H. Steele, and A.T. Howe, J. Solid
    State Chem. 18, (June) 1976.
34. W.L. Roth, General Electric (Schenectady, New York)
    Report No. 74CRD054 (March 1974).
35. T.L. Markin and R.J. Bones, U.K.A.E.A. Report A.E.R.E.
    R4042, 1962.
36. B.C.H. Steele and C.C. Riccardi, Proc. Int. Symp. on
    Metallurgical Chemistry, ed. O. Jubaschewski (H.M.S.O.
    London, 1972)

37. T. Kudo, H. Obayashi, and T. Gejo, J. Electrochem. Soc. 122, 159, 1975.
38. U.S. Patent 3,804,674 (16/4/70)
39. Y. Ebisuyatu and M. O'Keeffe, p.187, Progress in Solid State Chem. Vol. 4., ed. H. Reiss (Pergamon, Oxford, 1967).
40. B.E.F. Fender p.243, M.T.P. International Review of Sci. Series One, Inorganic Chem. (London, Butterworths, 1972)
41. L.M. Atlas, p.425, The Chemistry of Extended Defects in Non-Metalluc Solids (Amsterda, North Holland 1970).
42. P.G. Dickens and P.J. Wiseman, p.211, M.T.P. International Review of Science, Series Two, Inorganic Chem. (London, Butterworths 1975).
43. B.C.H. Steele, p.135, 'Heterogeneous Kinetics at Elevated Temperatures' ed. G.R. Belton and W.L. Worrell (Plenum Press, New York, 1970).
44. H. Sato and R. Kikuchi, J. Chem. Phys. 55, 677, 702, 1971.
45. W.J. Pardee and G.D. Mahan, J. Solid State Chem. 15, 310, 1975.
46. R.T. Foley, p.959, 'Physics of Electrolytes' ed. by J. Hladik (Academic Press, New York, 1972).
47. C.C. Lsaing, p.19, 'Fast Ion Transport in Solids' ed by W. Van Gool (North Holland, Amsterdam, 1973)
48. J.H. Kennedy, p.931, 'Physics of Electrolytes' ed. by J. Hladik (Academic Press, New York, 1972)
49. M. Vainov, p.431, 'Electrode Processes in Solid State Ionics' ed. M. Kleitz and J. Dupay, (Reidel, Dordrecht, 1976)
50. T. Takahashi and O. Yamamoto, J. Electrochem. Soc. 118, 1051, 1971.
51. B.W. Faughan, R.S. Crandall, and P.M. Heyman, R.C.A. Review, 36, 177, 1975.
52. I.F. Chang, B.C. Gilbert, and T.I. Sun, J. Electrochem. Soc. 122, 955, 1975.

THE ELECTROCHEMICAL PROPERTIES OF SOME SOLID ELECTROLYTES

R. D. Armstrong and T. Dickinson

Electrochemistry Research Laboratories, Department of Physical Chemistry, The University, Newcastle-upon-Tyne NE1 7RU, England

INTRODUCTION

This paper reviews many of our studies of solid electrolyte systems. The emphasis has been toward studying systems which could be used as power sources. Three main topics are covered. The first deals with the kinetics and mechanisms of metal/solid electrolyte systems, including Ag/Ag_4RbI_5, $Ag/Ag_6(WO_4)I_4$, Cu/complex cuprous bromides and Na/β-alumina.* Studies of solid electrolyte/cathode systems have involved I_2/Ag_4RbI_5 and Na_2S_3/β-alumina. Finally cell degradation processes are discussed including iodine diffusion through Ag_4RbI_5, the anodic decomposition of a complex cuprous bromide and sodium penetration into β-alumina.

EXPERIMENTAL TECHNIQUES

When the electrolyte was easily compressible, the electrolysis cell was a Perspex cylinder with a 5mm diameter axial hole. A 1mm diameter radial hole mid-way along the cylinder accommodated a silver- or copper-wire reference electrode. This electrode was embedded in powdered electrolyte located between two silver or copper electrodes backed by short steel rods. The assembly was compressed in a vice and the applied pressure measured by a load gauge. The working electrodes were normally short cylinders with polished end faces. With vitreous $Ag_6WO_4I_4$, high pressures (>4000 kg cm^{-2})

* In all our studies, the β-alumina was in the β'' form.

were needed before intergranular effects became unimportant.
Under these conditions the Perspex cell was enclosed in a
stainless steel tube and the reference electrode was an in-
sulated silver wire, which passed through a small axial hole
in the working electrode.

The β-alumina was generally a sintered disc (2mm thick,
24mm diameter), sealed onto the end of an α-alumina tube with
a sealing glass. The α-alumina tube was attached to a glass
tube mounted within a glass cell. In studies of the system
Na/β-alumina/Na, purified molten sodium was admitted into
the evacuated cell and the α-alumina tube. Platinum electrodes were located in both compartments. A similar arrangement was used to study the Na_2S_3/β-alumina/Na_2S_3 system,
except that there were also two platinum probes within 1mm
of either side of the β-alumina disc. These cells were studied at 150-360°C, by using a fluidised sand bath.

Current-potential measurements were made using a pot-
entiostat in conjunction with a function generator and X-Y
recorder, or an oscilloscope. Impedance measurements were
made using a.c. bridges, phase-sensitive detection systems
or more recently an automated impedance measuring unit.

Studies of the silver and copper ion conductors were
normally performed at 20°C. With these materials, the un-
compensated resistance between the working and reference
electrodes was measured as described in reference (1).

METAL/SOLID ELECTROLYTE SYSTEMS

Silver and Copper Systems

In most of these systems the current at a given potential increased as the pressure was increased. This was
probably due to improved contact between the electrode and
electrolyte, resulting in an increase in the effective surface area of the electrode. With Ag/Ag_4RbI_5, the pressure-
independent value of the exchange current density (i_o) could
not be measured since at pressures >700 kg cm^{-2} the current-
potential relation was essentially determined by the IR drop
between the working and reference electrodes. Only a lower
limit to the exchange current density of 20 mA cm^{-2} could

Table 1. Variation of i_o and Tafel Slope with Pressure for $Cu/Cu_7(CH_3C_6H_{12}N_4)Br_8$.

Pressure/kg cm^{-2}	$i_o/\mu A$ cm^{-2}	Tafel slope/mV decade^{-1}
50	165	312
100	100	322
$\geqslant$ 150	55 $\pm$ 7	380 $\pm$ 10

be established (1). The system $Cu/Cu_7(CH_3C_6H_{12}N_4)Br_8$ was unusual (2) in that i_o decreased with increase in pressure (Table 1). No wholly satisfactory explanation of this be-haviour is available.

For some systems exchange current densities were meas-ured both by extrapolating Butler-Volmer plots, of $\log(i/[1-\exp(-F\eta/RT)])$ against η, to $\eta = 0$ and from impedance measurements. In the latter case the low frequency inter-cept on the R_s axis, of a plot of $1/wC_s$ against R_s at $\eta = 0$, equals R_{ct} the charge transfer resistance. Knowing this quantity, i_o may be calculated from the relation,

$$i_o = RT/nFR_{ct}$$

The values of i_o obtained by these two methods and the Tafel slopes are presented in Table 2. The values of i_o from im-pedance measurements are shown in parenthesis.

For a simple electron transfer reaction, the Butler-Volmer plot should be linear over its whole length and have a slope of $2.303RT/\beta F$ where β is the symmetry factor of the reaction. Linear plots were obtained at $\eta > 50mV$ and the slopes, although rather large with the copper compounds, were acceptable. At low overpotentials marked deviation from linearity occurred (Fig. 1). This curvature and the much lower i_o values obtained from the impedance measurements suggest that a process, other than a simple charge transfer reaction, controls the rate of low overpotentials.

Table 2. Values of i_o and Tafel slope.

System	Pressure /kg cm^{-2}	i_o /mA cm^{-2}	Tafel slope /mV decade^{-1}	Ref.
Ag/Ag$_4$RbI$_5$	>700	>20.0		1
	200	2.8	200	1
	200	7.8(0.6)	200	3
Ag/Ag$_6$WO$_4$I$_4$ polycryst.	⩾1000	1.1	110	4
vitreous	⩾4000	3.7^a	110	4
Cu/CuBr(I)^b	100	0.10(0.02)	320	2
Cu/CuBr(II)^c	⩾100	0.11	320	2

(a) Small particles of the electrolyte become embedded in
    the electrode thus increasing the effective area of
    contact.
(b) CuBr(I) refers to the compound formed between 87.5^m/$_o$
    CuBr and 12.5^m/$_o$ N-methyl hexamethylene tetramine bro-
    mide (CH$_3$C$_6$H$_{12}$N$_4$Br).
(c) CuBr(II) refers to the compound formed between 94^m/$_o$
    CuBr and 6^m/$_o$ 1,4-dimethyl 1,4-diazobicyclo[2,2,2]
    octyl dibromide ((CH$_3$)$_2$C$_6$H$_{12}$N$_2$Br$_2$).

This process was investigated for Ag/Ag$_4$RbI$_5$, by attempting to fit the i-E relation at low overpotentials to the
theoretical equation for the theories concerning metal dissolution reactions. Only when the rate determining step
was the nucleation and growth of 2-dimensional holes, was
agreement between theory and experiment obtained (3). Also
in accord with experiment, this model is expected to break
down at higher overpotentials when a 2-dimensional nucleus
becomes comparable in size to an atom. The rate of dissolution will then be charge-transfer controlled and the Butler-
Volmer equation should be obeyed.

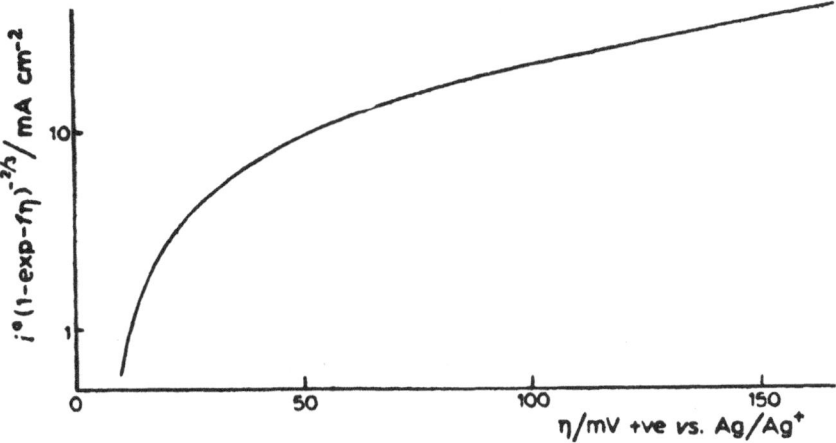

Fig. 1. Butler-Volmer plot for the dissolution of Ag into Ag_4RbI_5, pressure 200 kg cm^{-2}.

Recently, further evidence for this mechanism was obtained (4). With $Ag/Ag_6WO_4I_4$ (p.c.) the complex impedance plots at anodic potentials show an inductive semi-circle at low frequencies, its radius decreasing with increasing η (Fig. 2). This behaviour has been predicted(5) for 2-dimensional nucleation and growth of holes.

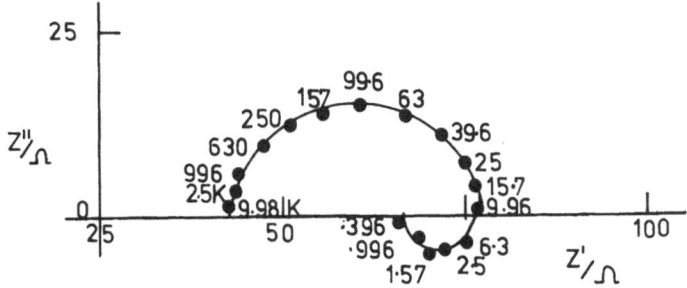

Fig. 2. Impedance spectrum for the dissolution of Ag into $Ag_6WO_4I_4$ (p.c.) at +40mV. Pressure 1000 kg cm^{-2}. Electrode area 0.2 cm^2. Frequencies in Hz.

With all these systems, electrodeposited metals showed a much higher dissolution rate than bulk metal. Thus with the former (1,2), dissolution began at underpotentials of 11 to 69mV. This enhanced activity is probably due to the small particle size of the electrodeposited material (1).

With vitreous $Ag_6WO_4I_4$, intergranular impedances were detected at pressures below 4000 kg cm^{-2}, especially at high overpotentials (400mV). Here the charge transfer semi-circle, in the complex plane plot, is followed by a second semi-circle at low frequencies. Increasing the pressure removes this feature and the impedance spectra become similar to those with the polycrystalline material.

Deposition of silver from Ag_4RbI_5 onto platinum, vitreous carbon, silver and pyrolytic graphite was investigated (6) by potential step techniques (Fig. 3). With the first three substrates the current at short times (t) was proportional to t^2. This is consistent with either growth of instantaneously nucleated 3-dimensional centres, or layer by layer growth by progressive 2-dimensional centres. In the latter case, however, damped oscillations of the current should occur at longer times (7), which was not observed. With pyrolytic graphite, the increase in current was proportional to t^3. This is consistent with growth of 3-dimensional centres with a constant nucleation rate.

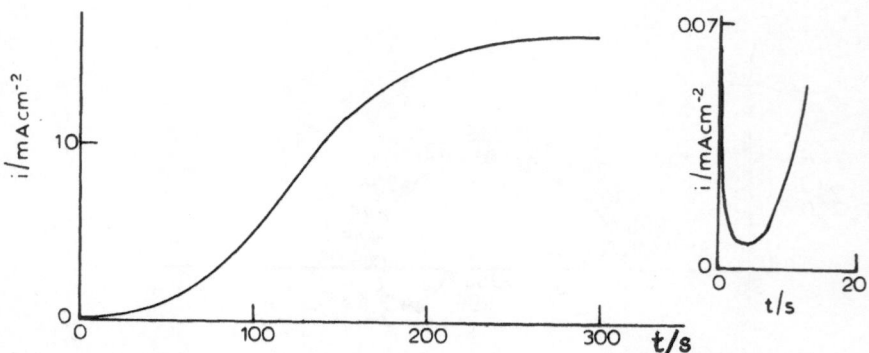

Fig. 3. Current-time transient for Ag deposition onto Pt from Ag_4RbI_5 at -25.6mV (w.r.t. Ag/Ag^+).

The Sodium/β-Alumina System

This system showed no evidence of interfacial polarisation even at 150°C, providing the ceramic had polished,
dry surfaces (8). Thus the current-voltage curve, for the
cell Na/β-alumina/Na, was linear and the impedance was almost purely resistive at frequencies up to 10k Hz (Fig. 4).
The resistance values from these two methods agreed and equ-
alled that of the β-alumina. From the frequency independence of the impedance and assuming the double layer capacitance to be 20μF cm^{-2} the minimum value of i_o was calculated as about 500mA cm^{-2} at 150°C.

Unpolished β-alumina displayed two new phenomena. The
first was that the a.c. resistance of the cell Na/β-alumina
/Na fell with time. At 350°C it fell over 4-5 hours from
6ohm to 0.6ohm. The latter value agrees with the accepted
resistivity of β-alumina. Since the D.C. resistance also
fell by a similar factor, this strongly suggests that a resistive film is present initially but is removed or modified
by the molten sodium. The impedance spectrum obtained with
an unpolished disc, after about 6 hours immersion in molten
sodium, was dominated by surface roughness effects (9-11).

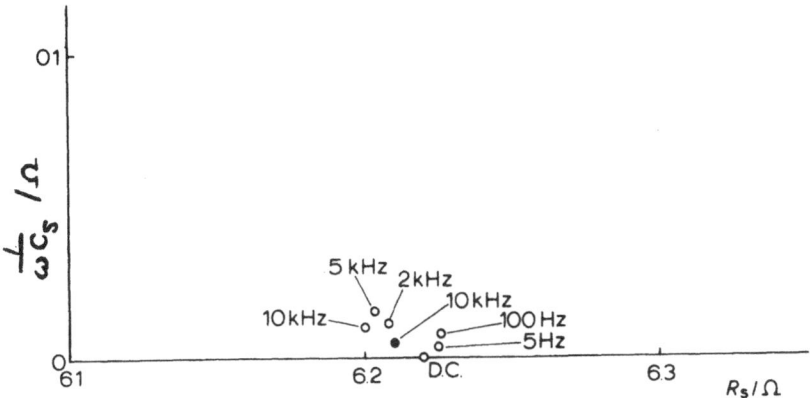

Fig. 4. Complex impedance at 150°C of the cell Na/β-alumina
(polished)/Na. ● Wheatstone bridge; O P.S.D.

SOLID ELECTROLYTE/CATHODE SYSTEMS

$$Ag_4 Rb I_5 / I_2$$

The stability of $Ag_4 Rb I_5$ toward iodine was determined by making e.m.f. measurements on cells of the type

$$Ag | Ag_4 Rb I_5 | Ag_4 Rb I_5 (x_1) + I_2 (x_2) | Pt.$$

At mole fractions of iodine (x_2) between 0.004 and 0.162, the e.m.f. was constant at $666 \pm 2mV$ (12), implying that over this composition range, at least a two phase system is formed. Topol and Owens (13) have shown that the following reaction occurs spontaneously,

$$I_2 + Ag_4 Rb I_5 = 4AgI + Rb I_3$$

Consequently when $x_2 > 0.004$, the mixture probably contains AgI, $Rb I_3$ and $Ag_4 Rb I_5$. Unfortunately, the e.m.f. was unstable at lower values of x_2 so that the composition corresponding to the two phase/one phase boundary could not be established; it must lie at $x_2 < 0.004$.

The rate of iodine reduction at a platinum electrode was obtained from impedance measurements (14). Since the couple $Ag/Rb I_3$,C had an e.m.f. of 666mV, the exchange current density was measured at this potential; it was about $36mA \ cm^{-2}$. Similar measurements for the $C, Rb I_3/Ag_4 Rb I_5$ interface gave i_o as about $60mA \ cm^{-2}$. This suggests that the graphite, in the electrolyte mix, catalyses the electrode reaction.

$$Na_2 S_3 / \beta\text{-alumina}$$

To measure any polarisation at the $Na_2 S_3 / \beta$-alumina interface, the potential difference (ΔV) between Pt probes on opposite sides of the β-alumina was measured as a function of the current flowing through the cell

$$Pt | Na_2 S_3 | \beta\text{-alumina} | Na_2 S_3 | Pt$$

At $360^{\circ}C$, ΔV varied linearly with the current (15) over a range of $\pm 280mA \ cm^{-2}$ $(\Delta V = \pm 160mV)$. This strongly suggests

that the bulk resistivity of the β-alumina determines the
i-ΔV relationship and that there is no measurable interfacial
polarisation. This was confirmed by measuring the resistance of the disc using molten sodium nitrate, when interfacial effects are negligible (16). The resistance determined in this way was 0.72 ± 0.03 ohm, in good agreement with
that from the i-ΔV curve $(0.74 \pm 0.005$ ohm).

CELL DEGREDATION PROCESSES

Solubility and Diffusion of Iodine in Ag_4RbI_5

This study was undertaken because the main factor determining the shelf-life of cells such as those based on the
Ag/RbI_3,C couple is diffusion of iodine to the Ag electrode.
In our experiments the iodine was produced by anodic decomposition of a small amount of electrolyte at Pt or vitreous
carbon electrodes.

Initially (12), current-time transients following a
potential step were recorded. For applied potentials (E_2)
between 650 and 676mV, the current decreased monotonically
with time. With $E_2 > 676$mV a maximum appeared in the transient, moving to higher currents and shorter times at more
anodic potentials. The transients at $E_2 < 676$mV (Type I) were
attributed to the formation of an oxidised species (presumably I_2) which diffuses into the electrolyte. When $E_2 > 676$mV
a thin layer of another phase, probably RbI_3, is formed on
the electrode surface. This explanation is consistent with
the e.m.f. measurements, described earlier.

The current-time relation for Type I transients obeyed
the equation for a 2-electron oxidation reaction, a first
order reduction process and diffusion of the oxidised species into the electrolyte. The logarithm of the intial current density varied linearly with potential with a slope of
44mV decade^{-1}. This is consistent with the mechanism

$$I^- \rightleftharpoons I + e$$

$$I + I^- \longrightarrow I_2 + e$$

The same conclusions were reached from impedance measurements (17). At potentials between 640 and 670mV the

complex plane plot displayed a high frequency semi-circle,
coupled to a Warburg impedance at lower frequencies (Fig. 5).
This shape is consistent with the mechanism suggested previously. Apparent values of i_o were obtained from such
graphs. Log i_o varied linearly with potential giving a slope
of 44mV decade^{-1}, in agreement with the results from pulse
measurements.

Assuming the iodide concentration adjacent to the electrode remains essentially constant and that the stoichiometry of the anodic decomposition is

$$2I^- = I_2 + 2e \qquad (1)$$

then the Warburg coefficient (σ) is given by

$$\sigma = RT/n^2 F^2 C_c \sqrt{2D_o} \qquad (2)$$

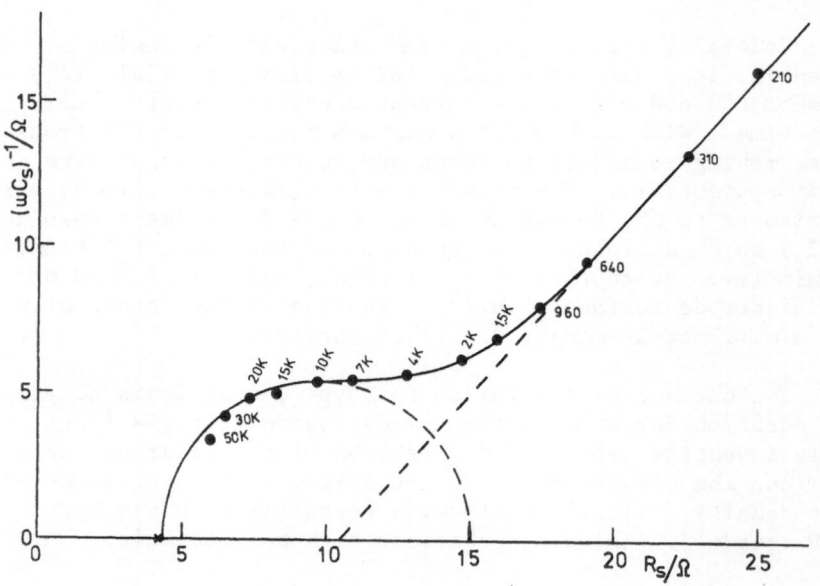

Fig. 5. Complex impedance display for the Pt/Ag, Rb I$_5$ interface at + 650mV (w.r.t. Ag/Ag$^+$). Pressure 1000 kg cm^{-2}.
Ungraded electrolyte.

where C is the apparent surface concentration of iodine and
D_o its diffusion coefficient. If reaction (1) is in a state
of pseudo-equilibrium $\log \sigma$ should be a linear function of
E with a slope of -29.5mV decade^{-1}. The experimental graph
was linear with slope -33mV decade^{-1} in good agreement with
theory.

At potentials above 670mV, the impedance plot was a
flattened semi-circle intersecting the R_s axis at about 8000
ohm. This is consistent with the formation of a highly resistant layer on the electrode, as deduced from both the
e.m.f. and potential step results.

To obtain C_o and D_o, the steady-state current in a thin-
layer cell, with a Pt anode and Ag cathode, was measured at
various cell voltages. The logarithm of the current increased linearly with potential (the slope being 33mV decade^{-1},
theoretical slope 29.5mV decade^{-1}) until the potential was
about 675mV. Above this value, the current was independent of potential. Since the steady-state current is given by

$$i = nFC_o D_o / \delta \qquad (3)$$

where δ is the thichness of electrolyte, the limiting current presumably corresponds to the limiting surface concentration of iodine i.e. its effective solubility (C_s). Equ-
ations(2) and (3) therefore allow calculation of C_s and D_o.
The diffusion coefficient decreased rapidly with increase
in pressure, becoming almost constant at pressures above
1000 kg cm^{-2}; C_s increased with pressure. At 1000 kg cm^{-2},
D_o decreased and C_s increased as the particle size of the
electrolyte was increased (Table 3).

Table 3. Solubility and Diffusion Coefficients at 1000kg cm^{-2}

Particle size	$10^6 \cdot D_o / cm^2 s^{-1}$	$10^6 \cdot C_s / mole\ cm^{-3}$
< 50μm	2.8	0.6
ungraded	1.4	1.6
> 50μm	0.8	2.2

The density of electrolyte pellets increase sharply as the pressure of formation is increased (12), up to 1000kg cm^{-2}. It then rises more slowly, becoming essentially constant at pressures above 3000 kg cm^{-2}. The decrease in D_o with increasing pressure therefore suggests that one diffusion path is through voids in the compacted solid. Since, at 1000 kg cm^{-2}, the voidage was essentially independent of particle size (14), the variation of D_o shown in Table 3 indicates that diffusion must also occur long the grain boundaries.

The increase in c_o and decrease in voidage with pressure suggests that the iodine is adsorbed on the sides of the grains; the improved contact between electrode and electrolyte at higher pressures resulting in a larger surface area for adsorption. Presumably iodine vapour also exists in the voids.

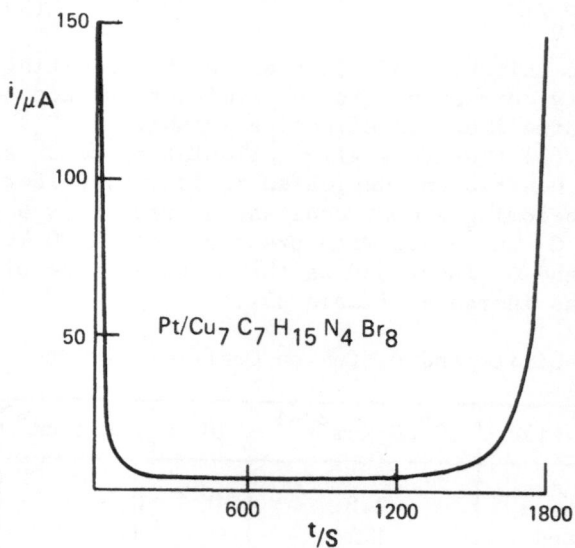

Fig. 6. Current-time transient for Pt/CuBr(I) at +700 mV (w.r.t. Cu/Cu$^+$). Pressure 200 kg cm^{-2}. Electrode area 0.2 cm^2.

The Anodic Decomposition of CuBr(I)

Cells based on complex cuprous bromide electrolytes and the Cu/Br$_2$ couple, appeared to offer a higher theoretical voltage and energy density (1.03V and 695 Ws g^{-1}) than those using the Ag/I$_2$ couple (0.69V and 282 Ws g^{-1}). Unfortunately the electrolytes are unstable toward bromine (19). We studied the oxidation of the CuBr(I) electrolyte by the behaviour observed at a platinum anode (19).

Oxidation of cuprous to cupric ions occurred at potentials +600mV. The electrolyte developed appreciable electronic conductivity, as a result of the mixture of valency states. Following a potential step into this region, the current first fell rapidly, reached a steady value and finally increased rapidly (Fig. 6). During this time the electrolyte near the anode changed from pale yellow to greeny-blue and the boundary between these zones moved into the electrolyte. The final increase in current occurred when this electronically conducting, blue material reached the reference electrode.

The steady-state current varied logarithmically with the electrode potential. The slope of this graph was 55mV decade^{-1} suggesting that the electron transfer process is rapid. During the initial fall in the current, its value was proportional to t$^{-1/2}$. Since this is typical of a diffusion controlled process, the initial process is probably the diffusion of freshly formed cupric ions away from the electrode. Finally, the ratio Q_c/Q_A, where Q_A is the charge passed in an anodic pulse and Q_c that passed in a pulse of equal length at zero volts, decreased as the pulse length increased. It approached a limiting value for zero pulse length of about 0.59; again characteristic of diffusion controlled processes.

These results were explained by assuming that diffusion of freshly generated cupric ions continued until they reached a suitable site. They are effectively 'locked' into these sites and have low mobility. This latter step is necessary to explain the existence of a steady-state current. For electronic conductivity to exist, however, a finite concentration of high mobility cupric ions must always be present. The theoretical current-time and current-potential relations for this mechanism were derived and found to be in agreement with the experimentally observed results.

The Breakdown of β-Alumina Ceramic

Until relatively recently the main factor determining the lives of sodium-sulphur cells was the development of electronic conductivity in the β-alumina electrolyte. This process was investigated (20) by recording the change in the impedance of the cell Na/β-Al$_2$O$_3$/Na when subjected to constant current electrolysis at 300°C.

The a.c. impedance and d.c. resistance of the electrolyte were initially almost constant but, after prolonged electrolysis, both decreased fairly rapidly to a much lower, nearly constant value. The impedance fell before the resistance. Moreover the impedance spectrum, recorded after the d.c. resistance had fallen somewhat, has the shape associated with a very uneven electrode-electrolyte interface (Fig. 7). Since prior to electrolysis the β-alumina disc was polished and showed no frequency dispersion of the impedance, this change strongly suggested that sodium was penetrating into the β-alumina.

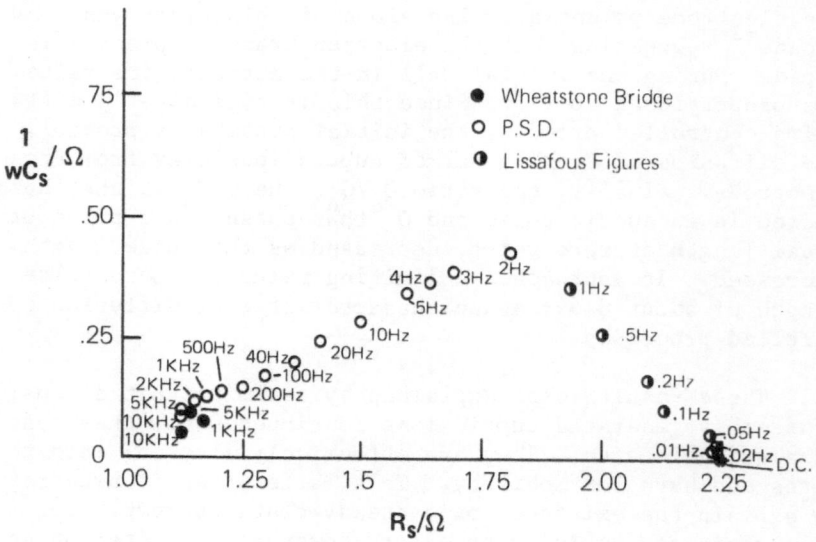

Fig. 7. Complex plane display for Na/β-alumina/Na after partial breakdown of the β-alumina. ● Wheatstone Bridge; ○ P.S.D.; ◑ Lissajous Figures.

To confirm that the observed behaviour was consistent with the growth of sodium dendrites, a model cell was constructed. This consisted of an 'electrolyte' of carbon granules, the electrodes were two perforated brass discs and steel needles passing through these represented the dendrites. The change of resistance, with increasing penetration of the needles into the electrolyte, closely followed the resistance-time curve obtained with β-alumina. Moreover the impedance spectrum, measured after a comparable fall in resistance had occurred, was very similar to that obtained with 'degrading' β-alumina.

The penetration was suggested as beginning with the gradual filling of small surface cracks by molten sodium, as a result of electrolysis. Since the current path to the tip of such a crack will be shorter than to the 'flat' surface of the β-alumina, continued deposition of sodium into the crack will occur. Flow of the molten sodium out of the crack requires the pressure difference, between the sodium inside and outside the crack, to equal that calculated from the Poiseuille equation. Under suitable conditions this pressure could be large enough to extend the crack further into the electrolyte. The whole process of crack filling and extension would then be repeated, leading to penetration of the ceramic by sodium dendrites.

REFERENCES

1. R. D. Armstrong, T. Dickinson, H. R. Thirsk and R. Whitfield, J. Electroanal. Chem., 29, 301 (1971).

2. R. D. Armstrong, T. Dickinson and K. Taylor, J. Electroanal Chem., 57, 157 (1974).

3. R. D. Armstrong, T. Dickinson and P. M. Willis, J. Electroanal. Chem., 57, 231 (1974).

4. R. D. Armstrong, T. Dickinson and K. Taylor, To be published.

5. R. D. Armstrong and A. A. Metcalfe, J. Electroanal, Chem., In Press.

6. R. D. Armstrong, T. Dickinson and P. M. Willis, J. Electroanal. Chem., 59, 281 (1975).

7.   R. D. Armstrong and A. A. Metcalfe, J. Electroanal.
     Chem., 63, 19 (1975).

8.   R. D. Armstrong, T. Dickinson and J. Turner, J. Elect-
     roanal. Chem., 44, 157 (1973).

9.   R. de Levie, in "Advances in Electrochem. and Electro-
     chem. Eng.," (P. Delahay, ed.,) p.329, Interscience,
     New York (1967).

10. R. D. Armstrong, T. Dickinson and P. M. Willis, J. El-
     ectroanal. Chem., 53, 389 (1974).

11. R. D. Armstrong, T. Dickinson and P. M. Willis, J. El-
     ectroanal. Chem., 67, 121 (1976).

12. R. D. Armstrong, T. Dickinson, H. R. Thirsk, R. Whit-
     field, J. Electroanal. Chem., 34, 47 (1972).

13. L. F. Topol and B. B. Owens, J. Phys. Chem., 72, 2106
     (1968).

14. R. D. Armstrong, T. Dickinson and P. M. Willis, J. El-
     ectroanal. Chem., 48, 47 (1973).

15. R. D. Armstrong, T. Dickinson and R. Whitfield, J. El-
     ectroanal. Chem., 32, App.9 (1971).

16. R. D. Armstrong, T. Dickinson and J. Turner, J. Elect-
     rochem. Soc., 118, 1135 (1971).

17. J. H. Sluyters, Rec. Trav. Chim., 79, 1092 (1960).

18. M. Lazzari, R. C. Pace and B. Scrosati, Electrochim.
     Acta, 20, 331 (1975).

19. R. D. Armstrong, T. Dickinson and K. Taylor, J. Elect-
     roanal. Chem., 64, 155 (1975).

20. R. D. Armstrong, T. Dickinson and J. Turner, Electro-
     chim. Acta, 19, 187 (1974).

INTERPRETATION OF AC IMPEDANCE MEASUREMENTS IN SOLIDS*

J. Ross Macdonald

Department of Physics and Astronomy
University of North Carolina
Chapel Hill, North Carolina 27514

INTRODUCTION

In order to use solid materials most effectively in solid electrolyte applications, they need to be characterized as fully as is practical. Here, one useful method of electrical response characterization will be discussed: measurement of the response when a small-signal sinusoidal potential difference is applied across a sample. A great deal more can be learned from the frequency and temperature dependence of the resulting impedance than from ordinary DC or single-frequency AC conductance measurements alone.

An exact solution of an idealized model of a system involving homogeneous material between two identical plane parallel electrodes separated by a distance ℓ has recently been given.[1,2] It incorporates the following five physical processes: charge separation, adsorption-desorption and charge transfer at electrodes, diffusion effects, and bulk generation-recombination (G/R). Many impedance results[3,4] and considerable background information[2] have already been presented; space limitation precludes their recapitulation here. Instead, the present work will concentrate on the use of earlier and current complex impedance plane results to illustrate methods of characterizing materials in terms of (a) pertinent equivalent circuits and (b) basic quantities such as mobilities, reaction rates, etc. derivable from equivalent circuit elements.

*Work supported by U. S. National Science Foundation
(Grant No. DMR 75-10739).

PERTINENT PARAMETERS

Figure 1 shows the equivalent circuit following most directly from the model.[1] All circuit elements will be given for unit electrode area. In general, $C_i(0) \equiv C_{i0}$ is finite and $R_i(\infty)$ is zero. Thus, the circuit necessarily incorporates an $\omega = 0$ resistive path, R_D, and as $\omega \to \infty$, ordinary bulk response involving the geometric capacitance $C_g \equiv \epsilon/4\pi\ell$, and the bulk resistance, $R_\infty \equiv \ell/[(ez_e)(\mu_n n_e + \mu_p p_e)]$. Here $R_\infty^{-1} \equiv R_E^{-1} + R_D^{-1}$; ϵ is the dielectric constant of the material; μ_n and μ_p are the mobilities of negative and positive charges of equilibrium concentrations n_e and p_e; and the valence numbers, z_n and z_p of these charges have, for simplicity, been taken equal to z_e here. For an intrinsic conduction situation, $n_e \equiv p_e \equiv c_i$. The dielectric relaxation time $\tau_D \equiv R_\infty C_g$ is also equal to $\epsilon/4\pi\sigma$, where $\sigma \equiv \ell/R_\infty$ is the high-frequency limiting conductivity of the material. It is often convenient to use the normalized frequency $\Omega \equiv \omega\tau_D$ and to normalize circuit elements with R_∞ and C_g. Such normalization will be denoted by a subscript "N." Thus, $R_{DN} \equiv R_D/R_\infty$, $C_{iN} \equiv C_i/C_g$, etc. Also, let $\pi_m \equiv \mu_n/\mu_p$.

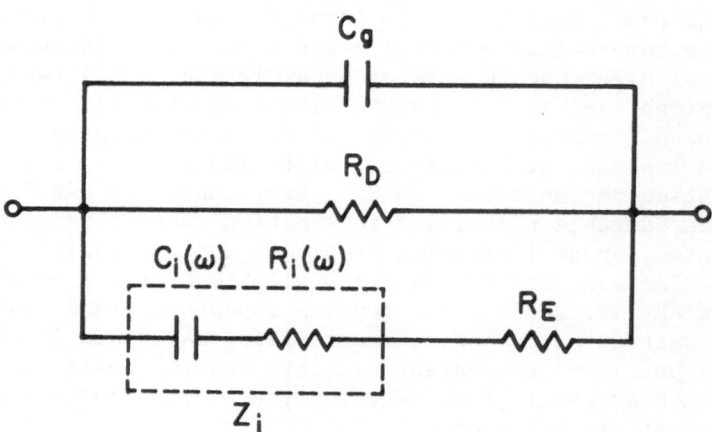

Fig. 1   General equivalent circuit following from theory.

For simplicity, the present analysis will deal primarily with complete blocking at the electrodes of mobile positive charges. Negative charges, however, may react at the
electrodes with a heterogeneous rate constant k_n, related
to a dimensionless boundary parameter r_n by $k_n \equiv (D_n/\ell)r_n =$
$(kT\mu_n/\ell e z_e)r_n$. Here k is Boltzmann's constant, T the absolute temperature, and ℓ the proton charge. When $r_n = 0$,
negative charges are also completely blocked, a $(r_p, r_n) =$
$(0,0)$ situation. In the presence of adsorption, r_n becomes
complex,[5,6] and one must introduce both its low-frequency
limiting value r_{nq} and its high-frequency value $r_{n\infty}$ as well
as the internal adsorption relaxation time τ_{An}. Generation/
recombination effects[7,8] will be largely neglected in the
present work. Define N_e as the net concentration of extrinsic centers, assumed fully dissociated, and N_i as the concentration of neutral intrinsic centers before any dissociation. Then $\chi \equiv N_e/2z_e c_i$ will determine the degree of extrinsic conduction and c_i/N_i the amount of intrinsic dissociation present.

EQUIVALENT CIRCUITS

Although the model[1] yields relatively simple expressions for the R_{EN} and R_{DN} of Fig. 1, that for $Z_{iN} \equiv$
$R_{iN} + (i\Omega C_{iN})^{-1}$ is very complicated, making it often difficult to use. Further, many of the effects of the five principal processes incorporated in the model are buried in the
frequency dependence of $C_i(\omega)$ and $R_i(\omega)$. Therefore, it is
worthwhile to consider other kinds of circuits which may
still represent the physical processes present adequately.

All linear circuits not containing inductance can be
represented by either the Maxwell model, the left circuit
of Fig. 2, or the Voigt model, the right circuit. By proper
choice of element values, these two circuits can be made to
have the same total impedance Z_T at all frequencies. The
right-hand circuit of Fig. 2 has been used to represent N-
layer Maxwell-Wagner (M-W) interfacial polarization.[9,10]
In this theory, it is assumed that the material involves N
separate layers, each with its specific dielectric constant
and conductivity. The two-layer model, being simplest, has
often been used. It could be appropriate for a two-phase
material such as a pressed and/or sintered compact in which
somewhat conducting particles were dispersed in a matrix of

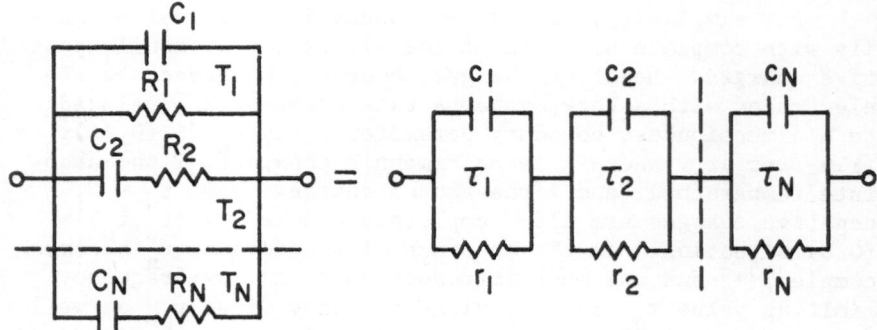

Fig. 2   Two equivalent circuits which may have the same
         overall impedance.

different conductivity or, in the absence of a matrix, the
particles themselves might have a more or less homogeneous
central region surrounded by oxygen-rich surface layers.
Alternatively, the two layers might represent bulk and surface regions of a macroscopic sample. The two-layer model
has been modified in various ways: for example by considering it as a two-layer capacitor with a continuous distribution of relaxation times[11] or by assuming[12] Schottky barriers are present in one of the layers and may cause the resistance and capacitance of the layer to depend upon the potential difference across it.

     Notice that the M-W model even with the above generali-
zation is rather simplistic. In a real N-layer M-W material,
one might expect that each layer or phase could separately
involve many of the physical effects discussed above.

     Since the present homogeneous-material model can often
lead to $Z_{T}(\omega)$ response of essentially just the kind predicted
by the inhomogeneous-material M-W model with N = 2 or 3, unless it is known that the material is strongly inhomogeneous,
it seems wise first to try to analyze its response on the
basis of a homogeneous model. If this is insufficient, then
two such homogeneous models connected in series and involving
different parameters might be considered. While it may sometimes be easy to mistake the response of a homogeneous system for an inhomogeneous one, distinction between them can
frequently be made on the basis of their possibly different
responses to a change of electrode separation distance, ℓ.

Although the two circuits of Fig. 2 may be made entirely equivalent electrically, they may not be equivalent
in terms of interpretive power for characterization purposes. When these circuits are thought to be appropriate,
which one should be used? If each of the main processes
which lead to the electrical response considered separately
produces one of the series RC branches of the left-hand circuit, then this circuit will generally be more appropriate
than the right-hand one, which will then characteristically
involve mixtures of processes for each parallel RC section.
From this point of view, dielectric relaxation with a distribution of relaxation times (N finite or infinite) seems
better represented, as is customary, by the left than the
right circuit. For a M-W system, involving layers physically in series, however, clearly the right hand one should
be employed.

Further, when impedances or admittances (or complex dielectric constants[13]) are to be plotted as parametric functions of frequency in the complex plane, admittance plane
(or complex dielectric constant) plots will be more appropriate when the left-hand circuit is the preferred one and
impedance plane plots will be more appropriate for the right
one.[2] For the present homogeneous model where, at least in
the limit of very loosely coupled processes, each parallel
RC section is associated with a separate process or physical region, impedance plane plotting is therefore preferred
and will result in a connected series of arcs, each again
associated with a single process and a single RC section.
Analysis will then be greatly simplified since mixture of
processes and effects will be avoided.

While straightforward transformation equations exist
which lead from known values of the elements of the right-
hand circuit of Fig. 2 to values of the left-hand circuit
elements, direct solutions for the reverse transformation
are not very practical for N > 2. It would therefore be
desirable if the present model led directly to a Voigt type
of circuit instead of that of Fig. 1. Although it is not
self-evident that the model can often be well represented by
a Voigt circuit, it indeed can be.[3,4]

The resulting approximate circuit is shown in Fig. 3-a.
This is an N = 5 Voigt system except that the last two

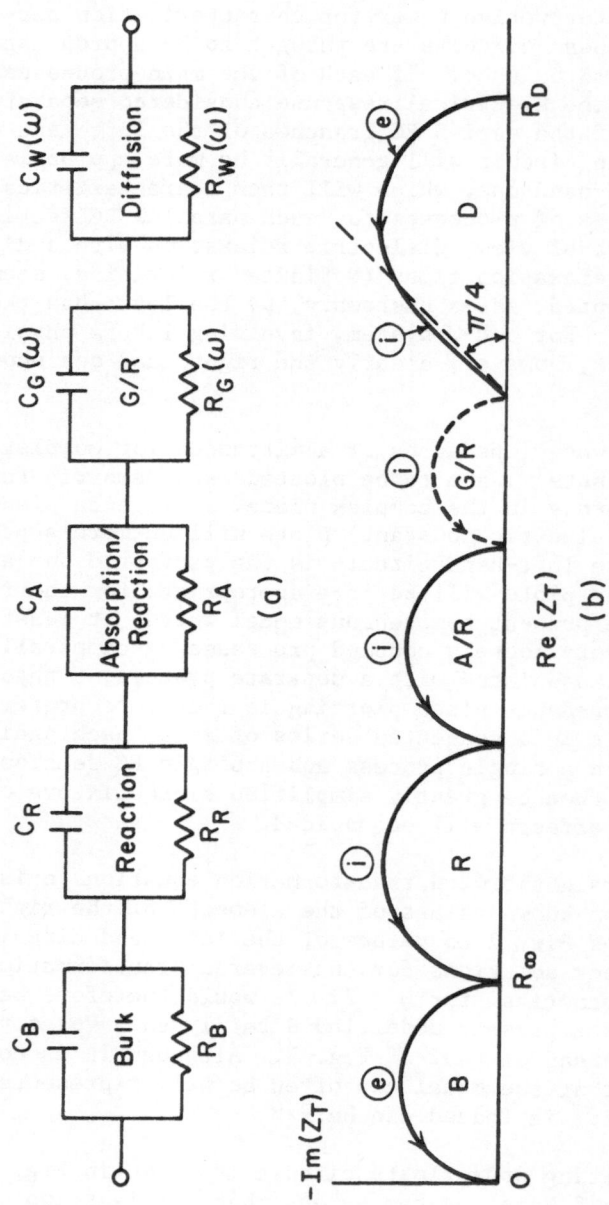

Fig. 3 (a) Equivalent circuit following from the present model. (b) Possible $Z_T^*(\omega)$ complex plane plot consistent with the circuit in (a).

sections may involve frequency dependent elements. The des-
ignations of the sections are appropriate only for very
loose coupling; i.e., each time constant must differ from
all others by at least a factor of 10^2. Surprisingly, however, it turns out, based on least squares analysis of exact model results, that the circuit shown here is often an
excellent approximation even in the close-coupled case.
Then we must designate the elements as C_k and R_k, with k =
1,2,...,5. It is hoped that future work will reveal how
these parameters, each of which may in the close-coupled
case depend on all processes, can be related to the basic
material characterization parameters.

For simplicity, most of the following results will apply only to the intrinsic-conduction loosely coupled case.
Unless the recombination internal relaxation time is much
greater than τ_D, contrary to the intrinsic-conduction prediction of the Langevin diffusion theory of the bimolecular
recombination constant,[2] it turns out that the G/R contribu-
tion, Z_G, to $Z_T \cong \sum_{k=1}^{5} Z_k$, is negligible compared to the other
contributions to Z_T. When it is not negligible, C_G and R_G
may or may not be frequency independent,[2] depending on the
value of π_m. Possible generation/recombination effects will
be considered in detail elsewhere.

Fig. 3-b is a plot of $Z_T^* \equiv \text{Re}(Z_T) - i[\text{Im}(Z_T)]$ in the
complex plane for a loosely coupled situation. Arrows in-
dicate the direction of increasing frequency. Frequently,
not all these arcs will be significant at the same time.
The G/R arc is shown dotted since it may be negligible under most conditions and may be either semicircular or similar in shape to the finite-length Warburg diffusion arc[3]
shown at the right. The sizes of the arcs depend on the
process parameters; they may thus vary widely in relative
magnitude. They have only been shown roughly similar here
in order to fit all five on the same diagram. Further, the
order in which specific arcs appear may vary greatly depending on the time constants involved. The theory does not
allow the A/R arc to occur at higher frequencies than the R
arc, however. Further, in cases of practical interest, the
B arc will always lie at the left.

The present model, when well approximated by the circuit·

of Fig. 3-a, leads to

$$Z_{TN} \equiv \sum_{k=1}^{5} Z_{kN} \cong \sum_{k=1}^{5} \left\{ \frac{(1 - \delta_{k5})R_{kN}}{(1 + i\Omega\tau_{kN})} + \left(\frac{\delta_{k5}}{\pi_m}\right)\left(\frac{\tanh(i\Omega\tau_{kN})^{1/2}}{(i\Omega\tau_{kN})^{1/2}}\right) \right\},$$

(1)

where δ_{kj} is Kronecker's delta and $\tau_{kN} \equiv R_{kN}C_{kN}$. In the usual case where the normalized bulk time constant is well separated from the others ($\tau_{kN} \gg \tau_{1N} \cong \tau_{BN} \cong 1$, k = 2 to 5), $R_{1N} \cong R_{BN} \cong 1$ and $C_{1N} \cong C_{BN} \cong 1$. Approximate expressions for the other normalized parameters will be discussed in the next section. Note that each semicircle of Fig. 3-b is associated with one of the k = 1 to 4 terms, while the k = 5 term accounts approximately for finite-length Warburg response.

Sometimes, experimental arcs are depressed[4] (less than a semicircle), possibly arising from a continuous distribution in one or more of the elements of the given process. Under some circumstances, the present model can yield some depression for the reaction arc without distributed elements, however.[4] Depression can be accounted for heuristically in the present approximate approach by writing[4,14] each k = 1 to 4 denominator in (1) in the Cole-Cole[13] form [1 + $(i\Omega\tau_{kN})^{1-\alpha_k}$], where α_k is a distribution parameter which is zero when there is no distribution.[13] It is sometimes found experimentally that the straight line portion of the diffusion arc lies at an angle different from the theoretical Warburg value of $\pi/4$. One possible way of accounting heuristically for such a result is to raise the k = 5 [tanh( )$^{1/2}$]/( )$^{1/2}$ quotient to the (1 - α_5) power, with -1 $\leq \alpha_5 < 1$. This modification is quite similar in effect to the skewed-arc Davidson-Cole heuristic relaxation distribution modification.[15] It is likely to be more appropriate here, however, since the approximate k = 5 term in Eq. (1) follows directly from the detailed model when $\alpha_5 = 0$.

Finally, the circled "e's" and "i's" in Fig. 3-b stand for extensive and intensive. Here "intensive" means that the R and C parameters of the arc are independent of ℓ. Arcs or portions of arcs which are extensive will, in contrast, change with change in ℓ, helping somewhat to identify

an experimental arc as being associated with a specific
process in the loosely coupled case.

CHARACTERIZATION RESULTS AND POSSIBILITIES

There are many different ways to plot frequency response results. Some of them have been discussed elsewhere.[2] It has been found that the present model (with
$\alpha_k = 0$ for all k) can lead to frequency dependence of C_p,
the overall parallel capacitance of the system, of the form
ω^{-m} over an appreciable range with m = 0.5, 1, 1.5, and 2
and even some intermediate values. For characterization,
however, only a complex-plane $Z_T^*(\omega)$ plot is needed. Then,
the data may be analyzed with as many terms of Eq. (1) as
required, preferably by nonlinear least squares fitting (of,
e.g., $Im[Z_T(\omega)]$), to determine the pertinent R and C parameters. Equation (1) may be transformed to unnormalized
form by omitting N's, multiplying the last term by R_∞, and
changing Ω to ω. Once $R_1 \cong R_\infty$ and $C_1 \cong C_g$ values have been
found, the remaining parameter estimates may be normalized
with them.

The remaining figures illustrate various response possibilities for the intrinsic-conduction, loose-coupled situation with G/R effects negligible. Figure 4-a shows Z_T^*,
equivalent circuit, and $Y_T \equiv Z_T^{-1}$ results in the simplest
situation, that of bulk response only: $(r_p, r_n) = (\infty, \infty)$,
implying ohmic electrodes. Here, $C_1 = C_B \cong C_g$, $R_1 = R_B =$
R_∞, $\omega_B = \tau_D^{-1} = (R_\infty C_g)^{-1}$, and $G_\infty \equiv R_\infty^{-1} = (ez_e^2/\ell)(\mu_n + \mu_p)$.
Values of R_∞ (= R_D here) and C_g can only yield, in this
ordinary conduction case, estimates of ε and $(\mu_n + \mu_p)c_i$.

Somewhat more can be learned from the completely
blocking (0,0) case of Fig. 4-b. Here ω_B is as above but
ω' involves a mixture of bulk and interface parameters.
Define the Debye length for the present $z_n = z_p \equiv z_e$ case
as $L_D = [\varepsilon kT/4\pi(ez_e)^2(n_e + p_e)]^{1/2}$ in the extrinsic-
intrinsic situation. The effective Debye length, L_e, is
in general a frequency-dependent function of χ, recombination,[7,8] and mobility parameters. For most practical
purposes, one may use $L_e = L_D$, however. In the intrinsic
case one should use $L_e \cong \sqrt{2}L_D$ for slow or zero $(c_i = N_i)$ recombination under the following conditions: (a) $\mu_p = 0$, all

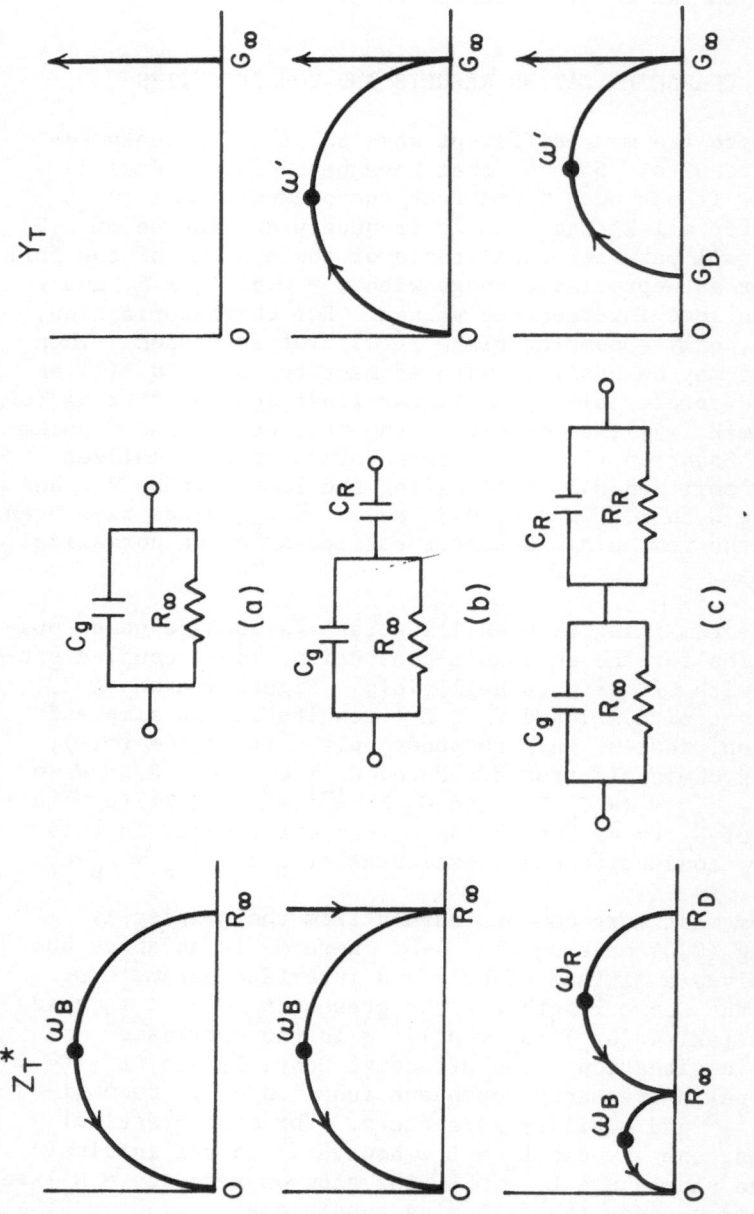

Figure 4. Response and equivalent circuits for (a) $(r_p, r_n) = (\infty, \infty)$; (b) $(0,0)$; and (c) $(0, r_n)$, $\pi_m \gg 1$, and $r_n \gtrsim 1$.

Ω, and (b) $\mu_p \ll \mu_n$, $\Omega \overset{\sim}{>} \pi_m^{-1}$. In this latter case, the frequency is too high for positive charges to move appreciably in a half cycle and they thus act as if they were immobile. Let $M \equiv \ell/2L_D$ and $M_e \equiv \ell/2L_e$, always taken $\gg 1$ here. Then it is found[2,4] in the present case that $C_R \overset{\sim}{=} M_e C_g = \varepsilon/8\pi L_e$, just the double-layer capacitance of two layers (one at each electrode) in series. Here, where a value of C_R is available, ε, c_i, and $(\mu_n + \mu_p)$ can generally be found.

For the $(0, r_n)$ situation of Fig. 4-c, the material can be even better characterized. Here $C_{RN} \overset{\sim}{=} (M_e + r_n)$, $R_{RN} \overset{\sim}{=} 2/r_n$, and $\omega_R = \tau_R^{-1}$. One obtains the 4-b results when $r_n = 0$, of course. Here, where $\mu_n \gg \mu_p$, analysis yields estimates of ε, c_i, μ_n, and $k_n \equiv k_{no} \equiv {}^P k_{n\infty}$. This is the situation where an electrode reaction is the limiting chemical step. Note that $R_{DN} \equiv (1 + \pi_m^{-1})[1 + (2/r_n)] \overset{\sim}{=} 1 + R_{RN}$ in the present case.

Figure 5-a shows the complementary situation, that where diffusion is the limiting step. Note that here μ_n, the mobility of the discharging carrier, may be much smaller than that of the blocked carrier. The low-frequency arc of the Z_T^* diagram arises from the $k = 5$ term of Eq. (1). Its initial (high-frequency) slope is unity when $\alpha_5 = 0$, the only case considered here. The height of this arc is about $0.42 \pi_m^{-1} R_\infty$ and its width about $\pi_m^{-1} R_\infty$. Here $R_{DN} \overset{\sim}{=} 1 + \pi_m^{-1}$. Thus for $\pi_m \ll 1$, $\pi_m^{-1} R_\infty$, which is proportional to $[\mu_n(1 + \pi_m)]^{-1}$, involves essentially only μ_n, the smaller mobility. In the present case, τ_{5N} may be approximated by $bM^2[(2N_i - c_i)/c_i] \equiv [(2 + \pi_m + \pi_m^{-1})/4][(2N_i - c_i)/c_i]M^2$, which takes incomplete intrinsic dissociation ($c_i \overset{\sim}{<} N_i$) into account. Thus, $\tau_5 \equiv \tau_{5N}\tau_D \overset{\sim}{=} (b\ell^2/2)(D_n + D_p)^{-1} \cdot [(2N_i - c_i)/c_i]$. Clearly, τ_5 can become very large and $\omega_W \simeq 2.5\tau_5^{-1}$ very small when $c_i \ll N_i$. The present $k = 5$ term needs some modification[12] when $\pi_m \overset{\sim}{<} M^{-1}$. When $\pi_m \sim 1$, analysis of data of the present type can yield estimates of ε, c_i, μ_n, and μ_p if N_i is known independently or ε, N_i, μ_n, and μ_p if M (and thus c_i) is obtained independently. The reaction rate constant, k_n, will be too large here ($r_n \gg 1$) to be determinable from the data (reaction arc too small to measure).

The results of Fig. 5-b can appear when adsorption of a charged species on the electrode occurs.[5,6] The boundary parameter r_n is then complex, r_n^*, and may involve r_{no}, $r_{n\infty}$,

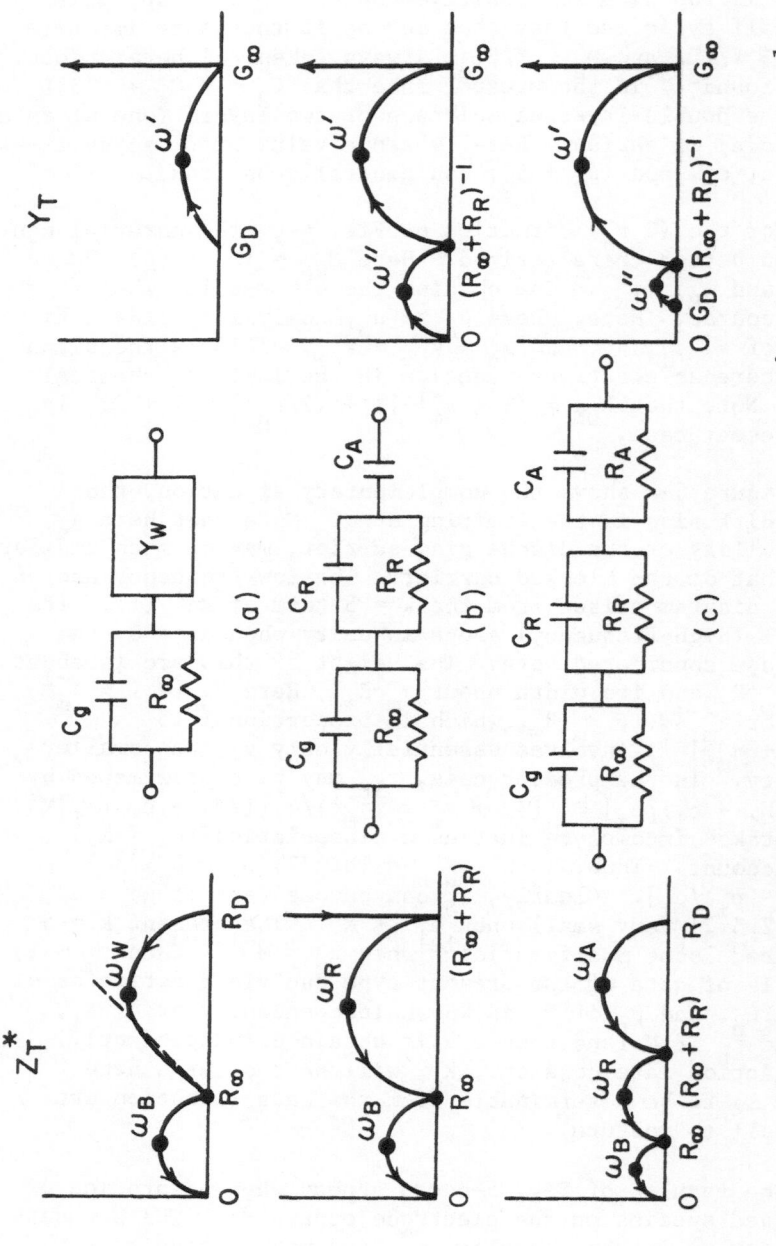

Fig. 5 Response and equivalent circuits for (a) $(0, r_n)$, $\pi_m \overset{\sim}{<} 1$, and $r_n \gg 1 + \pi_m^{-1}$; (b) $(0, r_n^*)$, $\pi_m \gg 1$, $r_{n\infty} \overset{\sim}{>} 1$, $r_{no} = 0$; and (c) like (b) but $r_{no} > 0$.

and $\xi_{An} \equiv \tau_{AN}/\tau_D$. In the loosely coupled case, $\xi_{An} \overset{\sim}{>}$ 100 M_e, and for $\pi_m \gg 1$, $C_{AN} \overset{\sim}{=} \xi_{An} r_{n\infty}/2$. Then, data analysis can yield estimates of ϵ, c_i, μ_n, $k_{n\infty}$, and τ_{AN}. The latter two quantities may themselves be expressed in terms of more basic properties of the system, as discussed elsewhere.[5] Note that when $R_R \gg R_\infty$ and frequency response measurements don't extend to high enough frequencies to delineate the bulk semicircle, the present adsorption-reaction results can be easily confused with the bulk-double layer response of Fig. 4-b. The two similar responses can be readily distinguished, however, on the basis of their extensive or intensive dependence. Further, note that 5-b shows that the electrode reaction semicircle may appear even in a kind of completely blocking situation, one where $R_D = \infty$. Unless frequency response measurements are extended to sufficiently low frequencies to show the beginning of the C_A capacitative rise, the presence of adsorption may not even be recognized.

The conditions appropriate for the results shown in Fig. 5-c are similar to those discussed above except that $r_{no} \neq 0$ and $r_{n\infty} > r_{no} > 0$. In this case, loosely coupled results are $C_{RN} \overset{\sim}{=} (M_e + r_{n\infty})$, $R_{RN} \overset{\sim}{=} (2/r_{n\infty})$, $C_{AN} \overset{\sim}{=} \xi_{An} r_{n\infty}^2/2r_{nm}$, and $R_{AN} \overset{\sim}{=} 2r_{nm}/r_{n\infty}r_{no}$, where $r_{nm} \equiv r_{n\infty} \overset{\sim}{=} r_{no}$. Here, $R_{DN} \equiv (1 + {}^{AN-1}\pi_m)[1 + (2/r_{no})] \equiv 1 + R_{RN} + R_{AN} \overset{\sim}{=} 1 + (2/r_{no})$. The radial frequency at the peak of the A/R arc, ω_A, becomes, in the present case, $\omega_A = (\tau_D R_{AN} C_{AN})^{-1} \overset{\sim}{=} \tau_{AN}^{-1}(r_{no}/r_{n\infty})$. It turns out for the situation of Fig. 5-c that as coupling increases and ξ_{AN} decreases below $\sim 100M_e$, $\omega_3 \equiv \tau_3^{-1}$ always remains larger than ω_2. In fact, C_{3N} decreases from the above $C_{AN} \gg C_{RN}$ value and approaches the smaller uncoupled C_{RN} value of M_e, while C_{2N} simultaneously increases above this value. At the same time, R_{2N} decreases toward zero and R_{3N} increases toward $R_{DN} - 1 \overset{\sim}{=} (2/r_{no})$. Thus, for sufficiently small ξ_{AN}, $Z_{2N} \to 0$, and Z_{3N} then plays the reaction role, involving as before the double-layer capacitance and a reaction resistance of $(2/r_{no})R_\infty$ instead of the uncoupled R_R value of $(2/r_{n\infty})R_\infty$. The original reaction arc disappears and its place is filled by what was, in the loosely coupled region, the adsorption-reaction arc. Therefore, if only a single reaction arc is present, it will usually be impossible to establish which of $r_{n\infty}$ or r_{no} is involved, although they are of course identical in the absence of adsorption.

There are numerous further curve-shapes which can arise
from $Z_3 \cong Z_A$ in the loosely-coupled situation. Some of them
appear in Fig. 6-a. Note that when $r_{no} > r_{n\infty}$, $r_{nm} < 0$ and
both C_{AN} and R_{AN} are negative, leading to the negative-
capacitance arcs shown below the real axis. The largest
such semicircle, appearing for $r_{no} \to \infty$, is the mirror image
of the reaction semicircle. Further when $-2 < r_{no} < 0$,
$R_{DN} < 0$, and thus the differential resistance of the system
is negative in the low frequency limit. The result of
Fig. 6-b shows that a diffusion arc may appear at even lower
frequencies than the reaction and A/R arcs. Experimental
results of just this form have been found by Gabrielli.[16]

Finally, Fig. 7 shows some further possible combinations of R, A/R, and D arcs. Typical parameter values
yielding results similar to those of Fig. 7-a and 7-b are
$r_{n\infty} = 10^9$, $r_{no} = 0$ for 7-a and 2 for 7-b, $\xi_{An} = 2$, $\pi_m = 10^{-1}$, and $M = M_e = 10^3$. Values leading to shapes similar
to 7-c and 7-d are $r_{n\infty} = 6$, $r_{no} = 0$ for 7-c and 2 for 7-d,
$\xi_{An} = 4 \times 10^8$, $\pi_m = 10^{-1}$, and $M = M_e = 10^3$. Notice that
the large decrease in $r_{n\infty}$ has been compensated by a large
increase in ξ_{An} so that C_{AN} is roughly the same in the a, b
and c, d cases. Many other arc combinations are possible.
For example, any of the A/R arc shapes of Fig. 6-a could
occur at the right of the 7-c and 7-d diagrams, not just
those shown.

In conclusion, it is worthwhile to summarize some general expressions which should be useful for data analysis in
arbitrary $(0,r_n^*)$, π_m, and conduction-type situations. Let[1]
$\pi_f \equiv n^*/p^* \cong (\phi + \chi)/(\phi - \chi)$, where $\phi \equiv (1 + \chi^2)^{1/2}$. Here
$\chi >> 1$ for strong n-type doping; $\chi = 0$ for intrinsic condi-
tions; and $\chi << -1$ for strong p-type doping. Further, de-
fine $\delta_n \equiv \pi_f/(1 + \pi_f)$ and $\varepsilon_n \equiv \pi_e/(1 + \pi_e)$, where $\pi_e \equiv \pi_m \pi_f$.
Then $R_{DN} = \varepsilon_n^{-1}[1 + (2/r_{no})]$. For loose coupling one finds:
$R_{1N} \cong R_{BN} \cong 1$, $C_{1N} \cong C_{BN} \cong 1$; $R_{2N} \cong R_{RN} \cong 2/\varepsilon_n r_{n\infty}$, $C_{2N} \cong$
$C_{RN} \cong M_e + \varepsilon_n r_{n\infty}$; and $R_{3N} \cong R_{AN} \cong 2r_{nm}/\varepsilon_n r_{n\infty} r_{no}$, $C_{3N} \cong$
$C_{AN} \cong \xi_{An}(\varepsilon_n r_{n\infty})^2/2\varepsilon_n r_{nm}$. As already mentioned, Z_{4N} will
often be negligible compared with the other contributions
to Z_{TN}. The π_m appearing in the fifth term of Eq. (1)
should be replaced by the more general π_e. The approximate
expression for τ_{5N} already given above may be used (with the
factor b given more generally[1]) in the Z_{5N} part of Eq. (1)

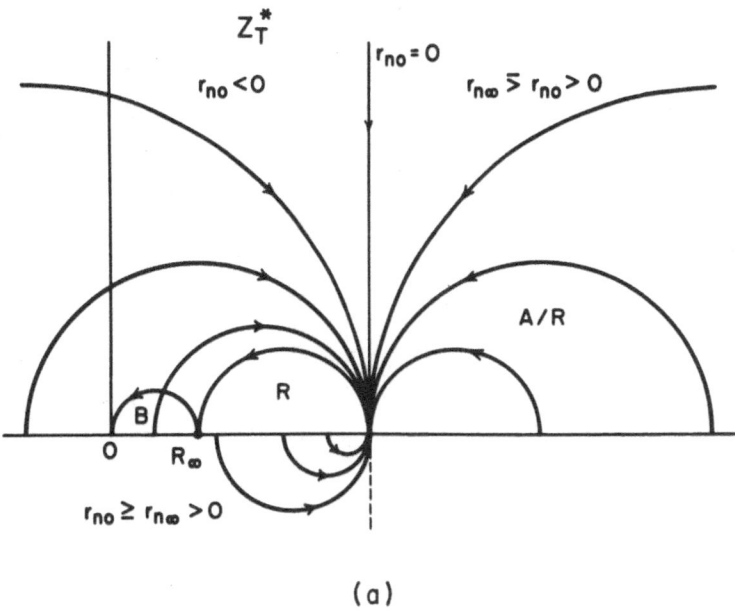

(a)

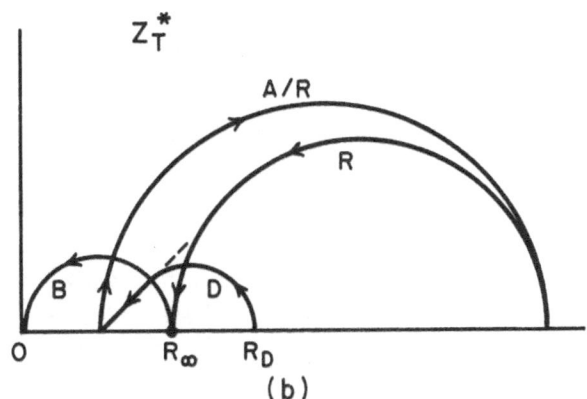

(b)

Fig. 6   Further $(0, r_n^*)$ response for (a) $\pi_m \gg 1$, $r_{n\infty} \lesssim 1$, and various values of r_{no}. (b) As in (a) but $\pi_m \lesssim 1$ and $-\infty < r_{no} < -2$.

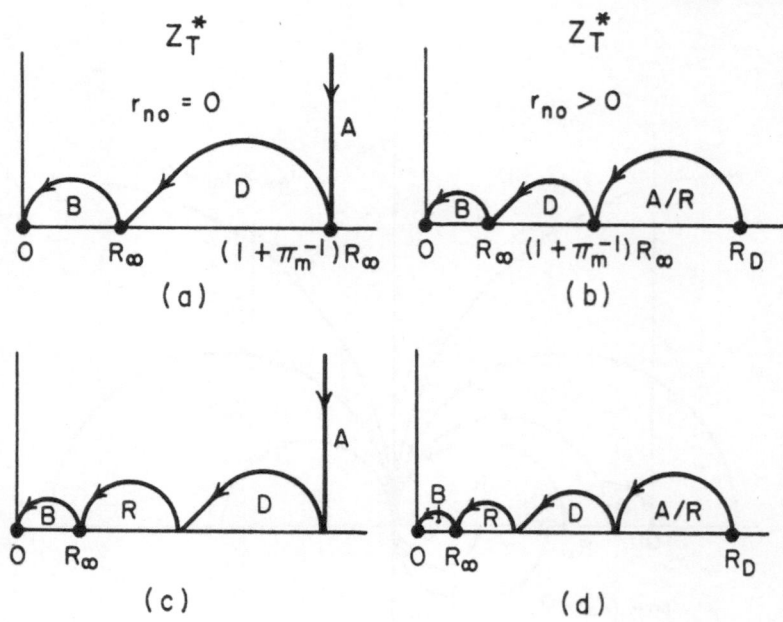

Fig. 7 Additional $(0,r_n^*)$ response possibilities.

or its $\alpha_5 \neq 0$ generalization. The quantity M_e usually may be taken as M multiplied by a factor dependent on π_m, recombination, and δ_n. This factor reduces to $\delta_n^{1/2}$ in the $\pi_m \gg 1$, $\approx$ full dissociation situation.

REFERENCES

1. J. R. Macdonald, J. Chem. Phys. <u>58</u>, 4982 (1973).
2. J. R. Macdonald in <u>Electrode Processes in Solid State Ionics</u>, edited by M. Kleitz and J. Dupuy (D. Reidel Publishing Co., Dordrecht-Holland, 1976).
3. J. R. Macdonald, J. Electroanal. Chem. <u>53</u>, 1 (1974).
4. J. R. Macdonald, J. Chem. Phys. <u>61</u>, 3977 (1974).
5. J. R. Macdonald, J. Electroanal. Chem. <u>70</u>, 17 (1976).
6. J. R. Macdonald and P.W.M. Jacobs, J. Phys. Chem. Solids, to be published.
7. J. R. Macdonald, J. Chem. Phys. <u>29</u>, 1346 (1958).
8. J. R. Macdonald, J. Phys. C: Solid State Phys. <u>7</u>, L327 (1974); <u>8</u>, L63 (1975).

9. A. R. von Hippel, Dielectrics and Waves, (John Wiley, New York, 1954), pp. 228-234.

10. J. Volger, Prog. in Semiconductors 4, 207 (1960).

11. R. B. Hilborn, Jr., J. Appl. Phys. 36, 1553 (1965).

12. G. S. Nadkarni and J. G. Simmons, J. Appl. Phys. 47, 114 (1976).

13. K. S. Cole and R. H. Cole, J. Chem. Phys. 9, 341 (1941).

14. J. R. Sandifer and R. P. Buck, J. Electroanal. Chem. 56, 385 (1974).

15. D. W. Davidson and R. H. Cole, J. Chem. Phys. 23, 493 (1955).

16. G. Gabrielli, Ph.D. thesis, University of Paris, 1973; Métaux, Corrosion, Industrie Nos. 573, 574, 577, 578 (1973).

Use of Pulse Methods in the Study of Solid Electrolytes. P. H. BOTTELBERGHS, National Energy Research Steer-
ing Committee.--The complex impedance behavior of solid
cell systems can lead to insight into transport phenomena
and other fundamental properties of solid electrolytes.
This complex impedance behavior can be studied mono-
chromatically, (i.e., by ac methods), but also polychro-
matically, (i.e., by use of pulse methods). Pulse methods
have some advantages over ac methods: (1) They are much
faster. (2) They can be used to distinguish between anodic
and cathodic processes. (3) They can be used under dynamic
conditions, (i.e., during electrolysis). Though in
principle they are less accurate than ac methods, this disadvantage often vanishes when one measures on solid cell
systems at elevated temperatures, because of the temperature
inhomogeneity (over time). The main disadvantage may lie
in the difficulties in interpretation, but in many cases
this problem can be dealt with, as will be shown. In the
interpretations, it is often convenient to make use of
Laplace transforms: $\mathcal{L}\{f(t)\} = {}_0\!\int^\infty f(t)\, e^{-st}\, dt$, with s =
jω. Another convenient tool is the description in terms
of a so-called Constant Phase Angle impedance (CPA): Z_p =
$K_p(j\omega)^{-p}$, in which p is independent of ω and has a value
between 0 and 1.

A Study of the Electrical Properties of Single
Crystal and Polycrystalline β-Alumina using Complex Plane
Methods. A. HOOPER, P. McGEEHIN AND A. E. HUGHES, Materials
Development Division, AERE, Harwell.--The now well
established technique of complex plane representation is
used to analyze frequency dependent a c conductivity data
from samples of single crystal and polycrystalline β-
Alumina. Measurements were made from room temperature to
around 300°C in vacuum. Wayne Kerr Bridges B. 221 and
B. 601 with external source (Krohn-Hite 4200 Oscillator)
and detector (Marconi TF1100 Valve Voltmeter) gave a
frequency range of 500 HZ to 2 MHZ. Electrical contact was
made via nickel or platinum foils in pressure contact with
evaporated metal electrodes. Prior to measurement every
specimen was annealed to 300°C in vacuum, within the
conductivity rig, to eliminate moisture from the system.
In single crystal β-Alumina, the as-received mate-
rial was part of a boule (~ 3 cm^2 cross section) grown by the

Union Carbide Co. X-ray analysis showed it to be single phase (β) material. The crystal was oriented with the mirror planes parallel to the long growth axis and clearly definable layers were visible perpendicular to the c-axis. Subsequent cutting with a diamond saw produced optically transparent specimens which contained interference fringes and became mechanically unstable; clearing easily along the mirror planes. Micron size cracks were observed by microscopy. This deterioration is believed to be due to internal crystal strains. Only mechanically sound samples were used for conductivity measurements and were typically 0.2 cm x 1·cm x 1 cm. The electrodes (Ag, Au, Al) were applied as evaporated layers to either as cut or 1 - 6 μm polished surfaces. Impedance, admittance and permittivity plots confirm the simple series R-C network expected for this situation of a single crystal with blocking electrodes. Conductivity values obtained from all types of electrode are similar; $\sigma(25^oC) = 0.035 \ \Omega^{-1}cm^{-1}$, $\sigma(300^oC) =$ 0.213 $\Omega^{-1}cm^{-1}$. The σT V 1/T plot appears to be slightly curved giving an estimated average activation energy of 0.13 ev.

In polycrystalline β-Alumina, specimens were cut in the form of discs (0.5 cm x 0.5 cm^2) from isostatically pressed and sintered ($\sim$1800^oC) rods of two phase (β & β'') material. Electrode surfaces were polished to 1 - 6 μm finish and evaporated layers of Ag or Al applied. A σT V 1/T plot leads to an activation energy (E_A) of 0.149 ev although once again the plot is slightly curved. $\sigma(25^oC) =$ 0.0078 $\Omega^{-1}cm^{-1}$. Below 200^oC a similar plot value gives $E_A = 0.268$ ev with $\sigma(25^oC) \sim 0.0012 \ \Omega^{-1}cm^{-1}$. Above 200$^oC$ the two plots merge with $\sigma(300^oC) = 0.065 \ \Omega^{-1}cm^{-1}$.

Theory

PHASE TRANSITIONS AND TRANSLATIONAL FREEDOM IN

SOLID ELECTROLYTES

M. O'Keeffe

Department of Chemistry
Arizona State University
Tempe, Arizona 85281

MELTING AND SOLID ELECTROLYTE TRANSITIONS

In 1834 Faraday (1) recorded an interesting phenomenon. In his words: "I formerly described a substance, sulphuret of silver, whose conducting power was increased by heat; and I have since then met with another as strongly affected in the same way; this is fluoride of lead. When a piece of that substance, which had been fused and cooled, was introduced into the circuit of a voltaic battery, it stopped the current. Being heated, it acquired conducting powers before it was visibly red hot in daylight; and even sparks could be taken against it whilst still solid."

This appears to be the first observations of the transition from the poorly conducting to the conducting states in ionically conducting materials that we now call solid electrolytes.

The picture that has emerged in recent years is that the solid electrolyte state is one of ionic crystals in which one subset of ions has a substantial measure of translational freedom and to this extent is fluidlike while the remainder of the ions execute motion (rotation and/or vibration) about a fixed center as in solids. It is this combination of solid-like and liquid-like properties that provide much both of the fascination and of the utility of solid electrolytes. This paper is concerned with firstly the evidence that there are indeed liquid-like degrees of freedom and secondly with the nature of the transitions from "solid-like" to "liquid-like" states. Finally the transition in crystals like lead fluoride

is discussed. Here the transition (which might appropriately
be termed the Faraday transition after its discoverer) from
solid to liquid-like behavior appears in every respect to be
continuous and spread out over a temperature range of typically a hundred degrees or more. The discussion is limited
mainly to binary salts which are simpler and better studied
than most ternary etc. compounds.

Elsewhere (2) attention has been drawn to certain regu-
larities in the thermodynamic properties and d.c. conductivity
changes associated with the non-conducting to conducting transition in simple salts. Thus for what might be called normal
salts (typified by the alkali halides) the entropy change on
fusion is remarkably constant at about 10-12 JK^{-1}/gatom of
ions. The conductivity change is remarkably constant also.
The ionic conductivity just below the melting temperature is
typically ~$10^{-1}\Omega^{-1}m^{-1}$ and that of the melt ~$10^{2}\Omega^{-1}m^{-1}$. This
last observation is a restatement of the well-known fact (2)
that for most liquids the diffusion coefficient D is typically
~$10^{-9}m^{2}s^{-1}$.

Many solid electrolytes, notably Ag^{+} and Cu^{+} salts (3,4)
but also some halides (examples are given in Table 1) are
now known to exhibit a first order phase transition (which
we termed a class II transition) that exhibits features similar to the melting transition. In particular (2) the entropy
change at the solid-solid phase transition is comparable to
the entropy of fusion, and the ionic conductivities both just
below and just above the transition temperature are in the
same range as for normal salts just below and above the melting temperature. It is convenient, and in some ways accurate,
to consider this transition (the solid electrolyte transition)
as a first melting in which one subset of ions acquires
liquid-like properties. Subsequently there is a second
melting at which point the remaining ions disorder.

Recently it has become apparent that there is a number
of salts in a third category. For these salts, the first
"melting" is not a first order transition but is spread out
over a substantial temperature range. This the the Faraday
transition which we (2) earlier called class III transitions.
Through the temperature range of the transition the conductivity passes smoothly from values typical of normal salts
to values typical of ionic melts. The heat content H is
likewise a continuous function of temperature but in the

TABLE 1

Examples of materials exhibiting the various types of transition discussed in the text. For references not given see (2).

Class I   -normal melting

alkali halides, $PbCl_2$, $MgCl_2$, $CaBr_2$, YCl_3 etc.

Class II   -first order solid electrolyte transition

cation conductors:
Ag^+, Cu^+ salts e.g. CuBr, CuI, AgI, Ag_2S etc. (3,4)
Li_2SO_4, Li_2WO_4 (5)

anion conductors:
LuF_3, YF_3, $BaCl_2$, $SrBr_2$, Bi_2O_3

Class III   -Faraday transition (diffuse)

cation conductors:
Na_2S (6), Li_4SiO_4 (7)

anion conductors:
CaF_2, SrF_2, $SrCl_2$ (8), PbF_2 (9), LaF_3, CeF_3 (10)

transition region the heat capacity $C_p = (\partial H/\partial T)_p$ increases many-fold although there are no data as yet to indicate any singularity in the temperature dependence of C_p.

The Faraday transition is best documented for crystals with the fluorite structure. Indeed all ionic crystals with this structure for which relevant conductivity or thermody-namic data exist, show signs of the transition. Judging from conductivity data it appears that other crystals, notably those with the tysonite (LaF_3) structure, behave similarly: examples are included in Table 1.

There are of course many ternary etc. solid electrolytes notably the β-aluminas, pyrochlores etc. They differ from binary salts and compounds like Li_2SO_4 etc. in that there is a relatively rigid framework (11) of bonded anions and cations in which the mobile ions move. A simple example is provided by the pyrochlore $AgSbO_3$ which has a rigid framework of SbO_6 octahedra connected by sharing corners so that the stoichio-metry is SbO_3^-. The open framework leaves tunnels which allow

easy passage of Ag^+ ions and indeed $AgSbO_3$ has a conductivity
(12) at room temperature comparable to that of e.g. β-alumina.
Other examples are provided by Goodenough et al. (11). By
way of contrast networks of corner-connected tetrahedra (such
as occur in the silicates) are not rigid but can collapse
by relative rotations of the tetrahedra about their centers
(13) and reduce the size of the "tunnels." A simple example
is provided by the sodalites which have disappointingly small
ionic conductivities (14).

In passing it might be noted that the solid electrolyte
transition can be induced by pressure in materials for which
a solid electrolyte phase is not known at zero pressure.
Thus the pressure-induced "rutile" → "fluorite" transition
in e.g. MnF_2 and ZnF_2 (17) is probably of this type. Certainly the slope dP/dT of the phase boundary indicates a very
large entropy change in the phase transition. It seems likely
too that the so called "metallic" ice formed at pressures
above ~1 Mbar (18) is a solid electrolyte rather than a metal.

VELOCITY AUTOCORRELATION AND
FREQUENCY-DEPENDENT CONDUCTIVITY

Most discussions of these phase transitions to date have
been in terms of a disordering or redistribution of atoms on
crystallographic sites (e.g. 19). Although conceptually
helpful, this approach ignores the essence of the solid electrolyte phase which is one in which there is a substantial
measure of translational freedom (as most obviously manifest
in the ionic conductivity).

An elementary discussion of the distribution of degrees
of freedom among translational (or better diffusive) modes and
lattice vibrations is conveniently carried out in terms of
the velocity autocorrelation function (20): $\tilde{Z}(\tau) \equiv <v(t)v
(t + \tau)>$ and its Fourier transform, or spectral density,

$$\tilde{Z}(\omega) = (1/2\pi) \int_{-\infty}^{\infty} Z(\tau) \exp(-i\omega\tau)d\tau. \qquad (1)$$

This is simple related to the diffusion coefficient by

$$\tilde{Z}(0) = (1/\pi) \int_{0}^{\infty} Z(\tau)d\tau = D/\pi. \qquad (2)$$

Note that the equipartition of energy requires that

$$\int_{-\infty}^{\infty} \tilde{Z}(\omega)\, d\omega = Z(0) = kT/m. \tag{3}$$

To the extent that one can neglect correlation effects in an ionic conductor, $\tilde{Z}(\omega)$ is proportional to the spectral density of the current autocorrelation function and hence to the ionic conductivity $\sigma(\omega)$ (cf. 21,22). It is instructive for this reason to compare in a qualitative way how $\tilde{Z}(\omega)$ differs for different states of matter.

For a solid with just one zone center optical mode of frequency Ω_0, $\tilde{Z}(\omega)$ is just $(kT/m)\delta(\Omega_0)$. Or with damping one has the well-known expression:

$$\tilde{Z}(\omega) = (kT/m)\Gamma_0^2\omega^2/[(\Omega_0^2 - \omega^2)^2 + \Gamma_0^2\omega^2] \tag{4}$$

where Γ_0 is the damping constant.

The other extreme is provided by a gas of particles undergoing Brownian motion ("free" diffusion). Again the result is well-known (20)

$$\tilde{Z}(\omega) = (kT/m)\tau(1 + \omega^2\tau^2)^{-1} \tag{5}$$

here $1/\tau$ is the friction constant related to the diffusion coefficient by the classical expression (cf eqs (2) and (3))

$$\tau = mD/kT. \tag{6}$$

Note that $\tilde{Z}(\omega)$ is a Lorentzian curve with half-width $\Delta\omega = \tau^{-1} = kT/mD$.

Many attempts have been made (23) to approximate a liquid (or solid electrolyte) with both oscillatory and diffuse degrees of freedom by incorporating the equation of motion for a damped oscillator (or distribution of oscillators) with that for diffusion by Brownian motion. These include using a memory function formalism which allows the high frequency oscillator restoring force to transform at low frequencies to a friction force. Bruesch, Strässler and Zeller (22) show how with a particular form of memory function one regains the form of eq (4) with Γ_0, Ω_0 replaced by $\Gamma(\omega)$, $\Omega(\omega)$ such that $\Gamma(0) = 1/\tau$ and $\Omega(0) = 0$. Thus one transforms from a response that is oscillatory at high frequencies to one that is diffusive at low frequencies as sketched in fig (1). This approach appears to provide a fair

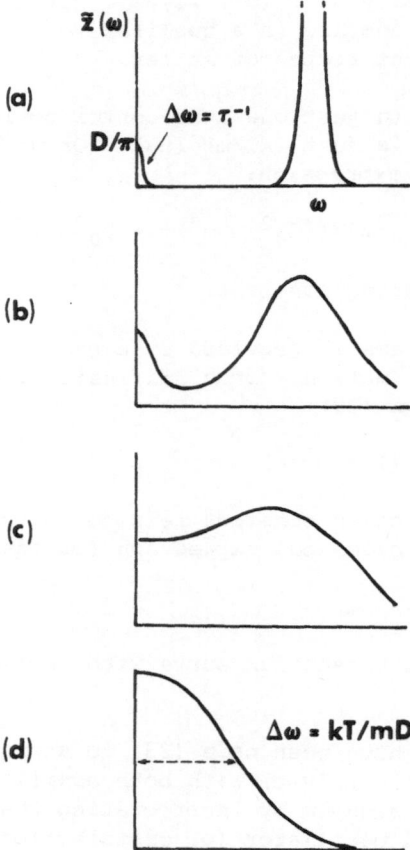

Figure 1   Sketch of the evolution of the spectral density,
           $\tilde{Z}(\omega)$ from (a) an idealised solid with jump diffu-
           sion through (b) and (c) liquids to (d) Brownian
           motion ("free" diffusion) in a dense gas.

description of $\sigma(\omega)$ for a "good"* ionic conductor such as AgI in the high frequency region ($>10^{12}$Hz) but it should be mentioned that the conductivity of AgI shows considerable structure in the region of 10^{10}-10^{12}Hz (24).

A note of caution is necessary at this point. Egelstaff (25) has pointed out that eq (6) holds only if all degrees of freedom contribute to diffusion, in practice only a fraction of $\Delta N/N$ will so contribute and τ (which is the time constant of the diffusive response of the system) will instead by replaced by:

$$\tau_1 = (mD/kT)/(\Delta N/N). \tag{7}$$

τ_1 as given by eq (7) is closely related to the mean time between "collisions" (changes of direction of flight) i.e.

$$\tau_1 \approx <\ell>/(kT/m)^{\frac{1}{2}} \tag{8}$$

where $<\ell>$ is a mean free path and $(kT/m)^{\frac{1}{2}}$ is the average velocity. Egelstaff (25) argues that $\Delta N/N$ should not be a very strong function of temperature in liquids just above their melting temperatures.

Alternatively one might try to approach a liquid starting from a solid. This has not often been attempted owing to the discontinuity between a solid and a liquid (contrast the continuity between a liquid and a gas) but as discussed below might be appropriate to the Faraday transition in e.g. PbF$_2$.

If one allows a small number of degrees of freedom in a solid to spill over into "jump" diffusion there will be a small contribution to $\tilde{Z}(\omega)$ at $\omega \sim 0$ which will again be of the form (compare eqs (5), (6))

$$\tilde{Z}(\omega) = (D/\pi)(1 + \omega^2\tau_1^2)^{-1}. \tag{9}$$

Where, as before, τ_1 is the time in transit, but now D is given by

$$D = \ell^2/2(\tau_0 + \tau_1) \tag{10}$$

---

*Bruesch, Strässler and Zeller suggest a measure of the quality of a solid electrolyte might be the ratio of the number of degrees of freedom in diffusive modes to that in oscillatory modes.

where τ_0 is the time between jumps (along the axis of v)*
and ℓ is the jump length. $\tilde{Z}(\omega)$ is again a Lorentzian of
half-width comparable to that for free diffusion.

The fraction of the total degrees of freedom involved
in jump diffusion may be estimated as the ratio of the integral of $\tilde{Z}(\omega)$ given by (9) to kT/m i.e.

$$\Delta N/N = Dm/2\tau_1 kT = \tau_1/(\tau_0 + \tau_1) \tag{11}$$

where the second equation is obtained using eqs (8) and (10).
Thus for jump diffusion the result is the intuitively obvious
one that the fraction of the degrees of freedom involved in
diffusion is the fraction of the total time spent in free
flight. In normal solids $\Delta N/N < \sim 10^{-5}$, but in PbF_2 at high
temperatures, to the extent that the jump diffusion is
appropriate, it may readily be shown (9) that this ratio
must be of the order of unity. Funke and his collaborators
(e.g. 24,26 and this volume) have indeed successfully anal-
ysed the low frequency part of $\sigma(\omega)$ for some Cu and Ag salts
using a jump diffusion model involving approximately equal
transit times and residence times.

We discuss below some of the evidence for considering
that in some materials there is a continuous transition from
jump diffusion at low temperatures to free diffusion at high
temperatures. How properly to treat the intermediate region
is not at all clear at present (but see (26)).

THE FARADAY TRANSITION

Class II solid electrolyte transitions are always accompanied by a change in structure of the immobile ions (although
in one or two instances such as Ag_2S or Cu_2S this is relatively small) suggesting that there is a strong coupling between the mobile ions and the "fixed" framework. In contrast,
in the Faraday transition (class III transitions) there is no
discernible discontinuity in the structure of the framework
which is a face-centered cubic array in the fluorite structure.
PbF_2 and $SrCl_2$ have been most studied in this respect and
data relative to the transition in these compounds are reviewed here.

---

*v is the component of the velocity in one direction
(cf eq (3)).

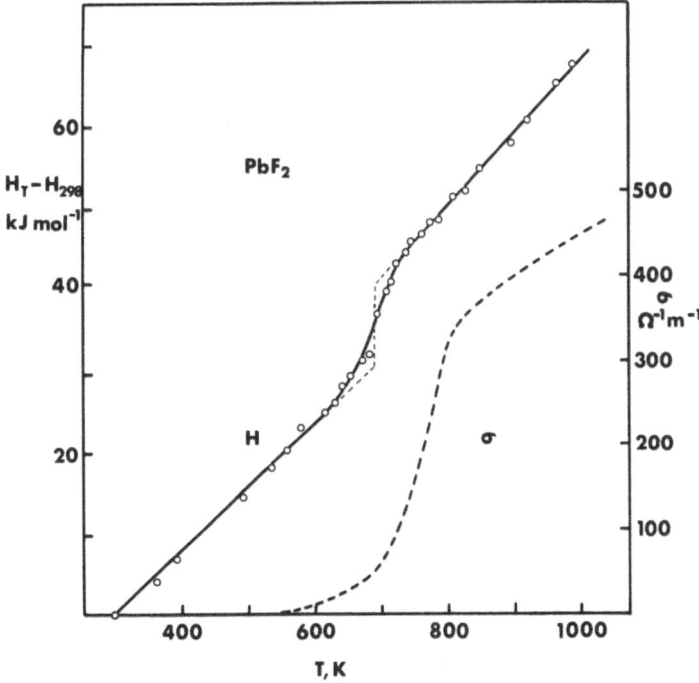

Figure 2   The heat content and ionic conductivity of PbF$_2$ (27)

Figure 2 shows the heat content and conductivity of
PbF$_2$ (9,27). The data are typical and compare closely to
those for CaF$_2$ and SrCl$_2$ (2), although it should be re-
marked that PbF$_2$ has a much lower transition temperature
(T$_t$ $\approx$ 0.6 T$_m$) compared with other crystals with the fluorite
structure (for which typically T$_t$ $\approx$ 0.8 T$_m$). Notable is the
rapid rise in heat content over the range 650-750 K. In
this same temperature region the conductivity changes from
a value typical of normal ionic crystals ($<10^{-1}\Omega^{-1}m^{-1}$) to
that typical of ionic melts ($\sim$100 $\Omega^{-1}m^{-1}$).

From the thermodynamic data one can calculate C$_p$ directly and using standard thermodynamic relationships estimate
the heat capacity at constant volume, C$_v$ (28); see fig (3).
Although there is some measure of uncertainty in the last
quantity it falls significantly below the high temperature
limit of 9R to be expected for a normal solid with 9N$_0$ (N$_0$
is Avagadro's number) vibrational modes. Indeed, if one

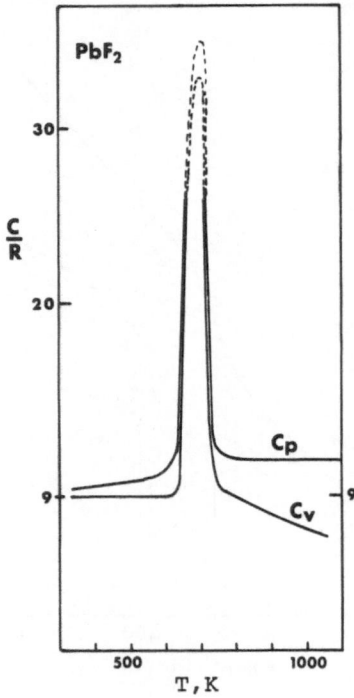

Figure 3 C_p and C_v for PbF_2 as a function of temperature
(28).   Note that C_v falls below 9R just below the
melting temperature.   The maximum values of C_p
and C_v are very approximate.

subtracts 3R as the contribution of the lead ions it may be
seen that per mole of fluorine ions C_v is approaching 3R/2
providing further support for the notion that a substantial
fraction of the degrees of freedom are diffusive (i.e. not
subject to a restoring force).   The limited data available
do not allow any statement to be made as to whether or not
there is a singularity in $C_v(T)$.

In the case of $SrCl_2$ (29) a careful study has been made
of lattice parameter as a function of temperature.   The
coefficient of expansion is continuous through the transition region with no sign of any anomaly either in V(T) or
in the temperature coefficient of $(\partial V/\partial T)_p$.   Likewise the
adiabatic compressibility of PbF_2, as inferred from Brillouin

scattering (30), appears to be a continuous function of
temperature although changing fairly substantially (by a
factor ~2) in the transition region. Both of these observations argue against there being a second order phase transition in the thermodynamic sense (contrast (16)).

Raman scattering studies of solid electrolytes also
provide informative results. In compounds with the fluorite
structure (29,30) there is a gradual evolution from a sharp
peak at low temperatures to a broad peak centered at $\Delta\omega = 0$,
much like the Raman spectrum of the solid electrolyte phase
of AgI (31). Figure 4 shows data (29) for $SrCl_2$. To the
extent that in the disordered material the scattered intensity reflects the phonon density of states, these spectra
are in qualitative agreement with the ideas of fig (1) in
demonstrating a substantial fraction of diffusive modes.
The results reported for PbF_2 (30) are perhaps even more
striking. At 100 K below the melting temperature there is
only a central peak rather like that for molten $SrCl_2$.

A very complete study of ^{19}F spin-lattice relaxation
time T_1 has been carried out in PbF_2 (32). At high temperatures when the relaxation time, τ, for diffusive motion is
much shorter than the inverse of the Larmor frequency
$T_1 \alpha \tau^{-1}$. It is found that below the region of the Faraday
transition τ calculated from T_1 is very close to that calculated from the conductivity using the jump diffusion
approach, eq (10). At temperatures above the transition
region T_1 varies inversely as σ (and hence D) as expected
for free diffusion, eq (7).

The bulk of the experimental evidence presented here,
particularly that pertaining to PbF_2, points to a gradual
transition from vibrational to translation degrees of freedom in the fluorites. In contrast to Class I and II transitions there is no evidence to date that this is a cooperative phenomenon at all. A simple model that has heuristic
value at least, is suggested by the Rice and Roth (33) model
of solid electrolytes. In an attempt to calculate the heat
capacity a simple model is considered (28) in which the
energy levels of an atom are described by harmonic oscillator
levels up to the rth level; at higher energies they are
translational levels appropriate to a one-dimensional box
of length ℓ. The partition function for an assembly of
such atoms, which is in fact that used for the "solid-like"

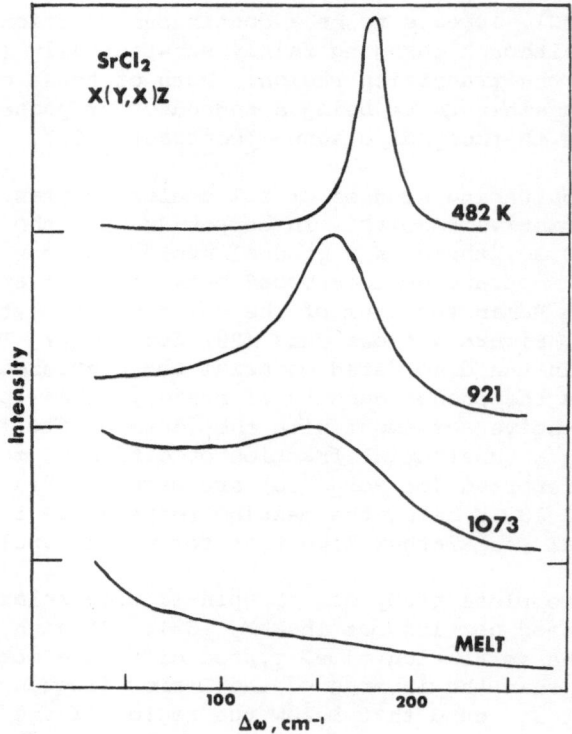

Figure 4   The Raman spectrum of SrCl$_2$ (29) close to room
            temperature, just below and just above the tran-
            sition region, and in the melt.

atoms in liquid Ar (34), is readily expressed in terms of
the dimensionless variable $x = \hbar\Omega_0/kT$, as

$$q = [\{1 - \exp(-rx)\}/\{1 - \exp(-x)\} + \exp(-rx)\alpha/\sqrt{x}]^{3N}$$

here Ω_0 is the oscillator frequency and $\alpha = (m\Omega_0/h)^{\frac{1}{2}}\ell$. Then

$$C_V/k = x^2(\partial^2 \ln q/\partial x^2).$$

For a choice of the parameters that might be appropriate
for PbF$_2$ (e.g. r = 15, α = 20) $C_V/3R$ varies from approximately
1 at x ≈ 1, has a maximum ~2.5 at x ≈ 1/4 and is <1 for 1/x > 9.

The peak in C_v is much wider than found experimentally as one would expect from a model of non-interacting atoms. The important point is however that this model, albeit crude, of the transition from "bound" to "free" states does predict a heat capacity anomaly of the general type that is observed.

This work was supported by a grant from the National Science Foundation.

REFERENCES

1. M. Faraday, Experimental Researches in Electricity, Art. 1339, Taylor and Francis, London (1839).
2. M. O'Keeffe and B.G. Hyde, Phil. Mag. 33, 219 (1976).
3. H. Wiedersich and S. Geller, in The Chemistry of Extended Defects in Nonmetallic Crystals (L. Eyring and M. O'Keeffe, eds.) North Holland, Amsterdam (1970).
4. T. Takahashi, J. Appl. Electrochem. 3, 79 (1973).
5. A Kvist and A. Lunden, Z. Naturforschg. 21a, 1509 (1966).
6. H.-H. Möbius, Z. Chem. 2, 100 (1962).
7. A.R. West, J. Appl. Electrochem. 3, 327 (1973).
8. C.E. Derrington, A. Lindner and M. O'Keeffe, J. Solid State Chem. 15, 171 (1975).
9. C.E. Derrington and M. O'Keeffe, Nature Phys. Sci. 246, 44 (1973).
10. L.E. Nagel and M. O'Keeffe in Fast Ion Transport in Solids (W. van Gool, ed.) North Holland, Amsterdam (1973).
11. J.B. Goodenough, H.V.-P. Hong and J.A. Kafalas, Mater. Res. Bull. 11, 203 (1976).
12. M. O'Keeffe, unpublished data.
13. D.K. Paul and I.F. Chang, J. Electronic Mater. 3, 709 (1974).
14. M. O'Keeffe and B.G. Hyde, Acta Cryst. in press.
15. W.V. Johnston, H. Wiedersich and G.W. Lindberg, J. Chem. Phys. 51, 3739 (1969).
16. W.J. Pardee and G.D. Mahan, J. Solid State Chem. 15, 310 (1975).
17. L.F. Vereshchagin, S.S. Kabalkina and A.A. Kotilevets, Sov. Phys. JETP 22, 1181 (1966); S.S. Kabalkina, L.F. Vereshchagin and L.M. Lityagina, Sov. Phys. Doklady 12, 946 (1968); Sov. Phys. JETP 29, 803 (1969).
18. N. Kawai, O. Mishima, M. Togaya and B. leNiendre, Proc. Japan Acad. 51, 627 (1975); L.V. Vereshchagin, E.N. Yakolev and Yu.A. Timofeev, JETP Lett. 21, 304 (1975).

19. B.A. Huberman, Phys. Rev. Lett. 32, 1000 (1974).

20. P.A. Egelstaff, An Introduction to the Liquid State, Academic Press, New York (1967).

21. B.A. Huberman and P.N. Sen, Phys. Rev. Lett. 33, 1379 (1974).

22. P. Bruesch, S. Strässler and H.R. Zeller, Phys. Stat. Sol. 31, 217 (1975).

23. C.A. Croxton, Liquid State Physics, Cambridge University Press (1974).

24. K. Funke and A. Jost, Ber. Bunsen-Gesell. phys. Chem. 75, 436 (1971).

25. P.A. Egelstaff, Adv. Phys. 11, 203 (1962).

26. C. Clemen and K. Funke, Ber. Bunsen-Gesell. phys. Chem. 79, 1119 (1975).

27. C.E. Derrington, A. Navrotsky and M. O'Keeffe, Solid State Comm. 18, 47 (1976).

28. M. O'Keeffe, to be published.

29. M. Shand, R.C. Hanson, C.E. Derrington and M. O'Keeffe, Solid State Comm. 18 769 (1976).

30. R.T. Harley, W. Hayes, A.J. Rushworth and J.F. Ryan, J. Phys. C 8, L530 (1975).

31. R.C. Hanson, T.A. Fjeldly and H.D. Hochheimer, Phys. Stat. Sol. (b) 70, 567 (1975).

32. J.B. Boyce, to be published.

33. M.J. Rice and W.L. Roth, J. Solid State Chem. 4, 294 (1972).

34. T.S. Ree, T. Ree and H. Eyring, Proc. Natl. Acad. Sci. (U.S.) 48, 501 (1962).

THEORETICAL ISSUES IN SUPERIONIC CONDUCTORS

G. D. Mahan

Physics Department
Indiana University
Bloomington, Indiana 47401

I.   INTRODUCTION

Superionic conductors[1] have been defined as solids with
ion conductances exceeding 0.01 ohm^{-1} cm^{-1}. Invariably
this occurs because an ion species in the solid begins
diffusing away from its normal lattice position. Usually
this ion species is one of the major constituents of the
solid. Thus in AgI, all of the Ag$^+$ ions are believed to
diffuse at high temperature, while the I$^-$ stay in position
to define a lattice. In CaF$_2$, it is the F$^-$ which move. In
usual ionic solids, the very small ionic conductivity is
provided by impurities or occasional defects. We are
interested in a quite different situation--where large
scale disorder appears to be an intrinsic process. Thus
the ionic conductivity is large, in part, because it is
proportional to the density of diffusing ions. This can be
very large--of the order of 10^{22} cm^{-3}.

The main theoretical issue is "Why does this occur?"

There is one necessary condition for the occurrence of
widespread disorder. The diffusing ion must be able to
move away from its normal lattice position, to another site,
without significantly disrupting the local arrangement of
neighboring ions.[2] This is often accomplished by having
more than one site per ion. Several examples: in β"-
alumina,[3] each Na$^+$ ion has nearly two identical sites

available to it; in α-AgI, there are 36 sites for 2 ions
per unit cube, if one just counts and d and h positions.[4]
Thus the ion often has an empty site next to it, to which
it may choose to move, thus making diffusion easy. Often
the ion moves to an interstitial site which has different
symmetry from the original one--as does F^- in CaF_2. In any
case, the lattice structure must be fairly open, and one
which allows the ions to move around.

This condition--of more than one site per ion--is
necessary but not sufficient for large scale disorder and
diffusion. There are lattices which have this feature,
which do not conduct ions well. There must be some other
conditions, and these have received extensive discussions.
Some ions obviously diffuse easier than others--even in
the same structure. So what conditions are put on the ion?
Empirically one finds that monovalent ions conduct best,
although there are expections: O^{--} in CSZ.[5] Ions diffuse
poorly if they are too small (Li^+ and Cu^+) or too large
(Rb^+, Cs^+, etc.).[6-7] The best ion conductors so far conduct
either Na^+ or Ag^+.[1] However, these two ions are not usually
regarded as being the same size if one adopts the Pauling
radii.[8] But they are the same size in Waddington's tables.[9]
Thus the criteria of ion size is not, by itself, adequate to
explain superionicity. These points will be discussed
further below.

II. PHASE TRANSITION

The ionic solids which show large ionic conductivity
often also have one or several phase transitions. Sometimes these phase transitions are from one crystal structure to another. There is no evidence for soft mode
transitions.[10] In other cases phase transitions arise
from the disordering of the ions themselves. Thus the
phase transition often marks the onset, with increasing
temperature, of sudden large scale disorder.

We will pause here to define an order-disorder phase
transition. This discussion will be brief, and you are
encouraged to read the classic discussion of Fowler and
Guggenheim for more details.[11] In Fig. 1 we show a two
dimensional square lattice, in which ions can sit at the
vertices. If half of the sites are occupied, and if the

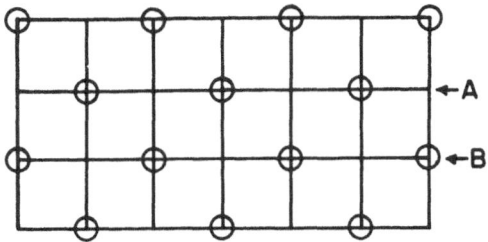

Fig. 1 The two dimensional lattice gas. The particles are circles, which may sit at any of the interstices. If the atoms repulse each other, then at half occupancy (n=1/2) the ions will have the arrangement which is shown.

ions repulse each other, then the lowest energy configuration is to have every other site occupied. Denote every other site A and B, and n_A and n_B are the fraction of these sites occupied at each temperature. Then we can define an order parameter

$$\xi(T) = (n_A - n_B)/(n_A + n_B)$$

It varies between zero and ±1. Fig. 2a, b, and c show the differing types of behavior of $\xi(T)$ as it increased from low temperatures.

In case 2a the transition to the disordered state (ξ= 0) is abrupt, and this is a first order phase transition. This is accompanied by a discontinuous increase in the ionic conductivity, often by several orders of magnitude. (Fig. 2d.) This occurs because the number of ions available for diffusion has suddenly increased, since they have left their ordered site. A model exhibiting this behavior, of sudden disordering, was first developed by Strassler and Kittel.[12] It was applied to superionic conductors by Huberman,[13] and also Rice, Strassler, and Toombs.[14] An example of a material which has a first order phase transition, of the order-disorder type, is Ag_2HgI_4.[15] Of course, structural phase transitions are usually first order. But here we mean phase transitions which are not structural, except for the disordering of the diffusing ion.

G.D. MAHAN

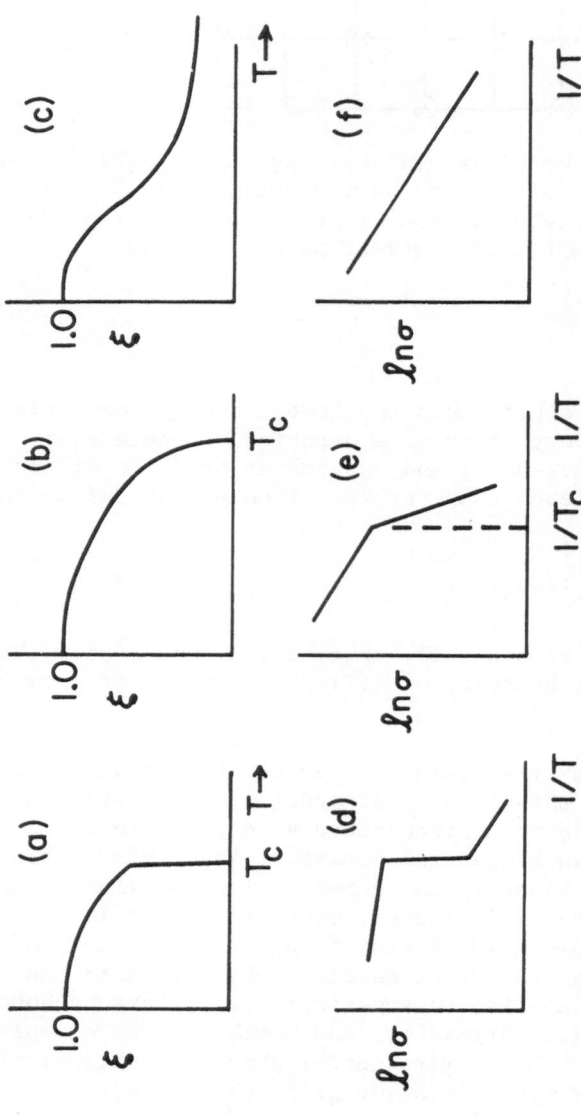

Fig. 2 The temperature dependent order parameter, which describes how the ions increasingly disorder with increasing temperature. (a) A first order phase transition, in which the disordering is abrupt. (b) A second order phase transition; (c) gradual disordering with no phase transition. (d), (e), and (f) show the ionic conductivity for each case. It is discontinuous in (d) for a first order phase transition. It is continuous in (e) for a second order phase transition, although the activation energy appears to change value.

A second order phase transition is described in Fig. 2b, where the order parameter goes continuously, albeit abruptly, to zero. It is experimentally distinguishable by the lack of latent heat, and a specific heat which diverges as a power law at the transition temperature $C_p \propto |T-T_c|^{-\alpha}$. Pardee and Mahan showed that the ionic conductivity was continuous through such a phase transition.[16] However, the activation energy for conductivity appeared to change, as in Fig. 2e. Specific heat measurements of Lederman and Salamon have identified such a transition in $RbAg_4I_5$ at 209°K.[17-19] Other systems in which the phase transitions are apparently second order are CaF_2, $SrCl_2$,[20] and $(C_5H_6N)Ag_5I_6$.[21]

The third possibility is where the order parameter gradually drops to zero, with increasing temperature, as in Fig. 2c. This seems to be the situation in Ag β-alumina as measured by Boilet et al.[22-24] This happens, for example, if there are significant energy differences between the sites A and B in Fig. 1. Here the ionic conductivity probably appears thermally activated, as in Fig. 2f, although detailed calculations are lacking.

There are also numerous structural phase changes in these materials, where the solid changes from one crystal structure to another. These are first order phase transitions, and are sometimes accompanied by a large discontinuous increase in the ionic conductivity. This is easily explained if ion disordering can occur in the high temperature phase, but not the low temperature phase. Often the entropy of disorder helps induce the phase transition. An excellent example of this behavior is provided by the fluorides, clorides, and bromides of Ca, Ba, and Sr. O'Keeffe and his coworkers showed that these nine materials provided three classes of behavior.[25-26] The four materials (CaF_2, BaF_2, SrF_2, $SrCl_2$) in the fluorite structure all had gradual disordering with increasing temperature. All compounds with this structure show such gradual disordering. (Why is disorder associated with this crystal structure?[21]) The three ($CaCl_2$, $CaBr_2$, $BaBr_2$) which do not have the fluorite structure only achieve high conductivity upon melting. The remaining two ($BaCl_2$, $SrBr_2$) do not have the fluorite structure at low temperatures, and have low conductivity. At high temperature they undergo a structural phase transition, probably to the fluorite structure,

and the conductivity increases to that of other fluorite-
structured materials. Thus the conductivity in these latter
two compounds increases because the crystal structure
changed to one which permitted large scale disorder.

Another quantity of theoretical interest is the
dependence $n(T_c)$, or perhaps of the transition temperature
T_c upon the concentration $\underline{n}$ of ions in the lattice. That
is, in Fig. 1, how does T_c change if the fraction of
occupied sites is decreased or increased. Nearly all
theoretical results reported so far have used a lattice gas
model with a Hamiltonian[11,16,27]

$$H = U \sum_{<ij>} n_i n_j - \mu \sum_\ell n_\ell \qquad (1)$$

Each site may have either $n_\ell = 0$ or 1 ions, and there is a
repulsive interaction $U > 0$ between ions on neighboring
sites. The free energy per site $F(U,\mu)$ is calculated, and
the concentration is obtained by differentiating with respect to the chemical potential

$$n = \frac{\partial}{\partial \mu} F(U,\mu)$$

Many versions of mean field theory have been developed,
including those of Bethe, Kirkwood,[17] and Kikuchi.[28]
Bienenstock and Lewis[29] have reported series results, at
high temperatures, for three different lattices. Recently
Subbaswamy and Mahan[30] reported results obtained from renormalization group methods. Their results are shown as
the solid line in Fig. 3, along with those of Sato and
Kikuchi,[27] for the two dimensional honeycomb lattice. This
lattice of Na^+ in β''-alumina. The renormalization group
results predict that the second order phase transition
occurs within a narrow concentration region $0.41 < n < 0.59$.
This is much narrower than that predicted by the mean field
theory of Sato and Kikuchi. Outside of this concentration
range the disordering is gradual, as shown in Fig. 2c. The
planar triangular lattice, which is the superlattice
structure of doubly occupied cells in Na β-alumina, was
found to not have a phase tranisiton at any concentration.[30]

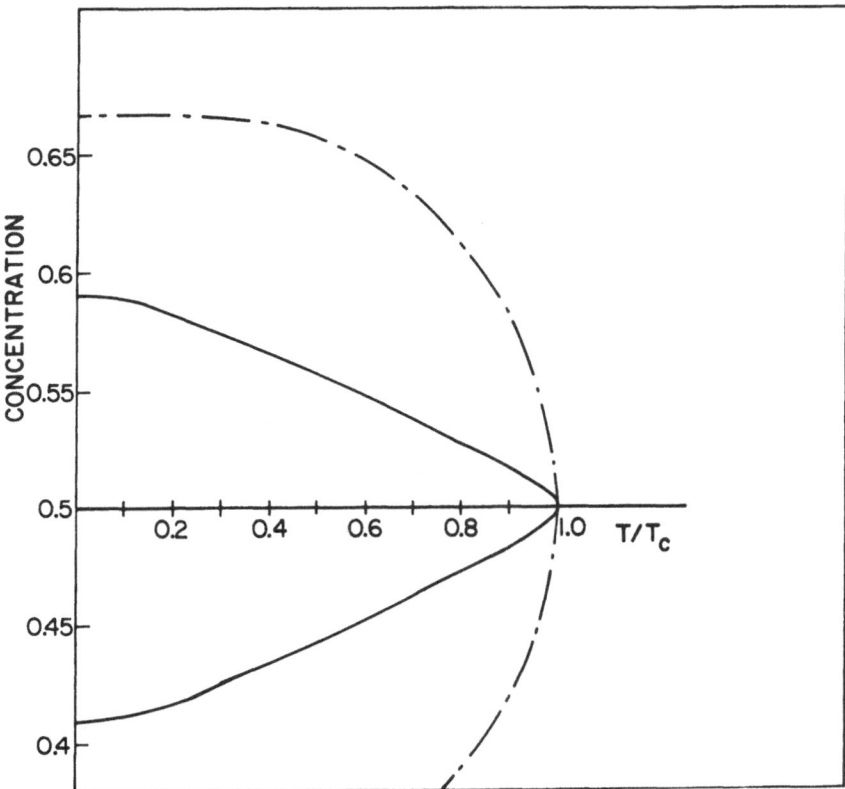

Fig. 3 Critical concentration versus temperature for the
two dimensional honeycomb lattice. Solid line are predictions of Subbaswamy and Mahan and dashed line that of Sato
and Kikuchi.

III.   THEORIES OF CONDUCTIVITY

Superionic conductors are technologically interesting
because of their high value of ionic conductivity. Indeed
it is this feature which is responsible for the name
"superionic".[1] Thus theories, and models, of conductivity
have played a central role in this field.

The ionic conductivity is often described by an Arrhenius equation

$$\sigma = \frac{C}{k_B T} \exp[-\Delta/K_B T] \tag{2}$$

where Δ is the activation energy, and C is the prefactor. This formula was originally applied to a low concentration of defect ions diffusing in a host media. Then C is[31]

$$C = \frac{1}{3}(Ze)^2 \, na^2 \, \omega_o \tag{3}$$

where a is the jump distance, and $\omega_o \sim$ Debye frequency is the attempt frequency. The energy Δ is the height of a barrier over which the particle must jump.

This formula must be modified if a significant fraction of the ions are disordering. The prefactor n might be interpreted as the fraction of ions which are disordered.[27] But this also depends upon temperature, as discussed above for lattice gas theories.[16-27] Indeed, the inclusion of an interaction term U between the disordering ions makes the theory of conductivity much more complicated.

Two quite different models have been proposed of collective disorder and conductivity--defects and domains. These will be discussed separately below. Keep in mind that each theory may apply to different materials, or to different temperature ranges of the same system. However, both theories are meant to apply to low temperatures, where the ion system is partially ordered.

There are some materials for which the prefactor C in (2) is anomalously large, at least if interpreted in terms of the factors in (3).[15] This led Rice and Roth[1] to suggest a "free ion" model, where a hopping ion could jump more than one site at a time. This suggestion has not been widely accepted. However, no other explanation has been proposed for the anomalously large prefactors.

3.1 Defect Models

The word defect is not meant to suggest impurity.

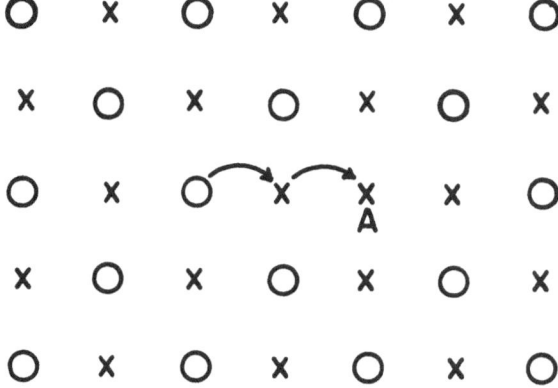

Fig. 4 A defect mechanism for d.c. ionic conductivity.
The caption of a vacancy at A means that an ion can move
into that spot with a double jump as shown. This is
equivalent to the movement of the vacancy, which provides
charge transport.

Rather, it means a local arrangement of host ions which
differs from ideal. A d.c. ionic conductivity is permitted
if this defect can move through the lattice. Several such
defects have been proposed by LeClaire.[32] One is shown in
Fig. 4, for a square planar lattice which is half occupied
with ions. An initial vacancy is at A. An ion can hop
twice to occupy its spot, which is equivalent to the
opposite movement of the vacancy. This gives charge transport. In the lattice gas model of equation (1), the energy
barrier for hopping is 2U. Also, the number of such defects
has an activation energy of 2U, so that the observed d.c.
conductivity from such defects would experimentally appear
to go as $\Delta = 4U$.

Mean field calculations can only include short range
order, and thus intrinsically are restricted to defect
models of ionic conduction. This applies to the results
obtained by Sato and Kikuchi,[27] and Pardee and Mahan.[16]
Both calculations used the Hamiltonian (1), and calculated
a conductivity dominated by single particle hops. Both
found a thermally activated conductivity, in which the
activation energy appeared to change slope at the second

order phase transition. The two calculations differed
mostly in the method of calculation, as one was via the
path probability method, and the other by the Kubo formula.
An exact calculation of single particle hops has shown that
the conductivity does not kink at the second order phase
transition, but gradually bends over and changes activation
energy as temperature is changed.[33]

There is a need for further development of this kind
of theory. The Kubo formula method has not been applied
to double jumps, of the kind shown in Fig. 4. The difficulty is that there are a great many terms, since there are
a great many possible double jumps! Thus present calculations do not include the interstitialcy mechanism of Fig. 4,
which has been proposed as the dominant one for low temperature conductivity.

3.2   Domain Models

There have been numerous suggestions, in the past, that
domains exist in ionic conductors. The domain walls are
charged, and conduction occurs by domain wall motion. Roth
suggested their application to β-alumina.[34] Formation
energies of small domains were calculated by van Gool and
Bottelberghs.[35-36] No one has yet calculated a dynamical
model of domain wall motion, and thus there is no actual
theory of conductivity.

The main features of domain walls are shown in Fig. 5,
where we use the example of β-alumina, as described by Roth,
van Gool and Bottelberghs. For Ag ions there is a planar
honeycomb lattice, with more than 50% of the sites occupied.
The excess charge was envisioned as being in domain walls,
which are the boundaries between ordered regions of the
plane. The sequence of arrows show the type of ion hopping
motion which leads to domain wall motion, and thus to
conductivity. If the hops are in a sequence along a wall,
like a zipper, each ion does not change its lattice gas
energy by its hop. Thus this wall motion should have a
low activation energy for motion.

The main question about domains is the probability of
their formation. In thermal equilibrium this must be
proportional to $\exp(-E_D/k_B T)$, where E_D is the energy needed

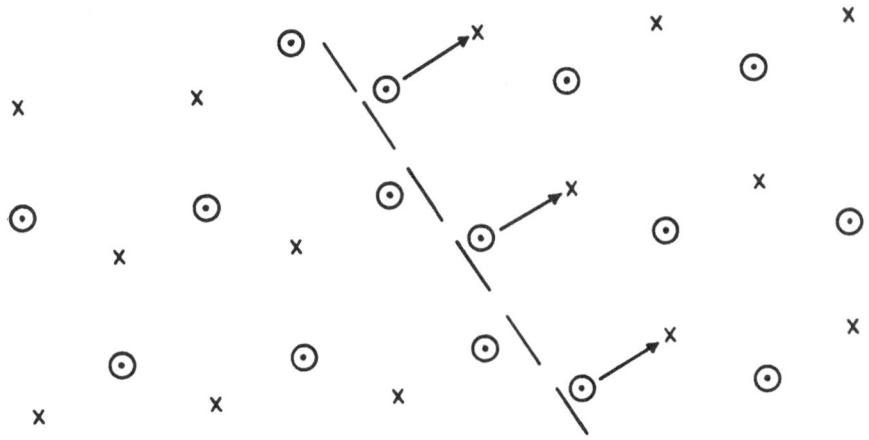

Fig. 5 Domain wall motion for d.c. conductivity. The ions in the wall hop in sequence, like a zipper. Since the wall is charged, wall movement leads to ionic conductivity.

to form the domain. Now if the domain walls are charged, then a macroscopic domain will have macroscopic electric fields. This takes a great deal of energy, and makes E_D very large. So large, that the probability of domains is negligible. For example, let the walls have a classical charge per area of σ_0 which has the approximate value of one electron charge per unit area of the crystal unit cell. From dimensional analysis, it is easy to show that $E_D \sim \sigma_0^2 V / \varepsilon_0$ where V is the volume of the domain and ε_0 is the static dielectric constant. But $\sigma_0^2 \sim 10^{23}$ eV/cm^3. Thus $E_D \gg k_B T$ if V has any size at all. The quantity E_D is small only if V is the size of a unit cell, but this limit is handled by the defect models. The same problem occurs for two dimensional domains. Here the coulomb energy is $S_0^2 L$ where S_0 is the charge per unit length of domain wall edge, and L is the dimension of the domain. One quickly estimates that $S_0^2 \sim 10^8$ eV/cm. Again the domains must be small to exist, and this situation is handled by defect models.

Recently it was discovered that the excess Ag^+ ions in β-alumina formed a regular super lattice at low temperature.[22-23] It now seems unlikely that domains play a significant role in this material.

Domains do occur in ferroelectrics. But here there are intrinsic internal electric fields, which the domains are trying to minimize. In superionic conductors the large scale electric fields do not exist, except if created by domains. Thus their existence appears improbable.

Theories of domain wall motion have recently been developed for ferroelectrics.[37] These could equally be applied to domains in superionic conductors, if they some-how existed. These theories employ the soliton concept, which emerges from solutions to the non-linear wave equa-tion.[38] The non-linearity introduces solutions which are not oscillatory, but instead propagate as a steady pulse shape, almost without dispersion or damping. They have been observed in one dimensional systems, and seem well established as a one-dimensional phenomena.[38] It is not certain whether the same types of phenomena occur in higher dimensions.

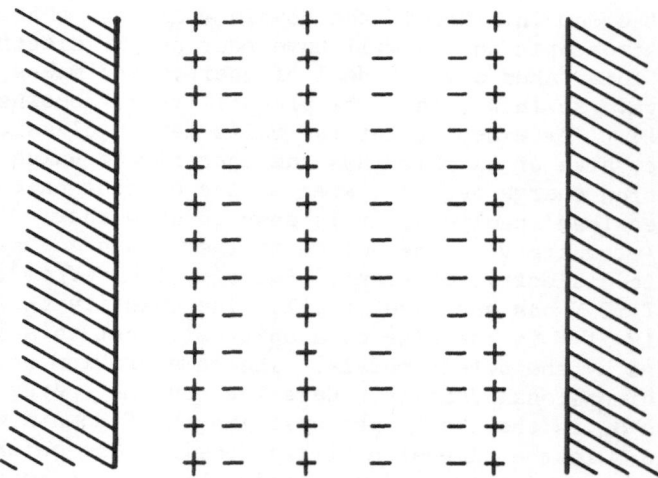

Fig. 6 A one dimensional model of domain wall conduction may be constructed by assuming that the domain walls are parallel charge sheets. They move to and from the elec-trodes. The statistical mechanics of the coulomb sheets have been solved exactly.

A quasi one-dimensional domain model could be constructed if the ionic conduction were assumed to occur between
the parallel planes of capacitor-like electrodes. The
domain walls would be charge sheets, either plus or minus,
which would drift in an electric field. This is shown
edgewise in Fig. 6. The effective mass of such sheets
could be obtained from soliton theory, as Krumhansel and
Schrieffer have done for ferroelectrics. This mass is
proportional to the area of the sheet and is macroscopically
large—thus such sheets are improbable. The statistical
mechanics for such classical sheets have been solved
exactly by Lenard.[39] In this case, there is indeed an
exact theory of domain walls.

3.3 Ionic Polaron

An ion, while sitting in a site, will either attract
or repel its ion neighbors. These will consequently move
away from their equilibrium positions, thus creating a
local polarization cloud about the central ion. When an
ion hops, it must carry this deformation cloud with it.
The theory of this was first developed for electron motion,
and is called "small polaron theory".[40] Flynn[41] suggested
that the same theory could apply to a diffusing ion, and
this we called the "ionic polaron". The simplest version of
the theory, which assumes that the local polarizations are
linearly coupled to the ion, predicts that the conductivity
has an activation energy of[42]

$$\Delta_p \simeq \frac{Z^2 e^2}{\pi a} \left(\frac{1}{\varepsilon_\infty} - \frac{1}{\varepsilon_o} \right) \tag{4}$$

where a is the jump distance and Z is the valence. A tabu-
lation of this quantity for Ag-, Cu-, and alkali halides show
that this quantity is only small for solids which are good
superionic conductors.[42] Thus a large amount of local
polarization about the diffusing ion will impede its
motion. Presumably this is why small ions diffuse so poorly.
In this picture, the activation energy Δ_p is thus caused by
the dynamic processes of polarization, rather than by the
static part of a simple barrier. Actual barriers, in real
materials, probably have a barrier Δ which has contributions from both the dynamic and static terms. And also the

lattice gas theories mentioned above. Unfortunately, one cannot just simply add these contributions to Δ.

IV. Ag^+ AND Cu^- CONDUCTORS

The best ion conductors at room temperatures are those which conduct Ag^+.[1] These compounds are often composed of Ag^+, I^-, and other ions or molecules. New ones are continually being discovered, many by Geller and his associates.[43]

A major theoretical problem, perhaps "The" problem, is to be able to predict in advance which crystal structures will become a superionic conductor. In fact it is a twofold problem. First to predict which crystal structures will be formed by a set of ions, and second to predict whether this crystal structure will permit fast diffusion. Theorists have not been very successful at either of these two tasks. For example, a systematic study by Bradley and Greene on Ag^+, Cu^+, and alkali halides has produced a wealth of data, which has not been explained.[44]

Another big question is why do Ag^+ ions diffuse so easily. The answer was provided by Armstrong, Bulmer, and Dickinson.[45] They make two points. First, low valence ions diffuse best, and monovalent best of all. A higher valence tends to increase all of the coulomb energies in the system, leading to higher activation energies for diffusion, as in (4). Second, Ag^+ ions prefer primarily covalent bonds, and are usually four-fold coordinated. Thus their sites are usually tetrahedra of I^-. To hop, they must go through the face of the tetrahedra, which is a three-fold coordinated site. This is also an acceptable coordination number for their bonding orbitals, which are d-states. Thus, they conclude, Ag^+ conducts well because it is a monovalent ion, with a closed d-shell providing the bonding orbitals. This conclusion, if correct, is discouraging because only Cu^+ and Au^+ also qualify. There are of course, good Cu^+ conductors also.[46]

Ag^+ ions move so easily through the I^- lattice, that the normal lattice gas models (1) do not apply. These models assume that the ion stays in a well defined site, except for a very rapid hop to a neighboring site. It is

assumed that the hopping time is very short compared to the
average dwell time on a site. Funke and his collaborators[47-49] have shown that this is not the case in AgI. In
the model they propose to explain their microwave and neutron scattering results, the hopping time is about twice
the average dwell time! The same result is found in α-CuI
and β-CuBr.

V. β-ALUMINA

This material is noteworthy because it is the best
conductor of Na^+ ions at moderate temperatures. As such,
it is in active development as a battery material. But all
monovalent ions can conduct in this material.[1]

For the alkali atoms Li^+, Na^+, K^+ and Rb^+, the
dominant trait, which determines ion motion, is the ion
size.[6-7] Apparently Na^+ moves through the lattice easiest,
because it fits well, and least disturbs the surroundings.
Li^+ is too small, and the others too large.

The crystals grow with an excess of cations in the
conducting plane. For alkali ions, this excess causes some
unit cells to have two cations instead of one.[24] X-ray
experiments show that the doubly occupied sites are about
one-third of the total.[24] They form a superlattice at low
temperature, which gradually disorders with increasing
temperature.[23] Wang, Gaffari, and Choi[50] have suggested a
cooperative motion of three ions, which lets one cation hop
from a doubly occupied cell to a neighboring singly occupied
one. This is equivalent to the motion of the doubly
occupied cell. Thus one can make a statistical mechanical
model, in which the "particles" are the doubly occupied
cells, and the "empty sites" are the singly occupied ones.
In the Wang, Gaffari, and Choi model, diffusion occurs because "particles" can move to "empty sites". As shown in
Fig. 7, for the triangular lattice found experimentally by
McWhan et al.,[24] the "particles" have a regular arrangement
at low temperature. This is to be expected if they have
the lattice-gas Hamiltonian (1), where U is the repulsive
interaction between "particles". This lattice also has the
peculiar feature that, for one-third occupancy, a particle
out of equilibrium position (e.g. at A) moves through the
lattice without changing its configuration energy. Thus
the conductivity from the lattice gas part gives an activa-

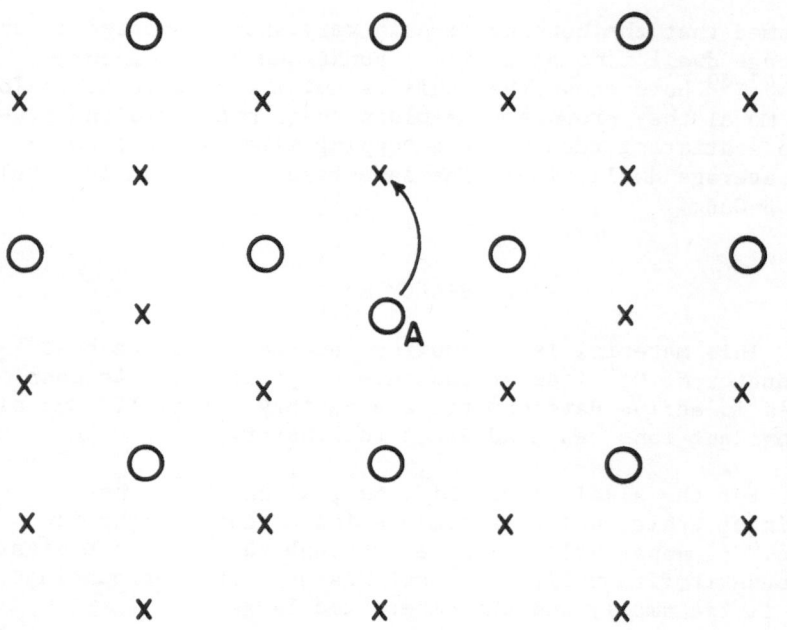

Fig. 7 If the planar triangular lattice is one third
occupied by ions, then the ordered arrangement is with each
third site occupied. This structure is quite open, and
permits other ions, such as A, to move through the lattice
without impediment. This is the conduction process in Na
β-alumina.

tion energy $3U/2$, which is just the energy required to
create the defect. A simple estimate is $U = e^2/\varepsilon_0 \alpha$. Taking
$\varepsilon_0 \sim 13$ and $d = \sqrt{3}a \simeq 9.7\text{Å}$, we get $3U/2 \simeq 0.17$ eV. This is
in exact agreement with experimental activation energies
for Na^+. This explanation of the experimental activation
energy requires that the "particle" hopping has a negligible energy barrier. Unfortunately, this latter quantity is
also estimated to be 0.17 eV.[50]

     This lattice seems exceptional, in this feature of
permitting diffusion of "particles" while not changing the
configuration energy. This, we suggest, is one of the
reasons for the high ionic conductivity of this material.

REFERENCES

1. M. J. Rice and W. L. Roth, J. Solid State Chem. 4, 294 (1972).

2. J. A. A. Ketelaar, Trans. Faraday Soc. 34, 874 (1938).

3. M. Bettman and C. R. Peters, J. Phys. Chem. 73, 1774 (1969).

4. G. Burley, Acta Cryst. 23, 1 (1967).

5. R. E. Carter and W. L. Roth, in Electromotive Force Measurements in High-Temperature Systems (Institute of Mining and Metallurgy, London, 1967) pg. 125-144.

6. Y. Y. Yao and J. T. Kummer, J. Inorg, Nucl. Chem. 29, 2453 (1967).

7. J. T. Kummer, Prog. Solid State Chem. 7, 141 (1974).

8. L. Pauling, The Nature of the Chemical Bond (Cornell University Press, Ithaca, 1960) pg. 514.

9. T. C. Waddington, Trans. Faraday Soc. 62, 1482 (1966).

10. J. F. Scott, Revs. Mod. Phys. 46, 83 (1974).

11. R. H. Fowler and E. A. Guggenheim, Statistical Thermodynamics, (Cambridge University Press, 1939) 541.

12. S. Strassler and C. Kittel, Phys. Rev. 139, A758 (1965).

13. B. A. Huberman, Phys. Rev. Letters 32, 1000 (1974).

14. M. J. Rice, S. Strassler, G. A. Toombs, Phys. Rev. Letters 32, 596 (1974).

15. K. W. Browall and J. S. Kasper, J. Solid State Chem. 15, 54 (1975); K. W. Browall, H. Wiedemeier, and J. S. Kasper, ibid 10, 20 (1974).

16. W. J. Pardee and G. D. Mahan, J. Solid State Chem. 15, 310 (1975).

17. S. Geller, Science 157, 308 (1967).

18. W. V. Johnston, H. Wiedersich, and G. W. Lindberg, J.
     Chem. Phys. 51, 3739 (1969); W. J. Pardee and G. D.
     Mahan, J. Chem. Phys. 61, 2173 (1974).

19. F. Lederman and M. B. Salamon, Bull. Am. Phys. Soc. 20,
     331 (March, 1975).

20. A. S. Dworkin and M. A. Bredig, J. Physical Chem. 72,
     1277 (1968).

21. S. Geller and B. B. Owens, J. Phys. Chem. Solids 33,
     1241 (1972).

22. Y. Lecars, R. Comes, L. Deschampes, and J. Thery, Acta
     Cryst. A30, 305 (1974).

23. J. P. Boilot, J. Thery, R. Collongues, R. Comes, and
     A. Guinier, Acta Cryst. (to be published).

24. D. B. McWhan, S. J. Allen, Jr., J. P. Remeika, and
     P. D. Dernier, Phys. Rev. Letters 35, 953 (1975).

25. C. E. Derrington, A. Lindner, and M. O'Keeffe, J. Solid
     State Chem. 15, 171 (1975).

26. C. E. Derrington and M. O'Keeffe, Solid State Comm. 15,
     1175 (1974).

27. H. Sato and R. Kikuchi, J. Chem. Phys. 55, 677 (1971).

28. R. Kikuchi, Prog. Theor., Phys. Suppl. 35, 1 (1966).

29. A. Bienenstock and J. Lewis, Phys. Rev. 160, 393 (1967).

30. K. R. Subbaswamy and G. D. Mahan, submitted to Phys.
     Rev. Letters.

31. A. B. Lidiard, Handbuch der Physik, Vol. 20, edited
     S. Flugge (Springer-Verlag, 1957) pg. 246-349.

32. A. D. Le Claire, Fast Ion Transport in Solids, ed.
     W. van Gool (North-Holland, London, 1973) pg. 51.

33. G. D. Mahan, "Lattice Gas Theory of Ionic Conductivity", submitted to Phys. Rev.

34. W. L. Roth, J. Solid State Chem. $\underline{4}$, 60 (1972).

35. W. van Gool, J. Solid State Chem. $\underline{7}$, 55 (1973).

36. W. van Gool and P. H. Bottelberghs, J. Solid State Chem. $\underline{7}$, 59 (1973).

37. J. A. Krumhansl and J. R. Schrieffer, Phys. Rev. $\underline{B11}$, 3535 (1975).

38. G. B. Whitham, Linear and Nonlinear Waves, (J. Wiley & Sons, New York, 1974).

39. A. Lenard, J. Math. Phys. $\underline{2}$, 682 (1969); $\underline{4}$, 533 (1963).

40. H. Reik, Polarons in Ionic Crystals and Polar Semi-conductors, ed. J. T. Devreese (North-Holland, London, 1972) pg. 679-714.

41. C. P. Flynn, Point Defects and Diffusion (Oxford University Press, 1972).

42. G. D. Mahan and W. J. Pardee, Phys. Letters $\underline{49A}$, 325 (1974).

43. S. Geller and P. M. Skarstad, Phys. Rev. Letters $\underline{33}$, 1484 (1974).

44. J. N. Bradley and P. D. Greene, Trans. Faraday Soc. $\underline{62}$, 2069 (1966); ibid $\underline{63}$, 2516 (1967).

45. R. D. Armstrong, R. S. Bulmer, and T. Dickinson, J. of Solid State Chem. $\underline{8}$, 219 (1973).

46. T. Takahashi, O. Yamamoto, F. Matsuyama, and Y. Noda, J. Solid State Chem. $\underline{16}$, 35 (1976).

47. G. Eckold, K. Funke, J. Kalus, and R. E. Lechner, Phys. Letters $\underline{55A}$, 125 (1975).

48. K. Funke, Phys. Letters $\underline{A53}$, 215 (1975).

49. C. Clemen and K. Funke, Berichte der Bunsen-Gesell-
     schaft fur Physikalische Chemie, $\underline{79}$, 1119 (1975).

50. J. C. Wang, M. Gaffari, and S. Choi, J. Chem. Phys.
     $\underline{63}$, 772 (1975).

PATH PROBABILITY METHOD AS APPLIED TO PROBLEMS OF

SUPERIONIC CONDUCTION

Hiroshi Sato
Purdue University, West Lafayette, IN   47907
    and
Ryoichi Kikuchi
Hughes Research Laboratories, Malibu, CA   90265

Different from superelectronic conduction, superionic conduction in ionic solids is closely related to a peculiar structural characteristic of crystals known as a substantial sublattice disorder or a "liquid sublattice."[1] The large ionic conductivity is due to a mass motion of a large number of diffusing ions in the liquid sublattice. The problems of superionic conduction can therefore be reduced to elucidating the following two fundamental problems: the origin which leads to the formation of such substantial sublattice disordering and the transport mechanism of ions in the substantially disordered sublattice. What we try here is to describe the characteristics of the mass motion of ions statistical mechanically when the characteristics of the "liquid sublattice" are specified. As such, the treatment is a generalization of the random walk theory of diffusion which is only applicable when the number of defects through which ions can move is extremely small.

Based on the random walk approach, the ionic conductivity σ is described by an Arrhenius equation

$$\sigma = \frac{C}{kT} \exp \left[-E/kT \right] \tag{1}$$

where E is the activation energy and C is the prefactor. C is given by

$$C = \tfrac{1}{3} (Ze)^2 n \, d^2 \, w_o \tag{2}$$

135

where d is the jump distance and w_o is the attempt frequency.
The activation energy E is the height of a potential barrier
over which the particle must jump. n is the effective
number of ions per unit volume which participate in the
diffusion motion or the number of defects through which
ions can move, while, the diffusion coefficient D is
described by

$$D = \tfrac{1}{3} n \, d^2 \, w_o \, \exp \, [-E/kT] \tag{3}$$

In both (2) and (3), if the defects are created thermally,
n is expressed also by an Arrhenius equation. Superionic
conductors are characterized by extremely low activation
energies and unusually large prefactors equivalent to
those of liquid states. The existence of a substantial
sublattice disordering is intrinsically connected to an
unusually low activation energy and to a large number of
defects (or vacancies) into which ions can hop. The
problem is then to generalize the random walk approach to
systems which include a large number of available vacant
sites.

    Isotope diffusion or ionic conduction is essentially
a time dependent Ising problem.[2] In the stationary state
where the rate of flow does not change with time, the
distribution of ions (including isotope ions) and vacancies
is kept close to equilibrium. Determination of the equilibrium distribution of ions on available lattice sites with
pair interactions can be made utilizing a generalized Ising
model and the deviation from the equilibrium distribution
by the presence of the chemical potential gradient is then
derived. Ions flow along the field gradient or the concentration gradient of isotope ions in a way not to disturb
the overall distribution of ions. By calculating this flow
rate, one can obtain the ionic conductivity or the diffusion
coefficient of both ordinary ions and isotope ions and,
hence, the correlation factor. It is then easy to envisage
that the calculation of the flow rate can be made by a
technique of irreversible statistical mechanics which is
an extension of the treatment of cooperative phenomena.

    The path probability method[3] has been developed as an
extension of a well known method of treating cooperative
phenomena in equilibrium, the Cluster Variation method.[4]

In the equilibrium theory, the convenient way of evaluating the partition function is to represent it by the maximum term. For nonequilibrium problems, a function corresponding to the partition function is sought whose extremum value corresponds to the way the system changes. Here, a set of variables $\{A_i\}$, $i=1,2,\ldots\ldots$ which specify the change of state of the system in Δt is defined, and the probability $P(A_1, A_2,\ldots\ldots)$ that the change of $\{A_i\}$ occurs is written. The values of A_i which makes P a maximum gives the direction the system changes into. The function P is called the path probability function. The idea is a natural generalization of the maximum principle. The problem lies in actually writing the path probability function $P(A_1, A_2,\ldots\ldots)$. In the path probability method, this is done by generalizing the expression used in the Cluster Variation method combined with the theory of absolute reaction rates by introducing the time axis as the fourth-dimensional axis.[3]

The advantage of the path probability method in the calculation of ionic conductivity in the disordered sublattice over the Kubo formalism is that it is specifically designed for highly cooperative processes which are occurring in crystalline lattice structures, while the latter is formulated based on continuum space. Another advantage is that the Path Probability method allows us to calculate the time dependent processes with the same approximation as that known in the Cluster Variation method and hence the nature of approximation involved in dealing with irreversible processes is easy to understand. This is especially advantageous for problems of diffusion or ionic conductivity in complicated crystals based on microscopic models. However, the mathematical complexities so far limited us to the pair approximation in the problems of diffusion. The pair approximation in the Cluster Variation method is equivalent to the Bethe approximation.

The path probability method was applied to the description of the superionic conductivity in β-alumina type compounds.[2] Although the calculation was made on specific models, the results are general enough to understand the mass transport characteristics of interacting ions. If the occupancy rate of ions on available sites is denoted by ρ, the effective number of ions which contribute to the conduction in the assembly of non-interacting ions is $N\rho(1-\rho)$, where N is the total number of equivalent, available sites per unit volume. Interactions among ions change the

situation a great deal. The repulsive interaction enhances
the diffusion while the attractive interaction suppresses
it. The effect of enhancement is mostly due to the lowering
of the activation energy and, hence, to the increase in the
jump frequencies; the repulsive interaction tends to push
nearest neighboring ions away into vacancies. On the other
hand, the repulsive interaction tends to create a regular
distribution of ions. If the long range order is formed,
this creates preferential sites among otherwise equivalent
sites. This increases the activation energy and lowers
the ionic conductivity. The lowering of the conductivity
in the ordered state, however, is largely due to the
correlation effect by the presence of the preferential
sites.[2,5] When ions on the preferential sites jump out of
the potential wells, they immediately tend to go back into
the original preferential sites. The path probability
method, just like in the case of the random walk theory,
eventually leads to an expression of diffusion coefficient
in terms of the correlation factor, the effective vacancy
availability factor, the effective jump frequency factor
which depend on the interaction and this similarity makes
the physical interpretations of transport mechanisms easy.[2]
If the available sites for ions are not equivalent and
preferential sites exist, it is necessary to take into
account the effect of redistribution of ions with temperature. Otherwise, the situation is similar. The correlation
effect due to the existence of preferential sites, however,
plays an important role just like in a system with a long
range order. This correlation effect is the physical
correlation effect and should be distinguished from the
geometrical correlation effect in the self-diffusion.[6]
Interested readers should refer to our original paper[2] for
details.

     It is to be noticed that the path probability method,
in its original form, is applicable only to so-called hopping diffusion where the staying time of ions in the
potential well is long compared to the hopping time. This
is because we treated the problems as Markovian systems
where each unit jump of ions is not affected by the history
of its previous jumps. The hopping assumption seems to apply
to β-alumina type compounds. In the case of α-Ag I or of
similar compounds, the applicability of this assumption has
been questioned.[7] It is possible, however, to revise the
definition of the path in the path probability method to
take into account the memory of previous jumps.

The Path Probability method is versatile enough to take into account complicated crystallographic features in dealing with the ionic conduction. It has been demonstrated that the ionic conductivity is lowered by the formation of long range order among diffusing ions.[2] It is also possible to show that order-disorder transformations in the rigid sublattice can cause a similar change in the ionic conductivity even if no long range order exists among conducting ions in the liquid sublattice. Calcia Stabilized Zirconia (CSZ) can be taken as a possible example which shows this type of effect. CSZ has a stable range with the CaF_2 structure between ~10 and ~20 per cent CaO.[8,9] In the CaF_2 structure, cations form a face centered cubic lattice while anions exist at eight tetrahedral interstitial positions of the unit cell of the face centered cubic structure. When tetravalent Zr ions are replaced by divalent Ca ions, vacancies are created in the anion lattice whose concentration is just a half of the concentration of Ca ions in order to compensate for the difference in charges. The increase in the concentration of vacancies results in the increase in the ionic conductivity due to oxygen ions as expected. However, beyond a certain concentration of oxygen vacancies, a drop in the conductivity is noticed.[8,9] The temperature dependence of the conductivity indicates that this decrease is due to an order-disorder transformation in the cation sublattice.[8] A brief outline of our calculation made on this system will be given to show how the flow of oxygen ions can be affected by the ordering of Ca and Zr ions in the cation sublattice.[10]

If there should be a cation ordering in the face centered cubic sublattice around 15% Ca, the ordered structure should correspond to that of Cu_3Au type. The phase diagram of ordered structure in the face centered cubic lattice with nearest neighbor interactions has been calculated with the tetrahedron approximation of the Cluster Variation method[11] and is shown in Fig. 1. This indicates the onset of ordering around 15% Ca. Fig. 2 shows the ordered cation sublattice (circles) and the anion sublattice (crosses). Filled circles indicate Ca ions. In both the ordered and the disordered state, the unit cubes composed of cations in Fig. 2 include one Ca ion each on the average (at 25% Ca) and, hence, the occupation probability of O^{--} ions on the anion sublattice is not affected by the order-disorder transformation in the cation sublattice. Since the concentration of vacancies of anion sites is low, it is not likely to have a long range order among O^{--} ions through their repulsive interactions.[2]

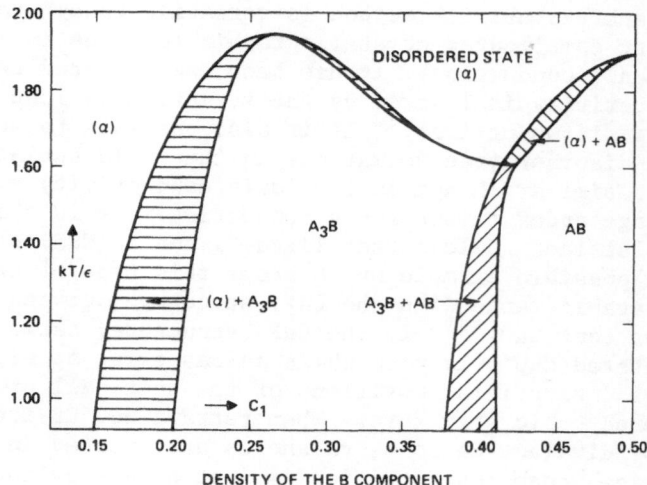

Fig. 1 — Phase diagram of ordered structures in the face
centered cubic lattice.

        Although there are no preferential sites for O^{--} ions,
two types of paths can be differentiated for oxygen ion
migration in the ordered state. If Ca-O interaction is
stronger than Zr-O interaction, the b-type jump in Fig. 2
is easier than the a-type jump and hence the a-type jump
tends to be avoided. This affects the efficiency of jumps
which contribute to the conductivity and the conductivity
in the ordered state is lowered. The result of the calculation is compared with experiments[7] in Fig. 3. The
agreement is quite favorable. The parameters adopted for
the calculation are the ordering energy ε_1 in the cation
sublattice, ε_2 which indicates the difference in the
interaction energies of oxygen ions and two kinds of cations,
respectively and ε_3 which indicates the pair interaction
energy among oxygen ions. These are defined as follows:

$$4\varepsilon_1 = \varepsilon_{ZrZr} + \varepsilon_{CaCa} - 2\varepsilon_{CaZr} > 0$$

$$\varepsilon_{Ca-O} = \varepsilon_{Zr-O} + 2\varepsilon_2$$

$$\varepsilon_{O-O} = \varepsilon_3 > 0$$

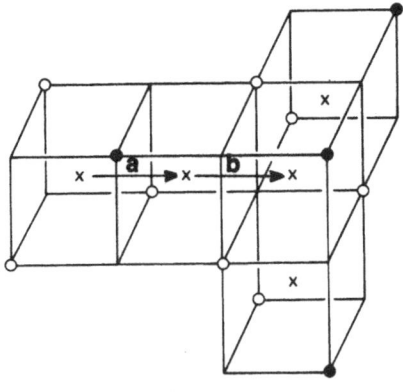

O AND ● :   CATIONS
X :   OXYGENS AND VACANCIES

Fig. 2 - Two types of jumps of oxygen ions.

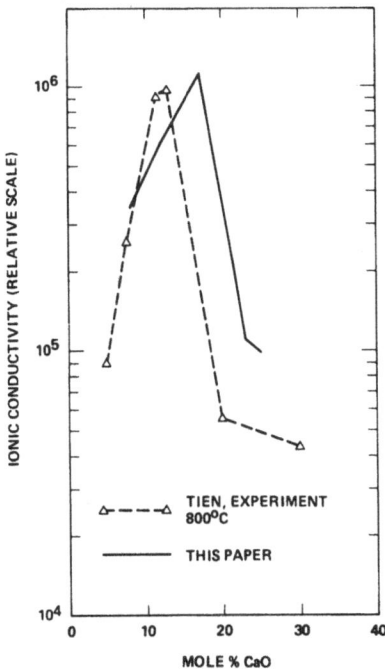

Fig. 3 - Dependence of ionic conductivity on composition.

with relative magnitudes

$$\varepsilon_2 = 5\varepsilon_1$$

$$\varepsilon_3 = 4\varepsilon_1$$

The details of the calculation will be published elsewhere.

REFERENCES

1. W. van Gool, "Fast Ion Conduction," Annual Review of Materials Science, vol. 4 (1974) p. 311.
2. H. Sato and R. Kikuchi, J. Chem. Phys. $\underline{55}$ 677 (1971).
3. R. Kikuchi, Prog. Theo. Phys. (Kyoto) Supplement No. 35, (1966) p. 1.
4. R. Kikuchi, Phys. Rev. $\underline{81}$ 988 (1951).
5. R. Kikuchi and H. Sato, J. Chem. Phys. $\underline{57}$ 4962 (1972).
6. H. Sato and R. Kikuchi, "Correlation Factor and Nernst-Einstein Relation in Solid Electrolytes," Mass Transport Phenomena in Ceramics, ed. A. R. Cooper and A. H. Heuer, Plenum, New York 1975, p. 149.
7. G. D. Mahan, Theoretical Issues in Superionic Conductors, this issue.
8. T. Y. Tien, J. Am. Cer. Soc. $\underline{47}$ 430 (1964).
9. R. E. Carter and W. L. Roth, Electromotive Force Measurements in High Temperature Systems, ed. C. B. Alcock, Inst. Min. Met., London (1968) p. 125.
10. R. Kikuchi and H. Sato, Superionic Conductivity and Order-Disorder Transition, this issue.
11. R. Kikuchi and H. Sato, Acta Met. $\underline{22}$ 1099 (1974).

This work was supported in part by the National Science Foundation grant DMR 7502959.

DOMAIN MODEL FOR SUPERIONIC CONDUCTORS

W. van Gool

Department of Inorganic Chemistry

State University, Utrecht, Netherlands

1. INTRODUCTION

Many years ago the liquid model has been proposed to explain the high ionic conductivity in solids such as α-AgI, Ag_2HgI_4, etc.[1] In the liquid model it is assumed that one or more constituant ion types remain in fixed positions and that the mobile ions move like a liquid through the frame. The liquid model is refined presently by both experimental and theoretical investigations[2].

The relationship between structure and superionic conduction (better: optimized ion-conduction) is important for the selection of new materials. Although the structures of superionic conductors differ widely, some of them have a common feature: they have more than the equivalent number of equal sites available for the mobile ion. It has often been suggested that this structural aspect is as such already and explanation for the high mobility of the ions, for they should move easily to the "other" sites.

Obviously, this suggestion is incorrect. There are many compounds which have more than the equivalent number of equal sites, but which are not superionic conductors, e.g. sphalerite. At least one additional condition must be fulfilled for good conductivity: the ions must be able to move easily to the other positions. This means that the energy barrier in the transition state must be low. This energy barrier can be interpreted in terms of space -available to squeeze the ion through- and potential energy due to the ions other than the mobile one under consideration.

The use of many sets of non-equivalent sites has no
sense for the explanation of the conductivity. Statements
that the two Ag^+-ions have available 42 positions in a
unit cell, are not meaningful. Including non-symmetric
points, one could equally well state that each Ag^+-ion has
available an infinite number of positions in one unit cell.
Nevertheless, the information -obtained from the interpretation of X-ray data- on statistical occupation of equivalent sites is a very helpful indication for selecting
possibly interesting new materials.

Assuming that the mobile ion has available more than
the equivalent number of equal positions and that a transition state has a low activation energy, one still has not
yet explained the good conductivity. The analysis of the
pre-exponential factor -used in the description of the
temperature dependence of the conductivity- shows that an
appreciable fraction (say 10^{-2} to 10^{-1}) of the ions of the
mobile species are at a certain moment migrating. This
excludes an explanation using isolated defects. Furthermore,
the energy necessary to create the defect is much larger
than the activation energy of the migration process[3]. These
problems can be solved by assuming domains in which the
ions occupy one corresponding position[4,5]. In an adjacent
domain another position is occupied. Considering the domain
as a "defect", the energy (per ion) necessary to create the
defect can be low enough to be in agreement with the experimental diffusion equation. This was demonstrated with an
idealized model of β-Al_2O_3[4]. Although in this model the
walls between the domains do represent an additional charge
compared to the completely undisturbed lattice, the energy
is already low enough to explain the low activation energy
of the conductivity data. It has already been pointed out
that other domain configurations can be used avoiding this
charged surfaces of the domains[4]. Obviously, the energy
must be even more favourable for these configurations.

Displacement of ions in the wall means migration
during which the wall moves. Kinetic considerations show
that wall movements are fast in good ionic conductors.
Therefore, direct observation by time averaging techniques
(X-ray, electron microscopy) will be impossible.

The local configuration within a domain might differ
from the time -and space- averaged models derived from the

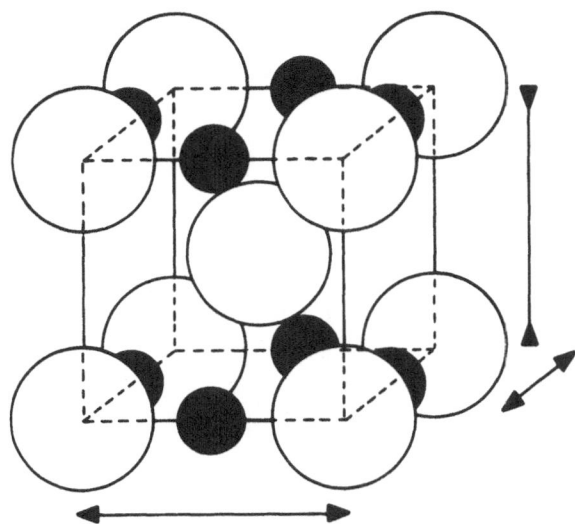

Figure 1. Placing two Ag-ions in the bcc unit cell leads
          to a tetragonal distortion.

X-ray data. See, for example, figure 1, in which two Ag-
ions have been placed in the bcc unit cell. It is immediately stressed that the differences will depend upon
the dimensionality of the diffusion paths[6]. For AgI where
Ag constitutes 50 % of the ions, the influence will be
larger than in β-Al$_2$O$_3$ (with Na in a special layer and
constituting less than 6 % of the ions). The problem arises
how to fit together the differently orientated domains
along the walls.

2. FITTING TOGETHER OF DOMAINS

The principal problem of fitting together domains
orientated differently will be illustrated with a two-
dimensional example. It is assumed that the space-averaged
configuration is a square. The local configuration will be
rectangular, with the long b-axis in different orientations
(see figure 2).

The diagonal has the same length and domains can be

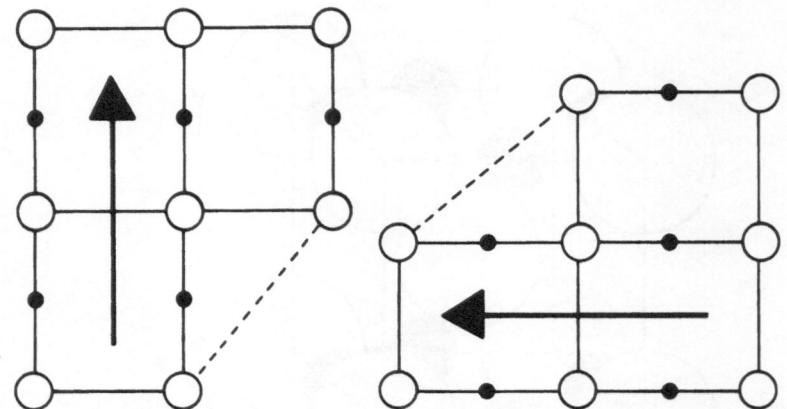

Figure 2. Rectangular lattices with the large axis in different directions can be fitted along the diagonal.

tied together along these diagonals (see fig. 3). The main
axis of the domains intersect under an angle δ, which is
related to the distortion of the unit cell according to

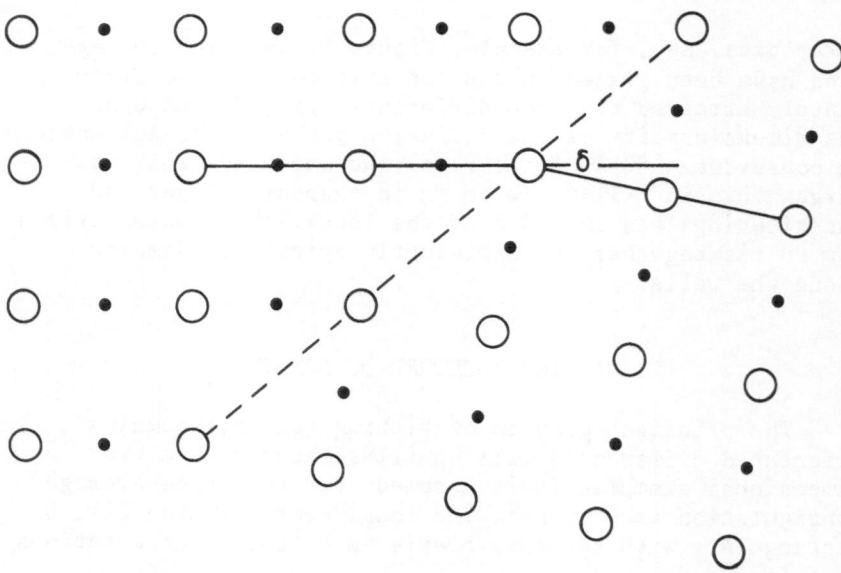

Figure 3. The directions of the main axis of the joint domains
differ the angle δ.

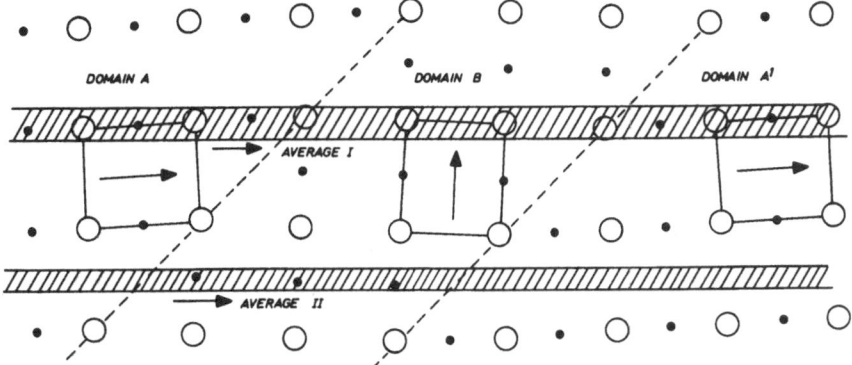

Figure 4. Joining several domains.

$tg\delta \approx p$, $b/a = 1+p$, in which a and b the dimensions of the unit cell and $p < 0.2$ (A deviation of 10 % from a square, thus $p = 0.1$ corresponds to $S \approx 6^{o}$).
The major aspects of the joining of many domains follow from figure 4. Space-averaging leads to an average square structure with a fractional occupation of the "equivalent" sites. Two conclusions are obvious from figure 4:
1) the space-averaged interpretation of X-ray data should lead to the conclusion that in the case of fractional (1/2) occupation, the charge density is not concentrated in the middle of the edge of the unit cell. This will be interpreted as a fractional occupation of near-by positions (see AVERAGE II).
2) the space-averaged interpretation of the fully occupied positions will lead to an unusual thermal vibration ellipsoide (see AVERAGE II).
Both aspects have been recognized in solid electrolytes when the normal procedures are applied to the X-ray data, assuming the material to be homogeneous. The present discussion demonstrates that another interpretation is possible.

3. FURTHER REMARKS AND CONCLUSIONS

Long ago it was recognized that the absolute intensity of X-ray reflections of single crystals is much higher than the value to be expected for ideal single crystals[7].

A mosaic structure can explain the experimental intensities[8] The orientations of the mosaic blocks make a small angle with each other. The reason for the occurrence of the mosaics was often obscure. A domain structure as discussed in the former section is a very natural example of mosaic structure in case that more than the equivalent number of sites is available. The suggestion is that domains do occur in that situation as static phenomena. In superionic conductors the domains have a dynamic character, in the sense that the walls and shapes are moving fast.

It must be stressed that a very limited amount of experimental data is available to analyze the proposed model with dynamic domains. Many interesting superionic conductors have to go through a phase transition in order to reach the highly conducting phase. Volume changes during the phase transition prevent the use of single crystals for the high temperature measurements. The powder diffraction methods give less information. It is interesting, however, that a tetragonal structure for α-AgI could not be excluded by Strock using the X-ray data only[1]. A tetragonal modification is to be expected when two Ag-ions are placed in a unit cell with the bcc structure. See figure 1.

It is also stressed here that the availability of more than the equivalent number of sites is not the only way to come to diffusion. Especially when rotating groups are present, the rotation together with a move of the mobile ion might create the new equivalent position. That position is not yet present when the move starts. Figure 5 illustrates the concept two dimensionally.

There is not much doubt that an instantaneous picture of solid electrolytes would have a domainlike character. The question is whether that observation would be of further importance. This brings us to the important question of time scales: flying time of the mobile ion versus residence time. When the ion is residing much longer at a certain spot than that it is on the move, the local configuration will be adjusted to the presence of the residing ion. Thus, the configuration will be more tetragonal than bcc cubic in α-AgI. With a slight generalization one can say that the local structure determines the high conductivity. If, however, the flying time is much larger than the residence time, the position of the I-ions will be more in a bcc con-

CONFIGURATION A

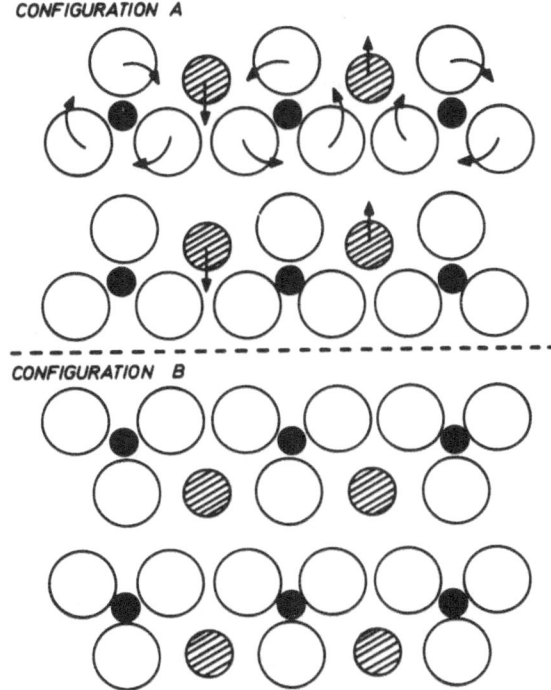

CONFIGURATION B

Figure 5. Creation of equivalent sites by the rotation of
        groups of ions.

figuration. In this case the conductivity determines the
structure. It is not to be expected that these time relationships will be correspondingly for all superionic conductors. The presently available information shows a slight
preference for a flying time larger than the residence time.
It must be pointed out, however, that the theoretical interpretation of the experimental data was performed by assuming
homogeneous crystals. It is not quite sure whether the
conclusions change by using a domain model, since interpretations have not yet be done using this approach.

    Summarizing, the conditions for the occurrence of super-
ionic conduction lead for certain electrolytes very naturally
to a domain structure in the description of an instantaneous

configuration of the mobile ions. Some results of the investigations of solid electrolytes are easily understood with
the domain model, such as unusual large thermal vibration
ellipsoids. The basic problem involved is dealing with time
scales, leading to the question whether the superionic
conduction determines the structure or the structure permits the superionic conduction. The value of the used concepts in selecting new materials has been discussed already[9,10].

Acknowledgement. The fruitful discussions with Mr.
H.A. Harwig are greatly appreciated.

REFERENCES

1. L.W. Strock, Z.Phys.Chem. B25, 441 (1934), B31, 132
   (1936).
2. See other contributions to this conference.
3. W. van Gool, J. Solid State Chem. 7, 55 (1973).
4. W. van Gool and P.H. Bottelberghs, J. Solid State Chem.
   7, 59 (1973).
5. W. van Gool in "Fast Ion Transport in Solids". W. van
   Gool, ed. North Holland, Amsterdam, 1973 p.201.
6. W. van Gool in "Mass Transport Phenomena in Ceramics",
   A.R. Cooper and A.H. Heuer, Eds., Plenum, New York,
   1975, p.139.
7. See, for example, H. Zachariasen, "Theory of X-ray
   Diffraction in Crystals" John Wiley, New York 1945,
   Chapter IV.
8. C.G. Darwin, Phil. Mag. 27, 315, 657 (1914), 43, 800
   (1922).
9. W. van Gool, Ann. Rev. Mater. Sci. 4, 311 (1974).
10. W. van Gool, in "Critical Material Problems in Energy
    Production", Ch. Stein Ed., Plenum Press, New York,
    to be published 1976.

REMARKS ON PHASE TRANSITIONS AND DYNAMICS IN

SUPERIONIC CONDUCTORS

B. A. Huberman*

Xerox Palo Alto Research Center

Palo Alto, CA 94304, USA

It is by now clear that the phase transitions one observes in superionic conductors can be described by a lattice gas model of one form or the other.[1-4] The recent developments in the field of critical phenomena, together with expansion techniques that allow for precise determinations of critical indices, give us a thorough understanding of Ising-like behavior in superionic conductors. One might wonder whether all what is left concerning their cooperative behavior is a systematic classification of phase transitions and thermodynamic data, with the remote possibility of using these systems to test dynamical scaling laws. The situation is far from being so settled. That there are still unresolved theoretical issues concerning phase transitions in superionic conductors is exemplified by the following question: What is the nature of the pseudoexchange coupling between ions? Although the possibilities have been discussed of its being Coulomb in nature[1,3,4], mediated via acoustic phonons[5] or optical ones as in the Jahn-Teller effect[2,6], the present experimental situation has not settled the question unequivocally. One of the reasons for this lack of precise answers can be traced directly to the very nature of the traditional measurements that are performed in superionic conductors; d-c conductivity and specific heat data are only able to provide clues to critical behavior that are independent of the nature of the pseudoexchange coupling between the mobile ions.

The purpose of these remarks is to point out that the dynamical response of superionic conductors can provide well-defined answers to the questions we posed above. Light scattering, diffuse X-ray scattering, ultrasonic attenuation and neutron scattering not only give valuable information concerning the excitations of the system, but can also probe the response functions near phase transitions. As we shall see, specific coupling mechanisms provide different signatures for the behavior of the correlation functions near critical points and hence there is a good possibility of sorting out the microscopic origin of the coupling between the excitations of superionic conductors.

In dealing with the excitations of the ionic liquid component of a superionic conductor one must recognize two types of processes that are relevant to the dynamics[5]: (i) charge density fluctuations of the mobile ions that behave like hydrodynamic modes with lifetimes of the order $(Dk^2)^{-1}$ where D is the ionic diffusion constant and k the wave vector of the fluctuation, and (ii) fluctuations in the local site population (pseudospin excitations) that leave the average density of the ions constant and which decay with lifetimes typical of elementary hopping times. Once the dynamics of the crystalline cage is considered (it is fully specified by giving its phonon spectrum) it is natural to ask for the coupled-mode behavior of a superionic conductor. In particular, since hopping frequencies of the mobile ions are of the same order of magnitude as phonon frequencies, one expects a rather strong interaction between the excitations of the ionic liquid and those of the cage.

The coupling between charged liquid fluctuations and acoustic phonons has been considered recently by Huberman and Martin.[5] By treating the effect of the phonons on the mobile ions as an effective frequency and wave vector dependent modulation of the chemical potential felt by the carriers it was possible to obtain the full response function of the system. In particular the case of phonon-pseudospin coupling (which is more effective than the phonon-density fluctuations interaction in damping certain phonon modes) leads to concrete predictions for the dynamics of a superionic conductor near a phase transition. In this case, the three pole structure of the response functions corresponds to a nonpropagating mode centered at imaginary frequency and two phonon modes at ω_p and $-\omega_p^*$ which are given by

$$\omega_p \simeq \omega_k - \frac{n_o A^2}{\rho V_s^2 k_B T} \frac{\omega_k \gamma}{(\omega_k^2 + \gamma^2)} \left(1 + i \frac{\omega_k}{\gamma}\right) \qquad [1]$$

with $\omega_k = V_s k$, V_s the speed of sound, n_o the density of mobile ions, γ their hopping frequency, ρ the mass density and A a coupling constant. As a phase transition is approached from below, Eq.[1] predicts two types of behavior, i.e., the real part of the phonon self energy is renormalized down in frequency (in other words, the phonon softens) and the imaginary part of the self energy, which determines its damping behaves like (assuming $\gamma = \gamma_0(T_c - T)$ and $\omega_k < \gamma$ at high temperatures)

$$I_m \omega_p \simeq \frac{n_o A^2}{\rho V_s^2 k_B T} \frac{\gamma_0(T_c - T)}{(1 + \gamma_0^2(T_c - T)^2/\omega_k^2)} \qquad [2]$$

which implies that as the phase transition is approached, the acoustic attenuation increases like $(T_c - T)^{-1}$ but close to T_c (i.e., when $\gamma_0(T_c - T)/\omega_k \approx 1$) it reaches a maximum and then starts decreasing. The softening of certain acoustic phonon modes in some fluorite crystals that are superionic conductors has been reported by the Oxford group[7] and although these systems do not ever reach critical behavior, these measurements seem to indicate that the coupling of the ionic degrees of freedom to acoustic phonons is not negligible. In the case of RbAg$_4$I$_5$ both Nagao and Kaneda[8] and Graham and Chang[9] have reported critical attenuation of the C_{44} phonon modes as the second order phase transitions at 208°K is approached. Therefore, in this particular system one is confident that the coupling of the liquid-like degrees of freedom to the acoustic modes of the cage plays an important role in the phase transition. Furthermore, the zero frequency pole of the response functions gives for the transition temperature

$$T_c \simeq \frac{n_o A^2}{\rho V_s^2 k_B} \qquad [3]$$

with k_B the Boltzmann constant. For $\rho/n_o \approx 290$ amv, $\rho = 5.4$ gm cm^{-3} $C_{44} = 0.49 \times 10^{11}$ dy and $A \approx 0.2$ eV one obtains for RbAg$_4$I$_5$ T$_c$≃170°K which is not far from the observed value of 208°K.

The coupling of the pseudospin excitations to the optical phonons of the crystalline cage also leads to specific predictions concerning the dynamics near a phase transition. In this case the poles of the response function give a nonpropagating mode and two optical phonons at ±Ω. If the coupling constant is given by δ, the zero frequency pole determines T$_c$ in the strong coupling limit through

$$T_c \simeq \frac{\delta^2}{\gamma_0 \Omega^2} \qquad [4]$$

As far as the dynamics of these coupled modes are concerned, Yamada, Takatera and Huber[10] have studied them near T$_c$ so again we have concrete predictions that can be tested by scattering experiments. So far, no soft optical phonon behavior has been reported in a superionic conductor.

To summarize, measurements that provide information on the dynamical response of superionic conductors are likely to provide us with answers as to the microscopic origin of the coupling between ions as well as the range of the interactions, which so far have been taken as phenomenological inputs into the theory.

References

1. H. Sato and R. Kikuchi, J. Chem. Phys. 55, 677 (1971).
2. H. J. Rice, S. Strassler and G. Toombs, Phys. Rev. Lett. 32, 596 (1974).
3. B. A. Huberman, Phys. Rev. Lett. 32, 1000 (1974).
4. W. J. Pardee and G. D. Mahan, J. Solid State Chem. 15, 310 (1975).
5. B. A. Huberman and R. M. Martin, Phys. Rev. B13 1498 (1976).
6. F. Lederman and M. B. Salamon, Bull. Am. Phys. Soc. 210, 331 (1976).
7. R. Harley, W. Hayes, A. Rushworth and J. F. Ryan, J. Phys. C, December 1975.
8. M. Nagao and T. Kaneda, Phys. Rev. B11, 2711 (1975).

9.   L. J. Graham and R. Chang, J. Appl. Phys. <u>46</u>, 2433
     (1975).
10. Y. Yamada, H. Takahera and D. L. Huber, J. Phys. Soc.
     Japan <u>36</u>, 641 (1974).

* Spring address: Racah Institute of Physics
                   Hebrew University of Jerusalem
                   Jerusalem, Israel

VIBRATION EFFECT IN THE ORDER-DISORDER TRANSITION THEORY OF $RbAg_4I_5$[†]

Sang-il Choi and W. M. Lee

Department of Physics & Astronomy
University of North Carolina
Chapel Hill, North Carolina 27514

ABSTRACT

Vibrational frequencies are expected to be quite diferent at different sites of the carrier ion in a superionic conductor. For a simple model of $RbAg_4I_5$, we have studied the effect of vibrational frequency differences on the equilibrium distribution of Ag^+ ions. It is found that one could introduce a considerable error by neglecting the above mentioned effect.

INTRODUCTION

In the theory of order-disorder transition it has been a common practice to neglect contribution from the vibrational motion of ions to the partition function[1,2]. This approximation is justified if the vibrational frequency distribution has a weak dependence on the spatial distribution of ions. One characteristic of superionic conductors is the availability of more than one site per carrier ion. Then it is natural to consider an order-disorder transition with respect to the arrangement of carrier ions over these sites. In applying a theory of order-disorder transition to this problem it is unlikely that one can justify neglecting the variation of vibrational frequencies over different sites. For example, previously we found that vibrational frequencies at the Beevers-Ross site of β- alumina is very different from that of the anti-Beevers-Ross site.[3]

Our objective of this study is to see the importance
of the vibrational partition function (or vibrational free
energy) through a relatively simple calculation. In 1969
Wiedersich and Johnston[4] reported an application of the quasichemical approximation of the theory of order-disorder
transition to $RbAg_4I_5$. Their work was based on the assumption that the vibrational frequencies are independent of the
distribution of Ag^+ ions. As a result of their analysis of
experimental data, they deduced the site energy differences
to be ~0.027 eV and ~0.042 eV. The mutual repulsion energy
between two silver ions on the adjacent II-sites was found
to be ~0.035 eV. In view of the existence of this interesting
work, we decided to study a simplified model of $RbAg_4I_5$.

MODEL AND CALCULATION

We consider the cubic phase of $RbAg_4I_5$ of which crystal
structure has been determined by Geller[5]. Iodide ions form
56 tetrahedra per unit cell which serve as sites for silver
ions. The four rubidium ions are surrounded by distorted
octahedra of iodide ions. The 56 sites are classified into
three types: 8 of I-type, 24 each of II-type and III-type.
Since there are only 16 silver ions, 3.5 sites are available
for each silver ion.

We assume that all the ions except silver ions form a
rigid crystal framework which produces the force field for
silver ions. Then, as far as our model is concerned, the
crystal is characterized by the site energy differences,
the interaction energy between silver ions, and the vibrational frequencies of silver ion at the three different sites.
The vibrational motion of silver ion at each site is assumed
to be represented by one isotropic three dimensional harmonic
oscillator. For interaction energies we consider only nearest neighbor pairs. Following Wiedersich and Johnston we
assume the pair interaction energy is inversely proportional
to the distance between the sites. Then our model is specified by seven parameters: 3 site energies, one pair interaction energy, and 3 vibrational frequencies. Some properties and notations of the three types of sites are listed
in Table 1. The fractional occupation (i.e. number of silver
ions on a given type site devided by the number of site of
this type) measured at room temperature[5] is given in the
sixth   column. The last column contains the fraction of

Table 1

Type	Site energy	Number of sites	Frequency	Number per unit cell	Frac. occup.	Frac. ion in type
I	e_1	N_1	ω_1	8	0.111	0.055
II	e_2	$N_2 = N$	ω_2	24	0.391	0.586
III	e_3	N_3	ω_3	24	0.229	0.344

silver ions in a given type site (number of silver ions on the given type site devided by the total number of silver ions). The notations used for pairs are summarized in Table 2.

In stead of the quasichemical approximation, we resort to Bragg-Williams approximation. For our purpose this approximation is expected to be adequate. In order to find the equilibrium distribution of silver ions, the expression for free energy, $F(T, \{n_i\})$, is obtained in terms of site energies (e_i), pair interaction energies (u_i), site vibrational frequencies (ω_i), and site occupation numbers (n_i). The free energy is, then, minimized with respect to the set of occupation numbers with the constraint that total number of silver ions is constant. In Bragg-Williams method, the number of each type of site pairs is simply related to the site occupation numbers:

$$m_1 = 3n_1n_2/N_2 , \quad m_2 = n_2^2/2N_2, m_3 = m_{\bar{3}} = n_2n_3/2N_2 \quad (1)$$

The free energy of the system, then, is given by

$$F = m_1u_1 + m_2u_2 + m_3u_3 + m_{\bar{3}}u_{\bar{3}} + E_H$$
$$- T[S_c(n_1,n_2,n_3) + S_H(n_1,n_2,n_3,\omega_1,\omega_2,\omega_3)] \quad (2)$$

where S_c is the configuration entropy while E_H and S_H are the energy and the entropy derived from the vibratonal partition function. The configurational entropy is given by

$$S_c = k\ell n \, W(n_1,n_2,n_3) \quad (3)$$

Table 2

Type of site pairs	Pair energy	Number of site pairs	Number of site pairs with occupation, Ag-Ag
II-I	u_1	$M_1 = N$	m_1
II-II	u_2	$M_2 = N/2$	m_2
(II-III)$_1$	u_3	$M_3 = N$	m_3
(II-III)$_2$	$u_{\bar{3}}$	$M_4 = N$	$m_{\bar{3}}$

where k is the Boltzman constant and W is the total number of distinguishable arrangement of silver ions for a given set of occupation numbers. In Bragg-Williams method

$$W = \frac{N_1!}{n_1!(N_1 - n_1)!} \frac{N_2!}{n_2!(N_2 - n_2)!} \frac{N_3!}{n_3!(N_3 - n_3)!} \quad (4)$$

With the help of Stirling's approximation, then,

$$S_c = k\left[\sum_{i=1}^{3} (n_i - N_i)\ln\left(1 - \frac{n_i}{N_i}\right) - \sum_{i=1}^{3} n_i \ln\frac{n_i}{N_i} \right] \quad (5)$$

In the high temperature approximation the partition function of a silver ion on i-site is given by

$$z_i = \left[kT/\hbar\omega_i \right]^3 \exp\left[\left(\tfrac{3}{2}\hbar\omega_i - e_i\right)/kT \right] \quad (6)$$

The total vibrational partition function of a given set of values of occupation numbers is

$$Z_T = \Pi_i\, z_i^{\,n_i} . \quad (7)$$

From this partition function we obtain

$$E_H = \sum_i n_i e_i + 3nkT - \frac{3}{2}\hbar \sum_i n_i\omega_i \quad , \quad (8)$$

$$S_H = 3k\left[n - \sum_i n_i \ln\left(\hbar\omega_i/kT\right)\right] \quad (9)$$

where $n = n_1 + n_2 + n_3$ and is the total number of silver ion. The free energy is given by the following equation.

$$F = \frac{3n_1 n_2}{N_2} u_1 + \frac{n_2^2}{2N_2} u_2 + \frac{n_2 n_3}{2N_2} u_3 + \frac{n_2 n_3}{2N_2} u_{\bar{3}}$$

$$+ \sum_i n_i e_i - \frac{3}{2} \hbar \sum_i n_i \omega_i \tag{10}$$

$$- kT \left[\sum_i (n_i - N_i) \ln \left(1 - \frac{n_i}{N_i}\right) - \sum_i n_i \ln \frac{n_i}{N_i} - 3 \sum_i n_i \ln \frac{\hbar \omega_i}{kT} \right]$$

The equilibrium value of occupation numbers are determined by minimizing F with respect to n_i. The two equations thus obtained are as follows.

$$- \frac{3u_1}{N_2} n_1 - \frac{(3u_1 - u_2)}{N_2} n_2 - \frac{(u_3 + u_{\bar{3}})}{2N_2} n_3 + (e_1 - e_2)$$

$$+ \frac{3}{2} \hbar (\omega_2 - \omega_1) + 3 \, kT \ln \frac{\omega_1}{\omega_2} + kT \ln \left[\frac{n_1}{n_2} \frac{N_2 - n_2}{N_1 - n_1} \right] = 0, \tag{11}$$

$$- \frac{3u_1}{N_2} n_1 - \frac{(2u_2 - u_3 - u_{\bar{3}})}{2N_2} n_2 - \frac{(u_3 + u_{\bar{3}})}{2N_2} n_3 + (e_3 - e_2)$$

$$+ \frac{3}{2} \hbar (\omega_2 - \omega_3) + 3 \, kT \ln \frac{\omega_3}{\omega_2} + kT \ln \left[\frac{n_3}{n_2} \frac{N_2 - n_2}{N_3 - n_3} \right] = 0, \tag{12}$$

Solution of these equations would yield the equilibrium silver ion distribution. Since our main objective is to see the effect of vibrational frequencies on the distribution of silver ions, we chose the parameter values determined by Wiedersich & Johnston, i.e. $e_1 - e_2 = 0.042$ eV, $e_3 - e_2 = 0.027$ eV, and $u_2 = 0.035$ eV. The values of other pair energies are also same as theirs. In order to facilitate solution of Equations (11,12) we assumed the following relationship among the frequencies.

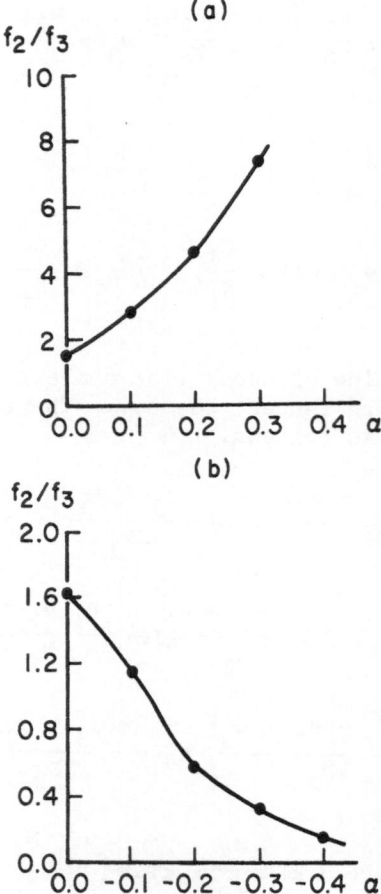

Figure 1. Ratio of number of silver ions in II-site and
           III-site.

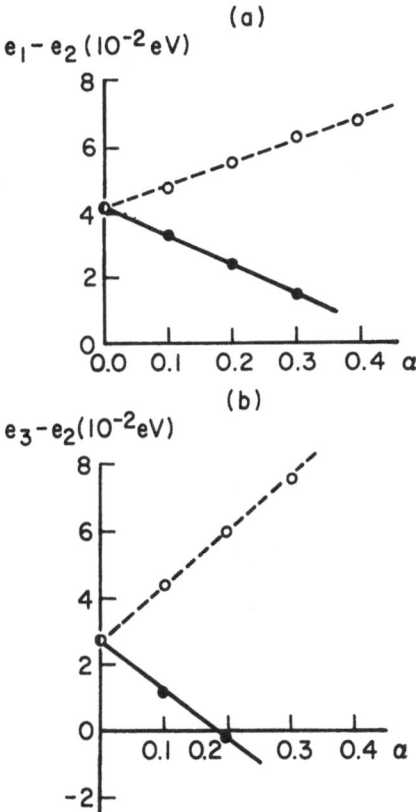

Figure 2.   Site energy difference (a) between I-site and
II-site, and (b) between III-site and II-site.

$$\omega_1 = \omega, \qquad \omega_2 = \omega(1 - \alpha), \qquad \omega_3 = \omega(1 + \alpha), \qquad (13)$$

where α is a parameter. At 300°K the values of $\{f_i = n_i/n\}$ are calculated for various values of α by numerical solution of Eqs. (11, 12). The results are shown in Fig. 1. The values of $\{f_i\}$ for $\alpha = 0$ agree quite well with experimental values as expected since the parameter values used were determined by Wiedersich & Johnston to reproduce the experimental values of $\{f_i\}$ when $\alpha = 0$, i.e. the vibrational effect is neglected. One must notice a large deviation from the experimental values as α is increased or decreased from 0.

The vibrational effect is demonstrated more effectively by evaluating the values of site energy differences which yield the experimental values of $\{f_i\}$. For various values of α, the site energy differences $e_1 - e_2$ and $e_3 - e_2$ are calculated and the results are shown in Figure 2. Set 2 (broken line) is for the negative values of α.

DISCUSSION

In this study a simple theory of order-disorder transition is applied to a simplified model of $RbAg_4I_5$ crystal in order to understand the effect of vibrational motion of silver ions at different sites. This work indicates that a large error may result in the estimation of site energies and pair interaction energies through application of a theory of order-disorder transition unless the vibration effect (kinetic effect) is properly included. Figure 2 shows that e_3 could be smaller than e_2 if ω_3 were large enough compared to ω_2 even though the experimentally observed occupation number of III-site is less than that of II-site. This is due to the larger entropy for a smaller frequency compensating the larger site energy.

Better theories of superionic conductors for equilibrium and transport properties with proper incorporation of vibration effect are desired for better understanding of this interesting class of materials.

[†]This work was supported by ARPA through UNC Materials Research Center and by the National Science Foundation.

REFERENCES

1. R. H. Fowler & E. A. Guggenheim, Statistical Thermodynamics, (Cambridge University Press, 1939) p.541.

2. E. A. Guggenheim, Mixturs(Oxford Press, 1952) p.101.

3. J. C. Wang, M. Gaffari, and S. Choi, J. Chem. Phys. 63. 772 (1975).

4. H. Wiedersich and W. V. Johnston, J. Phys. Chem. Solids, 30, 475 (1969).

5. S. Geller, Science 157, 310(1967).

ABSTRACTS

Superionic Conductivity and Order-Disorder Transition.*

R. KIKUCHI, Hughes Research Labs; H. SATO, Purdue University.--Superionic conductivity is calculated using the Path Probability method[1] to show how ionic conductivity of anions O^{--} (via the vacancy mechanism of migration) in the anion sublattice is influenced by an order-disorder phase change among non-migrating cations (A and B) in their sublattice. The A and B cations occupy an FCC sublattice, while O^{--} and vacancies occupy an interpenetrating SC sublattice. (The overall lattice structure is CaF_2 type). Possible examples include $CaO-ZrO_2$ and $CaO-CeO_2$. When the cation composition is varied, the FCC sublattice goes from the disordered phase to an A_3B type ordered phase. The calculation shows that this phase transition can reflect as a sharp change of the O^{--} ionic conductivity in the SC sublattice, although the distribution of anions and vacancies on the SC sublattice itself is not affected by the transition.

The basic cluster used in the Path Probability method is the double tetrahedron of the FCC sublattice in which a pair of the SC sublattice points sit.

*Work supported in part by NSF.
[1]H. Sato and R. Kikuchi, J. Chem. Phys. 55, 677 and 702 (1971)

A Microscopic Model for Sublattice Disorder in Ionic Crystals.*

D. O. WELCH AND G. J. DIENES, Brookhaven National Laboratory.--We have recently shown how the order dependence of the energy of sublattice disorder affects the form of the sublattice melting transition in ionic crystals.[1,2] A microscopic model for the energy of disorder has been derived by adapting an approximate theory for the energy of formation of isolated point defects to the case of large defect concentrations. The model, which is a modification of the Mott-Littleton theory of point defect energetics, takes account of electronic polarization and ionic displacements and is based on a realistic description of cohesion in ionic crystals. The approximate model agrees within about 30% with more elaborate calculations in the isolated defect limit. At large defect concentrations, in the molecular field approximation, relaxation of ionic positions and electronic polarization occur as a consequence of fluctuations of the local interaction potentials from the mean. The energy per added defect (disorder) decreases with increasing disorder, largely due to the volume dependence of elastic and dielectric constants. When this model is combined with an approximate description of the liquid state, conditions on the relative ionic charges, radii, and polarizabilities which permit the onset of sublattice melting prior to overall melting can be obtained. Such conditions are illustrated by model calculations for the zincblende structure which show that a large difference between the electronic polarizabilities of the two ionic species favors sublattice melting, while a large volume dependence of the total unit cell polarizability favors overall melting.

*Research supported by the Energy Research and Dev. Admin.
[1]D. O. Welch and G. J. Dienes, J. Electronic Materials $\underline{4}$, 973 (1975).
[2]D. O. Welch and G. J. Dienes, submitted to J. Phys. Chem. Solids.

Minimum Energy Path Model Calculation for Several Fast Ionic Conductors.* O. B. AJAYI, I. D. RAISTRICK, L. E. NAGEL, AND R. A. HUGGINS, Center for Materials Research, Stanford University.--The structurally-explicit minimum energy path model for fast ionic conductors, initially applied to the cubic silver iodide structure, has been extended to treat the tetragonal rutile and cubic fluorite structures. The results of these calculations and their relevance to experimental observations in a number of fast ionic conductors will be discussed.

*This work was supported by ARPA through ONR.

One-Dimensional Models for Superionic Conductors. J. C. WANG* AND DAVID F. PICKETT, Air Force Aero Propulsion Laboratory, Wright-Patterson Air Force Base.--Three one-dimensional models, interstitialcy-like, vacancy-like, and interstitial-like, are constructed to study general properties of superionic conductors. Potential energy curves along the diffusion path due to all ions other than conducting species are simulated by $(V_0/2) (1-\cos(2\pi x/a))$, where a is the lattice constant. By considering only the Coulomb and repulsive potentials among conducting ions, the activation energies, E_a, are calculated as functions of V_0. It is shown that due to local correlated motions of conducting ions, the values of E_a are much smaller than those of V_0. For instance, for the interstitialcy-like mechanism, E_a changes from 0.02 eV to only 0.4 eV while V_0 changes from

0.8 eV to 3.2 eV. For the vacancy-like mechanism, E_a is less than 0.2 eV for V_0 below 0.9 eV, and is less than 0.05 eV for V_0 less than 0.4 eV. Other properties such as attempt frequency, conductivity, ion distribution, and diffuse X-ray scattering pattern are discussed. The correspondences between the models and actual superionic conductors, such as β-alumina, β''-alumina, and α-AgI are also discussed. Comments on existing theories of superionic conductors are presented.
*NRC Resident Research Associate

A Quantum Mechanical Theory of Ionic Diffusion.*
E. GORHAM-BERGERON, Sandia Labs--The predictions of a quantum mechanical tunneling theory of low concentration ionic diffusion will be discussed. The theory uses the Kubo formula for the diffusion coefficient, harmonic wavefunctions for the ions and polaronic wavefunctions for the lattice. All direct tunneling processes (including those between states of different energies) are included, as well as processes that involve tunneling accompanied by transitions within a well. An exact expression for the diffusion coefficient has been found, which exhibits qualitatively different characteristics in different parameter regimes (predicting, for example, the differences in the temperature-dependent isotope effects for hydrogen diffusion in bcc and fcc metals). In addition, the theory predicts that even in the purely tunneling regime the diffusion coefficient may become large and exhibit an activated form. Isotope effects and deviations from an Arrhenius behavior for diffusing particles other than hydrogen will also be discussed.
*Work supported by the U.S. Energy Research and Development Administration, ERDA.

Silver Conductors

CRYSTAL STRUCTURE AND CONDUCTIVITY OF AgI-BASED SOLID ELECTROLYTES II

S. Geller

Department of Electrical Engineering

University of Colorado, Boulder, CO 80309

INTRODUCTION

Since the last review of the author's work in the field of solid electrolytes at a NATO conference,[1] two structures have been determined in his laboratory, those of $Py_5Ag_{18}I_{23}$ and the intermediate temperature form of $RbAg_4I_5$. Some difficulties have developed with the structure of the silver iodide tungstate solid electrolyte mentioned in the abstract for this conference. In particular, very recent conductivity measurements at elevated temperatures indicate that its formula might be $Ag_{51}I_{35}(W_4O_{16})_2$ instead of $Ag_{26}I_{18}(W_4O_{16})$. (Takahashi, Ikeda and Yamamoto[2] first reported this solid electrolyte to have the formula $Ag_6I_4WO_4$.) It will therefore be necessary to report on this material at a later date.

This paper will be divided into three parts, a brief review of the structure of $Py_5Ag_{18}I_{23}$, a general discussion of AgI-based solid electrolytes, and a brief report on the low temperature modifications of $RbAg_4I_5$.

THE STRUCTURE OF $Py_5Ag_{18}I_{23}$[3,4]

This material was discovered in the course of earlier work on $PyAg_5I_6$.[5,6] Crystals of this material belong to space group $P\bar{6}2m$, with one formula unit, $[C_5H_5NH]_5Ag_{18}I_{23}$ in the trigonal unit cell with lattice constants a=13.62, c=12.58Å. Fig. 1[3,4] depicts the structure and its pathways for Ag^+ ion diffusion. There are 55 iodide tetrahedra and

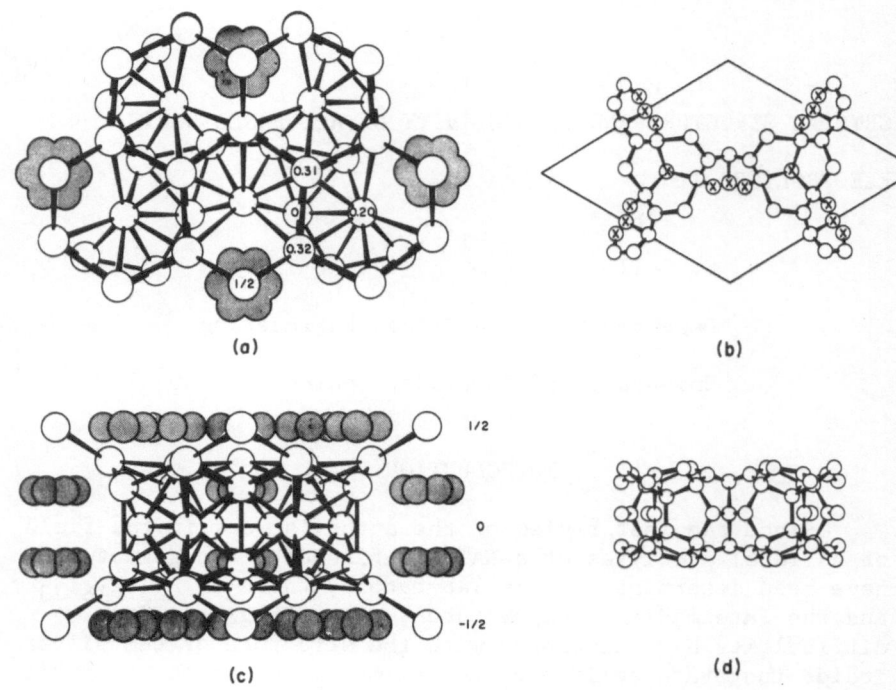

FIG. 1. (a) Top view of iodide-ion arrangement in $Py_5Ag_{18}I_{23}$.
The pyridinium ions (stippled) on the hexagonal axes are also
shown. (b) Top view of Ag^+-ion paths in $Py_5Ag_{18}I_{23}$. The
equilibrium Ag^+-ion sites (located at or near iodide tetrahedra centers) are shown. A cross indicates a connection between upper and lower halves of the Ag^+ path network within
the conducting layer. See also (d). (c) Side view of the
iodide arrangement in $Py_5Ag_{18}I_{23}$. The pyridinium ions (stippled) in the $\pm\frac{1}{2}c$ levels of the unit cell are shown. These
together with the I^- ions at $\pm(0,0,\frac{1}{2})$ block movement of Ag^+
ions in the c-axis direction. (d) Side view of the Ag^+-ion
paths in $Py_5Ag_{18}I_{23}$. (From ref. 3.)

18 Ag^+ ions per cell. This implies that if the silver ions
were uniformly distributed over the available sites, the
occupancy per site would be 18/55 or 0.327. There are, however, <u>seven</u> crystallographically nonequivalent sets of Ag^+

ion sites in the unit cell and their occupancies <u>per</u> <u>site</u>
are quite different: 0.60(0.12), 0.45(0.03), 0.18(0.06),
0.30(0.06), 0.42(0.02), 0.53(0.02) and 0.01 (by difference)
for 3-, 4-, 6-, 6-, 12-, 12- and 12-fold sites respectively.
(The numbers in parentheses are standard deviations as calculated by least squares.) This distribution is <u>consider-
ably</u> removed from a uniform one; note, for example, that
one set of 12-fold sites is essentially empty.

   That these sites are essentially empty does not mean,
of course, that no carriers pass through them. It <u>can</u> mean
that their site energies are relatively high and thus that
the carriers are ejected from them quite rapidly. In fact,
this <u>must</u> be the case for this crystal because[4] every Ag^+
ion passageway through the crystal includes at least two of
these sites in each unit cell. What is more, it <u>is</u>
possible to find passageways which bypass each of the six
other symmetry sets of equilibrium Ag^+ ion positions.

   The most important aspect of this material is that its
crystal structure shows that it must be a "two-dimensional"
solid electrolyte. This may readily be seen in Fig. 1.
Crystals of $PyAg_5I_6$[5,6] are three-dimensional conductors,
but the increased ratio of PyI to AgI in the $Py_5Ag_{18}I_{23}$
crystal results in layers of Py^+ ions surrounding iodide
ions in 3:1 ratio such that diffusion of Ag^+ in the c-axis
direction is precluded. The material has low conductivity
(see Fig. 3 of ref. 4) over a temperature range of about
90°K above room temperature. Over this range the conductivity is an order of magnitude lower than that of $PyAg_5I_6$.[6]
Part of the reason for this large difference must be the
restriction to essentially two dimensions for the diffusion
of Ag^+ ions. A substantial amount of the crystal space is
occupied by the pyridinium and iodide ions which block diffusion. One way of seeing this[4] is to calculate the total
volume of the iodide tetrahedra in $Py_5Ag_{18}I_{23}$ and compare
it with the total volume of iodide polyhedra in $PyAg_5I_6$. In
the former the number is 31.6% while in the latter, it is
52.5%, in $RbAg_4I_5$ it is 47.1%. Of course, such comparisons
do not give the whole story. It is not only that three-
dimensional networks are most favorable for high conductivity, because they provide more passageways for the
carriers, but simplicity of the pathways must also play a
role. (See Discussion.)

DISCUSSION

The property of electrolytic conductivity in ionic solids does not have specific symmetry requirements (as does ferroelectricity, for example) and there is no apparent reason that there should be. But there are indeed <u>structural</u> requirements for the existence of this property. We refer here to those materials called "true solid electrolytes" by the author. A detailed description of these relative to electrolytic conductivity that occurs because of deviation of all crystals from ideal structure has been given elsewhere[4]. This structural dependence was first enunciated almost a decade ago in the paper on the crystal structure of RbAg$_4$I$_5$,[7] the material that, to date, has the highest room temperature (electrolytic) conductivity, $0.27(\Omega cm)^{-1}$.[8,9] (The material was discovered independently by Owens and Argue[10] and by Bradley and Greene[11].) Since the first discussion of this structural dependence, the determination of several other structures of AgI-based solid electrolytes was considerably facilitated by utilizing the ideas expressed in the first paper, especially the requirement that there be present a network of passageways, formed by the face-sharing of iodide polyhedra.

In a paper written shortly after that[7] on RbAg$_4$I$_5$, Wiedersich and I[12] showed that it was possible to describe all the then known chalcogenide and halogenide true solid electrolytes in the same model. Because at that time very few structures of solid electrolytes were known, it was too early to rule out <u>completely</u> the type of distribution of Ag$^+$ ions suggested many years ago by Strock[13] to exist in α-AgI. Ultimately, the only way to rule out this type of distribution in α-AgI is to do a single crystal structure determination of that phase (which we intend). However the results of all present <u>definitive</u> structural work on AgI-based solid electrolytes are consistent. The equilibrium positions of the Ag$^+$ ions reside inside the iodide polyhedra. These iodide polyhedra are either tetrahedra or octahedra with the former <u>much</u> more favored. The results show that it is highly unlikely that the Ag$^+$ ions will stay anywhere else very long. Our model predicts 12 equivalent tetrahedral sites for Ag$^+$ ions in α-AgI.

In the first paper on the crystal structure of RbAg$_4$I$_5$,[7] it was strongly emphasized that there are in RbAg$_4$I$_5$ three crystallographically nonequivalent sets of sites over which the Ag^+ ions are <u>nonuniformly</u> distributed (see last section). Only over crystallographically equivalent sites should the distribution be expected to be uniform -- if it were not, the sites could not be crystallographically equivalent. Conversely, if sites are not crystallographically equivalent, the distribution should not be uniform over them. In the several structures for which these distributions have been determined, all are very far from uniform. Further they should not be expected to become uniform below the melting point of the material, unless a transition occurs making all sites that can be occupied by the current carriers crystallographically equivalent. This does not happen to any of the structures with which this author has thus far directly dealt.

Listed in Table 1 are several self-explanatory quantities for six AgI-based solid electrolytes including α-AgI itself. We wish to make comparisons with α-AgI and therefore have obtained values of conductivity for the other compounds by extrapolation. These extrapolations are reasonable; all the conductivities have been measured to at least 100°C. It should be pointed out that these conductivities are averages over the polycrystalline materials. For the cubic materials $RbAg_4I_5$ and α-AgI the conductivity is, of course, isotropic. For $PyAg_5I_6$, $Py_5Ag_{18}I_{23}$, and $[(CH_3)_4N]_2Ag_{13}I_{15}$ there are two independent components in the conductivity tensor. For $Py_5Ag_{18}I_{23}$ the crystal structure establishes that $\sigma_3=0$; therefore, $\sigma_1=1.5<\sigma>$. It is very likely that the conductivities of the other two crystals are also highly anisotropic. To date, no directional conductivity measurements have been made on any of the anisotropic materials, though we are hoping to do so.

The dramatic entry in Table 1 is the conductivity of $PyAg_5I_6$, at 145°C, which is comparable with that of α-AgI. If, as expected, the conductivity of $PyAg_5I_6$ is highly anisotropic, then one may also expect an even higher conductivity in one of the principal directions, most likely along the c-axis, at 145°C.

One might speculate about this high conductivity of $PyAg_5I_6$. $PyAg_5I_6$ has only about two-thirds the carrier concentration of α-AgI and little more than one-half its volume given to the passageways for the Ag^+ ions as opposed to 100%

for α-AgI. As shown elsewhere,[6] the passageways in $PyAg_5I_6$ that are skew to the c-axis are rather complicated, but the simplest passageway of any found in solid electrolytes of noteworthy conductivity is in $PyAg_5I_6$. It consists of face-sharing octahedra along the c-axis. These are <u>straight</u> channels as opposed to the "zig-zag" of the channels formed by tetrahedra as in $RbAg_4I_5$, $[(CH_3)_4N]_2Ag_{13}I_{15}$ and α-AgI. It appears[6] that at 145°C, about 55% of the sites in these simple channels of $PyAg_5I_6$ are occupied. There is no problem of nearest Ag^+ ion neighbor interference along these channels; i.e. all the sites in these channels can be occupied and are at -30°C.[5]

Now the enthalpies of activation of motion of the solid electrolytes have also been obtained from measurement of conductivity on polycrystalline materials and they are therefore an average. But even if they were obtained from directional measurements of conductivity, they would still be a directional average. This, in fact, is one of the outstanding problems in solid electrolytes; it would be most enlightening to know what the <u>individual</u> activation enthalpies of motion are. My guess is that they are much lower for the octahedral sites in $PyAg_5I_6$ than they are for the tetrahedral sites, not only in $PyAg_5I_6$, but also in all the other compounds including α-AgI and that this is an important reason that $PyAg_5I_6$ has such a high conductivity at 145°C. (Please note: I have chosen 145°C so that comparison with α-AgI could be made. Actually, the conductivity of $PyAg_5I_6$ is rather high at lower temperatures: at 23°C it is 0.077 (but see the discussion in ref. 6); at 55°C it is 0.29; at 129°C it is 0.88 $(\Omega\text{-cm})^{-1}$.)

Table 1 indicates that the dependence of the conductivity on carrier and site concentration is not simple. Compare for example, the cases of $PyAg_5I_6$ and $RbAg_4I_5$: The carrier concentrations are the same for both while the site concentration of the former is larger than that of the latter; yet at 145°C, the conductivity of the former is <u>much</u> larger than that of the latter. Again, compare α-AgI and $PyAg_5I_6$: Both the carrier and (especially) the site concentrations of the former are larger than those of the latter; yet at 145°C, the conductivity of the latter is comparable with that of the former. Compare $Py_5Ag_{18}I_{23}$ with $[(CH_3)_4N]_2Ag_{13}I_{15}$: the carrier and site concentrations of the latter are respectively 16 and 20% higher than those of the former, but the conductivity of the latter is five times

Table 1. Some specialized data
on AgI-based solid electrolytes.

Formula	Mobile Ag^+ ions/unit cell	Ag^+ ion sites/unit cell	$\dfrac{\text{Vol. of channels}}{\text{Vol. of unit cell}}$
α-AgI	2	12	1.00
$RbAg_4I_5$	16	56	0.471
$PyAg_5I_6$	10	34	0.525
$[(CH_3)_4N]_2Ag_{13}I_{15}$ [a]	13	41	0.39
$Py_5Ag_{18}I_{23}$	18	55	0.316

Formula	n_{Ag} $10^{22} cm^{-3}$	n_s $10^{22} cm^{-3}$	σ (ohm-cm)$^{-1}$ 296K	418K	Ref. for σ
α-AgI	1.57	9.42	--	1.31	14
$RbAg_4I_5$	1.13	3.94	0.27	0.66	9
$PyAg_5I_6$	1.07	3.65	0.077	1.1	6
$[(CH_3)_4N]_2Ag_{13}I_{15}$	1.04	3.28	0.04	0.26	15
$Py_5Ag_{18}I_{23}$	0.89	2.72	0.008	0.05	4

[a]For the crystal structure of this material, see ref. 15.

Note: All <u>calculated</u> quantities are room-temperature values.

that of the former. The lowest conductivity of all is that
of $Py_5Ag_{18}I_{23}$ which has the lowest site concentration (or
passageway volume) of all. There can be no doubt that the
low conductivity of $Py_5Ag_{18}I_{23}$ is a direct result of its
being only a two-dimensional conductor.

It should be clear that while carrier and site concentrations are important parameters, three-dimensional are
better than two-dimensional networks and that simplicity of
the passageways must enhance the conductivity of solid electrolytes. One should include in this "simplicity" the effects of ion-repulsions resulting from proximity to (nearest)
neighboring Ag^+ ion sites and to the structure-stabilizing

and immobile ions, such as Rb^+, Py^+, $(W_4O_{16})^{8-}$, etc., site energy differences, and correlations in the movement of ions.

PHASE TRANSITIONS IN $RbAg_4I_5$

The reader should be forewarned that there is always the danger that too brief an account will lead to misunderstanding; however, we hope to have a more detailed account ready for publication after the work on the lowest temperature phase is completed.

There are two low temperature transitions of $RbAg_4I_5$, a first order one at 121.8°K[10,16] and a second order one at 209°K[7,16]. In both cases the iodide superstructure persists, so that all three phases (i.e. including the high temperature cubic one) are closely related, despite the first order transformation at 121.8°K. (Actually, there are really two high temperature cubic phases, a right- and a left-handed one[7,17] belonging to space groups $P4_132$ and $P4_332$, but the difference between them is unimportant to this discussion.)

The transition at 209°K is to space group R32. The lattice constants of this rhombohedral phase at 130°K are a=11.16Å and α=90.0° within a few minutes. The unit cell contains four $RbAg_4I_5$ just as in the cubic case. In this case, the fourfold positions of the Rb^+ ions split into a threefold and a 1-fold set; each eightfold set of positions of the I^- and Ag^+ ions splits into a 6-fold and a 2-fold set, the 12-fold set of positions of the I^- ions splits into a 6-fold and two 3-fold sets and the general or 24-fold positions split into four sets of 6-fold positions. As in the cubic case, it is <u>impossible</u> for this rhombohedral phase to be ordered. There is simply a redistribution of the Ag^+ ions which is more preferential than in the cubic case. Thus the transition <u>is from one disordered state to another</u>!

The understanding of the intermediate phase is facilitated by an understanding first of the phase that exists below 121.8°K. This phase is trigonal, belonging to space group P321 and still very closely related to the rhombohedral phase. It is a <u>distortion</u> of the triply primitive cell of the rhombohedral phase to a primitive one. It therefore contains three times as many $RbAg_4I_5$ as the cubic or rhombohedral cells, or 12 $RbAg_4I_5$; its lattice constants are a=15.78, c=19.32Å. The space group in this case <u>does</u> allow the structure to be completely ordered (but see below).

In the trigonal cell of the lowest temperature form, the Rb^+ ions are in 4 sets of positions: a 1-fold, 2-fold, 3-fold and 6-fold. The iodides are in three 2-fold sets, two 3-fold sets and eight 6-fold sets. The original cubic 8-fold set for the Ag^+ ions goes into three 2-fold sets and three 6-fold sets and the original cubic 24-fold sets <u>each</u> go into twelve 6-fold sets. There are numerous ways in which the Ag^+ ions could order in these sets, but there are restrictions depending on the nearness of the Ag^+ ion sites. For example, in the cubic case, only half the Ag(II) sites can be filled. This is still the case for their analogs in the trigonal case (as well as in the rhombohedral case, of course). This eliminates two of the twelve 6-fold sets immediately because half of each of these are too close to the other half. Of the remaining ten 6-fold sets only certain combinations of five sets may be filled. The filling of these certain combinations of Ag(II) type sets puts constraints on the filling of the Ag(III) type sets. As to the Ag(c) type sets, there are also only certain ones that can be filled when certain combinations of the Ag(II) sets are filled.

Now when a single crystal of $RbAg_4I_5$ is cooled below the 209°K transition, there is no way of preventing its becoming multiply-twinned. The cubic fourfold axis becomes a rhombohedral pseudo-fourfold axis; this is now an <u>experimentally determined fact</u>. Thus each Bragg intensity for a general set of hkℓ indices is composed of contributions from four independent Bragg reflections. If two indices are equal, the contributions from two of the reflections are equal. If one of the indices is zero, two pairs are equal; if two of the indices are equal, all contributions are equal.

The intensity data were collected with an automatic diffractometer at 130°K. The task of determining the distribution of the Ag^+ ions was arduous mainly as a result of the twinning which reduced considerably the sensitivity of the data to the Ag^+ ion distribution. There are, however, three reflections which are especially sensitive to this distribution, namely the 320, 321 and 322, and especially the last. The 322 intensity could be seen to increase monotonically and substantially as the temperature was decreased toward that of the lower transition. Also, one could see the <u>onset</u> of the 209°K transition by the <u>onset</u> of the increase in the 322 intensity. Thus regardless of how good the overall agreement between calculated and observed structure factors might be it was especially necessary for the agreement of the 320, 321 and 322 cases to be very good. (The calculated "structure factor" is $\frac{1}{2}(F^2_{hk\ell}+F^2_{\bar{h}k\ell}+F^2_{h\bar{k}\ell}+F^2_{hk\bar{\ell}})^{\frac{1}{2}}$.)

A knowledge of the allowed <u>ordering</u> schemes in the lowest temperature phase gives insight into the possible <u>distributions</u> in the intermediate phase and all these were used in reaching the determination. The Ag^+ ion sites and their occupancies are shown in Table 2. (The agreement index, $\Sigma||F_o|-|F_c||/\Sigma|F_o|,=0.082$.) The sites Ag(1), Ag(3) and Ag(4) derive from the cubic Ag(II) set; the sites Ag(6) and Ag(7) from the cubic Ag(III) set, and the sites Ag(c2) from the cubic Ag(c) set.

In the cubic $RbAg_4I_5$, the distribution of the Ag^+ ions is as follows: Ag(c), 0.88; Ag(II), 9.38; Ag(III), 5.50. It is seen that the analogous numbers in the rhombohedral phase are 1.3, 10.5 and 4.2 respectively. This is so even though it would have been possible for all the Ag^+ ions to be distributed only over Ag(III) type sites without restriction. Ten Ag(II) type Ag^+ ions go into 30 Ag(II) type Ag^+ ions in the trigonal structure. If the latter is ordered, there is no case in which sites analogous to Ag(6) can be

Table 2. Ag^+ ion sites and their occupancies in the intermediate phase of $RbAg_4I_5$.

Atom designation	Site symmetry	Occupancy, Ag^+ ions	x	y	z
Ag(1)	1	3.0	0.160	0.886	0.428
Ag(3)	1	2.7	0.921	0.601	0.344
Ag(4)	1	4.8	0.397	0.327	0.096
Ag(6)	1	1.2	0.096	0.264	0.409
Ag(7)	1	3.0	0.313	0.127	0.786
Ag(c2)	3	1.3	-0.198	-0.198	-0.198

occupied in that phase. Yet the data were taken at a temperature very close to that transition, and though exhaus-
tive calculations were made, it appeared necessary to include these Ag(6) sites in the calculation to obtain the
agreement. In fact, this was very troublesome in the work,
because of the bias against inclusion in the calculations
of Ag^+ ions in these sites.

While the work on the lowest temperature form is not
yet completed, it already seems that the "extra" reflections
which occur come mainly from displacements of the iodide
ions away from the triply primitive arrangement. Thus,
presently, it seems that the first transition involves mainly a redistribution of the Ag^+ ions (to a still disordered
state), while the transition at 121.8° appears to involve
mainly a collection of small independent displacements of
the increased number of sets of iodide ions. One can conclude, at least tentatively, that the Ag^+ ions are still
disordered in the trigonal phase and it too may well be a
solid electrolyte. This has actually been predicted earlier
by Johnston, Wiedersich and Lindberg.[16]

Finally, it was pointed out in the paper[7] on the
structure of $RbAg_4I_5$ that even after cooling through the
121.8°K transition, the crystal became single again after
warming to room temperature. This low temperature structural investigation shows why this is so. Both low temperature phases are closely related to the cubic one and contain the same symmetry elements (32) which form a subgroup
of the cubic space group ($P4_132$ or $P4_332$). (There are no
3-fold screw axes in P321, however.) The domains therefore
go rather easily back to the rhombohedral and then to the
cubic structure. Needless to say, if the multidomain crystals, especially in the lowest temperature phase, were to
be highly anisotropically stressed, the resultant strains
could be such as to disrupt the domain walls, and then the
crystal would tend not to remain intact.

ACKNOWLEDGMENTS

A substantial part of the work reported here has been supported by the National Science Foundation. Dr. P. M. Skarstad worked with the author on the structure of $Py_5Ag_{18}I_{23}$.[3,4] I thank Dr. Lilian Chan for applying her expert skill to the programming of some of the calculations required in the work on the low temperature phases of $RbAg_4I_5$ and Mr. S. A. Wilber for carrying out conductivity measurements.

REFERENCES

[1] S. Geller, in "Fast Ion Transport in Solids," W. van Gool, Editor, pp. 607-616, North-Holland/American Elsevier, Amsterdam-London (1973).

[2] T. Takahashi, S. Ikeda and O. Yamamoto, J. Electrochem. Soc. 119, 477 (1972).

[3] S. Geller and P.M. Skarstad, Phys. Rev. Letters 33, 1484 (1974).

[4] S. Geller, P.M. Skarstad and S.A. Wilber, J. Electrochem. Soc. 122, 332 (1975).

[5] S. Geller, Science 176, 1016 (1972).

[6] S. Geller and B.B. Owens, J. Phys. Chem. Solids 33, 1241 (1972).

[7] S. Geller, Science 157, 310 (1967).

[8] D.O. Raleigh, J. Appl. Phys. 41, 1876 (1970).

[9] B.B. Owens in "Advances in Electrochemistry and Electrochemical Engineering," C.W. Tobias, Editor, Vol. 8, pp. 1-62, Wiley, N.Y. (1971).

[10] B.B. Owens and G.R. Argue, Science 157, 308 (1967).

[11] J.N. Bradley and P.D. Greene, Trans. Faraday Soc. 63, 424 (1967).

[12] H. Wiedersich and S. Geller in "The Chemistry of Extended Defects in Non-Metallic Solids," L. Eyring and M. O'Keefe, Editors, pp. 629-650, North Holland Publishing Co., Amsterdam (1970).

[13] L.W. Strock, Z. Physik. Chem. B25, 441 (1934); B31, 132 (1936).

[14] C. Tubandt and E. Lorenz, Z. Phys. Chem. 87, 513 (1914).

[15] S. Geller and M.D. Lind, J. Chem. Phys. 52, 5854 (1970).

[16] W.V. Johnston, H. Wiedersich and G.W. Lindberg, J. Chem. Phys. 51, 3739 (1969).

[17] L.D. Fullmer and M.A. Hiller, J. Cryst. Growth 5, 395 (1969).

AgI-TYPE SOLID ELECTROLYTES: PROPERTIES AT FREQUENCIES BETWEEN 10^9 AND 10^{13} Hz

K. Funke

Institut für Physikalische Chemie
der Universität Göttingen and
Sonderforschungsbereich 126
Göttingen, West Germany

Atoms in liquids, hydrogen in hydrogen-metal systems, and cations in AgI-type solid electrolytes have comparable coefficients of self-diffusion of the order of $10^{-5} cm^2 s^{-1}$ Yet the diffusion mechanisms in these systems are quite different from each other. The two extreme possibilities are presented on the left- and right-hand sides of Fig. 1. The atoms in a simple monoatomic liquid display a continuous type of motion obeying the laws of simple diffusion down to roughly 10^{-12} s and 10^{-8} cm. On the other hand, hydrogen in systems as H-Pd and H-Nb definitely performs a jump diffusion from one well-defined site to another. In this case it is generally assumed that the mean time of flight, τ_1, is negligible compared to the mean residence time, τ_0.

From our present microwave, far-infrared, and quasi-elastic neutron-scattering data, we conclude that the diffusion in AgI-type solid electrolytes is in a way intermediate between these extremes. As will be shown, a superposition of jump-diffusion and a local random motion can be regarded as a good candidate for a first approach, see Fig. 1. However, in AgI-type systems, the mean times of flight, τ_1, seem to be comparable to the mean residence times, τ_0.

Our high-frequency measurements were initiated by the following consideration. If a jump-diffusion model is valid in the highly conducting solid electrolytes, we can use the

Systems with $D \gtrsim 10^{-5} cm^2 s^{-1}$

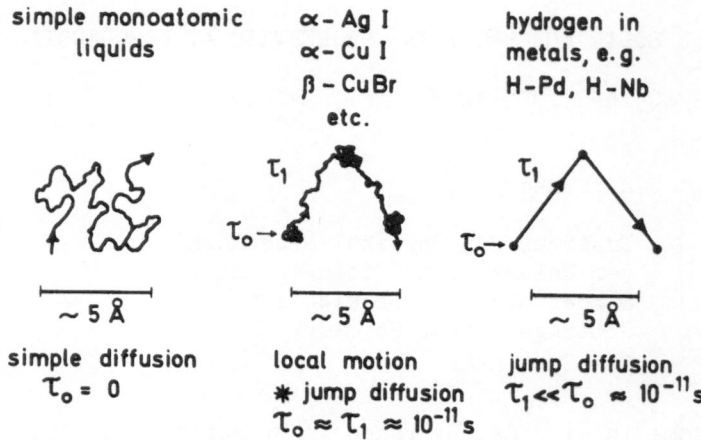

| simple monoatomic liquids | α - Ag I α - Cu I β - CuBr etc. | hydrogen in metals, e.g. H-Pd, H-Nb |

simple diffusion
$\tau_o = 0$

local motion
❋ jump diffusion
$\tau_o \approx \tau_1 \approx 10^{-11} s$

jump diffusion
$\tau_1 \ll \tau_o \approx 10^{-11} s$

FIG. 1. Diffusion in different systems.

equation

$$6D(\tau_o + \tau_1) = \ell^2 ,$$

where ℓ is the mean jump distance. Inserting for ℓ values of a few Å, we find that the characteristic times τ_o, τ_1 might be of the order of 10^{-11} s. In 1967, W. Jost pointed out that in this case a dispersion of the electrical and di-electric properties should be observed in the microwave range [1]. This dispersion, directly reflecting the process of jump-diffusion, could in the meantime be experimentally established in α-AgI and related compounds, at microwave frequencies between 2 and 40 GHz. At higher frequencies, up to 10^{13} Hz, conductivities and permittivities have been obtained by Fourier-transform far-infrared spectroscopy. These spectra give evidence for the large-amplitude local motion which is superimposed onto the jump diffusion of the cat-ions. Thirdly, since 1973, the quasielastic scattering of cold neutrons from α-AgI has been investigated at the high flux reactor of the Institut Laue-Langevin, Grenoble. The neutron spectra contain information on the cation motion in the entire frequency range below 10^{13} Hz. Moreover, they

also provide a simultaneous spatial resolution of this motion. In the following, the experimental results obtained
by the different techniques will be briefly presented and
discussed.

Microwave Spectra

In the microwave experiments, a special technique was
used which is appropriate for the measurement of high-loss
materials [2]. The solid electrolyte under investigation forms
the side-walls of a rectangular waveguide. In comparison to
a normal, nearly ideally conducting waveguide, two effects
are observed. Firstly, a TE_{10}-wave is attenuated because of
losses in the side-walls and secondly, there is a small
change of wavelength. From these data, the complex permitti-
vity $\hat{\varepsilon} = \varepsilon' - i\varepsilon''$ and the complex conductivity $\hat{\sigma} = i\omega\varepsilon_0\hat{\varepsilon}$ are
calculated by means of the proper continuous solutions of
the wave-equations within the waveguide and within the walls.
At low frequencies, i.e. at 2-8 GHz, a coaxial short-cir-
cuit technique was used.

So far, measurements have been performed on three AgI-

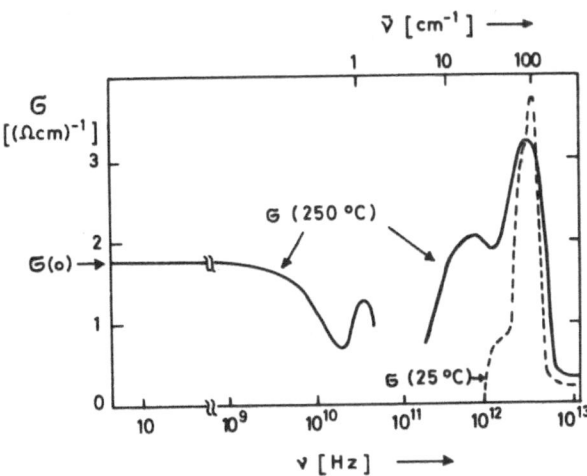

FIG. 2. Conductivity σ of α-AgI at 250 $^\circ$C and of β-AgI at
25 $^\circ$C in a broad frequency range.

type solid electrolytes with different structures of their
anion lattices, namely on α-AgI (b.c.c.)[3], α-CuI (f.c.c.)[4],
and ß-CuBr (h.c.p.)[5,6]. The results, $\sigma(\nu) = \text{Re}\,\hat{\sigma}(\nu)$ and
$\varepsilon'(\nu) = \text{Re}\,\hat{\varepsilon}(\nu)$, $\nu = \omega/2\pi$, are plotted in Figs. 2,3, and 4.
In Fig. 2, the electrical conductivity $\sigma(\nu)$ of α-AgI is
shown in a broad frequency range comprising both microwave
frequencies, $\nu < 10^{11}$ Hz, and far-infrared frequencies,
$\nu > 10^{11}$ Hz. The microwave conductivities and permittivities
of ß-CuBr and α-CuI in Figs. 3 and 4 are mutually consistent
according to the criterion of the Kramers-Kronig relations.
The curves drawn in the figures are calculated from a simple model which will be briefly discussed in this paper.
They automatically satisfy the Kramers-Kronig relations.

As a common feature, the $\sigma(\nu)$ spectra exhibit a decrease

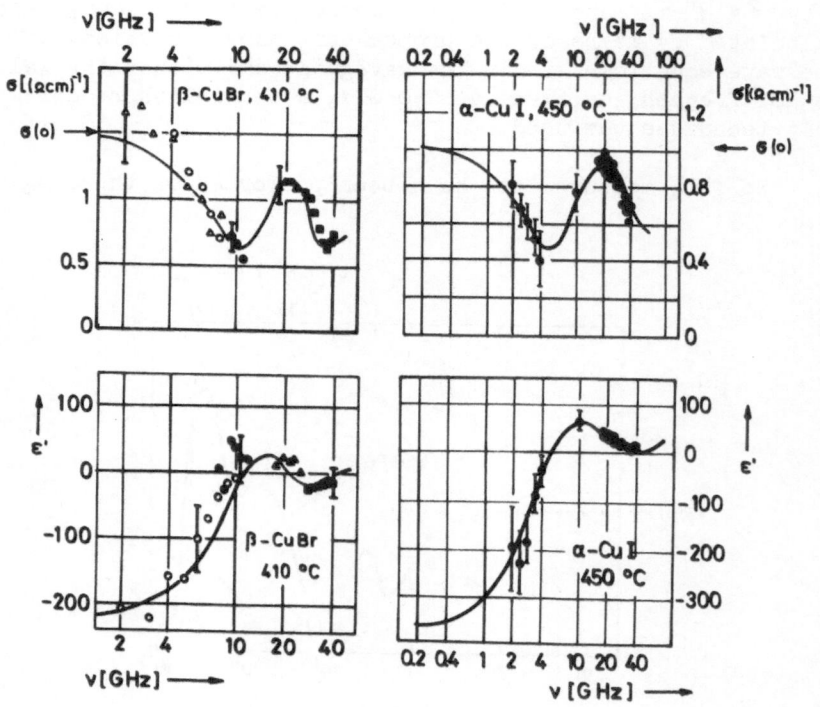

FIG. 3. Conductivity σ and   FIG. 4. Conductivity σ and
       permittivity ε' of   permittivity ε' of
       ß-CuBr at 410 $^\circ$C. α-CuI at 450 $^\circ$C.

with increasing frequency and an additional maximum in the
20-30 GHz range. At low frequencies, σ seems to approach
its known low-frequency value $\sigma(0)$. Correspondingly, according to the data and the Kramers-Kronig relations, ε' should
become constant and negative in the low-frequency limit. Indeed negative values of ε' have been observed in α-CuI and
in ß-CuBr, where data exist in the 2-8 GHz range.

In search of an explanation of the decrease in conductivity let us first disregard the conductivity maximum. The
remaining part of the spectrum looks much like the $\sigma(\omega)$-
curve predicted by the Drude model[7], and this model has actually been proposed by Huberman and Sen[8,9] and, more tentatively, by Armstrong and Taylor[10]. Applying the Drude model as a first attempt, we would find the momentary fraction
α of moving ions to be $\sigma(0) \cdot \omega_0 \cdot q^{-2} \cdot n^{-1}$, according to the sum
rule for the Drude-like part of the conductivity spectrum.
Here, ω_0 is defined by $\sigma(\omega_0) = \sigma(0)/2$, m and q are the ionic
mass and charge, and n is the number density of the cations.
In the case of α-CuI, the result would be $\alpha \approx 10^{-3}$. On the
other hand, the Drude model would give a mean time of flight
$\tau_1 \gtrapprox 2/\omega_0$, the $>$ sign being valid if there is friction during each individual flight. However, such friction has to be
expected because of the existence of the random local motion
which is in turn caused by the fluctuating local potentials
due to the motions of the neighbouring ions. Thus we would
find $\tau_1 \gtrsim 10^{-10}$ s in the case of α-CuI. From $\alpha = \tau_1/(\tau_0 + \tau_1)$ and
$6D(\tau_0 + \tau_1) = \ell^2$ we would then infer $\tau_0 \gtrsim 10^{-7}$ s and $\ell \gtrsim 400$ Å
which is quite unrealistic for an AgI-type solid electrolyte.

The difficulties in applying the Drude model are at once
resolved if we introduce one simple additional assumption[5,6].
It is assumed that the jump-lengths of the ions are fixed,
being given by the geometry of the anion lattice and not being affected by the electrical field. In this case, the effect
of acceleration or retardation of the moving ions by the electrical field changes their times of flight, but not their
jump rates. Therefore it cannot give any contribution to $\sigma(0)$.
However, as will be shown, it can explain the observed conductivity peaks at 20-30 GHz. This effect will be called the
"acceleration effect".

Assuming fixed jump lengths, we have to explain the low-
frequency conductivity and its decrease in the microwave-
range by another model. This can be done by a simple consider-

ation which is familiar in the case of "normal" ionic crystals: the electrical field slightly supports or hinders the
initiation of jumps in or against the instantaneously preferred direction, thus causing a constant value of $G(\omega)$ in
the low-frequency limit. In this limit, $\omega \ll \tau_1^{-1}$, the phase of
the field is practically constant while an ion is performing
a jump. When the frequency is increased the following situation will occur. A cation which has started for an "extra"
jump with the help of the electrical field is still on its
way while the field is already changing its sign. After this
change of sign the ion will move against the momentarily preferred direction and will give a negative contribution to the
overall conductivity, which will result in a decrease of G.
In the high-frequency limit, $\omega \gg \tau_1^{-1}$, the field will change
its sign many times during the flight of an ion and the overall conductivity will be zero.

We will now sketch the calculation of $\hat{G}(\omega)$ according to
our physical model, including the "start effect" and the
"acceleration effect". We start with the equation

$$\hat{G}(\omega) = q \cdot \hat{I}(\omega,t)/(E_o e^{-i\omega t}) , \qquad (1)$$

where $\hat{I}$ is the complex flux in the direction of the electric
field $E_o e^{-i\omega t}$. In order to formulate the problem in only one
dimension, let us introduce the flux $\hat{I}_{1D}$ which we would expect, if all jumps were being performed either directly in
or against the field direction. If in three dimensions the
field direction and the jump direction of an ion form an
angle ϑ, then the influence of the field on the motion of
the ion as well as the component of this motion in the field
direction are proportional to $\cos \vartheta$. Therefore we have

$$\hat{I} = \langle \cos^2 \vartheta \rangle_{sphere} \cdot \hat{I}_{1D} = \hat{I}_{1D}/3 . \qquad (2)$$

In our one dimensional calculation, $\dot{n}_+(t')dt'$ and $\dot{n}_-(t')dt'$
are the number densities of ions starting jumps in the +
and - direction, respectively, within the time interval dt'.
Secondly, let us consider an ion which has started a jump at
time t' and is still on its way at time $t > t'$. Depending on
the direction of its flight, its velocity at time t has the
positive value $v_+(t',t)$ or the negative value $v_-(t',t)$.
Thirdly, the time this ion needs for its jump of fixed length
ℓ is $\tau_+(t')$ or $\tau_-(t')$, respectively. Then the flux $\hat{I}_{1D}$ is
given by

$$\hat{I}_{1D} = \int_{-\infty}^{t} \dot{n}_+(t')v_+(t',t)\,\Theta(t'+\tau_+(t')-t)dt'$$

$$+ \int_{-\infty}^{t} \dot{n}_-(t')v_-(t',t)\,\Theta(t'+\tau_-(t')-t)dt' \qquad (3)$$

with

$$\Theta(x)=\left\{\begin{matrix}1\\0\end{matrix}\right\} \quad \text{for} \quad \left\{\begin{matrix}x>0\\x<0\end{matrix}\right\}.$$

The requisite start rates and velocities are written

$$\dot{n}_\pm(t')=\dot{n}_0\pm\Delta\dot{n}(t') \text{ and } v_\pm(t',t)=\pm v_0(t',t)+\Delta v(t',t), \quad (4)$$

where $\dot{n}_0$ and $\pm v_0(t',t)$ denote their values in the absence of an applied field. From equation (4), $\dot{n}_+(t')$ and $v_+(t',t)$ are inserted into equation (3). The jump times $\tau_\pm(t')$ are expressed by $v_+(t',t)$ under the constraint of fixed jump-length ℓ. Considering only first-order terms, we then obtain

$$\hat{I}_{1D} = 2 \int_{t-\tau_1}^{t} \Delta\dot{n}(t')v_0(t',t)dt'$$

$$+ 2\dot{n}_0\left\{ \int_{t-\tau_1}^{t} \Delta v(t',t)dt' - \int_{t-\tau_1}^{t} \Delta v(t-\tau_1,t'')dt'' \right\} \qquad (5)$$

$$= (\hat{I}_{1D})_{start} + (\hat{I}_{1D})_{accel.}$$

In equation (5), the first term involves $\Delta\dot{n}$ and therefore describes the "start effect", while the second term involves Δv and therefore describes the "acceleration effect".

For the calculation of the "start effect", it is supposed that $\Delta\dot{n}(t')$ is proportional to the electric field at time t', i.e. $E_0e^{-i\omega t}$. The fact that to a certain extent jumps of two ions in the same direction occur in a correlated manner has been taken into account, see Ref. 5.

In order to calculate the "acceleration effect", we have to formulate the influence of the applied field on the position and velocity of a moving ion. For an ion which has started a jump at time $t'<t$, the change in position due to the field at time t, $\Delta x(t',t)$, is obtained from

$$m\Delta\ddot{x}(t',t) + mb\Delta\dot{x}(t',t) + V''\Delta x(t',t) = qE_0e^{-i\omega t} \qquad (6)$$

with $\Delta x(t',t')=0$ and $\Delta \dot{x}(t',t')=\Delta v(t',t')=0$. The term b allows for the friction-like effect of rapidly changing local potentials, and V" is the (negative) second derivative of a smoothed potential barrier mainly due to the arrangement of the anions.

The derivation of the final equation for $\hat{\sigma}(\omega)$ is straightforward and is given in Ref. 5. The curves shown in Figs. 2, 3, and 4 have been calculated from this equation. They closely fit the experimental data.

The main parameter of the model is τ_1. The mean jump-time determines the position of the conductivity peak on the frequency scale as well as the frequency range where the decrease of $\sigma(\omega)$ is observed. In the case of α-AgI, at 250 °C, τ_1 is found to be ca. 15 ps. Somewhat larger values are obtained for ß-CuBr and α-CuI.

The height of the conductivity peak depends on the friction-term b and on the fraction of cations which are in flight at a given instant of time, $\tau_1/(\tau_0+\tau_1)$. In α-AgI at 250 °C, the neutron scattering results yield $\tau_0 \approx 8$ ps, see 11, 12. Then, b becomes approximately 15 ps^{-1} which enables us to give a first crude estimate of the "coefficient of self-diffusion for the random local motion", $D_{loc.}=kT/(mb)$. $D_{loc.}$ is thus found to be ca. $(0.5Å)^2$ps^{-1}.

In the case of α-AgI, the value of V" is approximately -15 meV/$Å^2$. Consequently the averaged potential barriers between the preferred residence regions of the cations can be estimated to be of the order of the thermal energy, in agreement with the formal activation energies derived from Arrhenius plots.

Far-Infrared Spectra

It is a common feature of AgI-type solid electrolytes that unusually large Debye-Waller exponents of the mobile cations are formally derived from X-ray and neutron structural refinements. They correspond to unusually high amplitudes of a local motion within flat potentials. However, not only within the voids provided by the anions, but also during flight from one void to another, some local kind of motion seems to be characteristic of the cations in these materials. This had to be taken into account by the introduction of the term b in the preceding section.

    In the present context, it is once more stressed that
the interactions of a cation with its moving neighbours cause
fluctuations of its local potential of the order of the thermal energy and on a time scale of ca. 10^{-12}s. Therefore, one
might tentatively describe the short-time behaviour of the
cations as that of highly damped, more or less stochastically driven oscillators. This view seems to be confirmed by
the far-infrared experiments so far performed on AgI-type
solid electrolytes.

    By the use of Fourier spectroscopy, the far-infrared
dispersion of the complex conductivity has been studied in
α-AgI [3] and α-RbAg$_4$I$_5$ [13]. The spectrometers used in these
experiments contain either a Michelson interferometer (Beck-
man FS 720) or a lamellar grating interferometer (Beckman
LR 100), the latter being preferable at frequencies below
ca. 10^{12} Hz. Transmittance and reflectivity measurements were
carried out on polycrystalline samples. The data were then
transformed into the $\sigma(\nu)$ and $\varepsilon'(\nu)$ representations, and

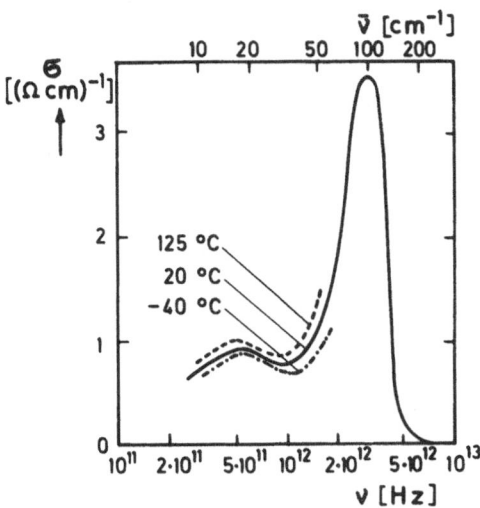

FIG. 5. Far-infrared conductivity of α-RbAg$_4$I$_5$ at different
    temperatures.

the mutual consistency of these results was checked by Kramers-Kronig analyses.

   From Fig. 2 it is seen that α-AgI exhibits a relatively
normal, though highly damped, lattice absorption spectrum
above 10^{12} Hz. This part of the spectrum does not differ
much from the one at room temperature. Below this frequency,
however, there is an intense, broad and rather structureless
absorption band, which is unusual in ionic crystals. The corresponding $\sigma(\nu)$ data from α-RbAg$_4$I$_5$ are given in Fig. 5. Qualitatively, the same features are observed as in α-AgI. Although the dc conductivity of α-RbAg$_4$I$_5$ at 20 $^\circ$C is lower
than that of α-AgI at 250 $^\circ$C by about one order of magnitude,
the far-infrared conductivity maxima near $5 \cdot 10^{11}$ Hz differ
only by a factor of two. The shapes and strengths of the observed far-infrared absorption bands hardly change with temperature; it is therefore concluded that these bands are not
mainly due to multiphonon difference processes [13].

   For an interpretation, one has to consider the fact that
Bloch's theorem is not valid in a crystal with structural disorder. Therefore the usual optical selection rule $Q \approx 0$ is
broken in the sense that all phonon-mode frequencies become
observable regardless of the values of their wave-vectors $\underline{Q}$.
So far, the situation is similar to that in polar liquids,
where a similar far-infrared behaviour has been found [14].
However, an interpretation in terms of "smeared-out liquid
lattice bands" [14] would imply that the phonon concept is applicable for the type of motion which is responsible for the
far-infrared absorption. In the limiting case of very high
damping one would rather prefer the description given at the
beginning of this section, treating the cations as an assembly of stochastically driven oscillators. Actually, our quasielastic neutron-scattering experiments on α-AgI demonstrate
that the local motion of the silver ions is of an overdamped
type. For an overdamped oscillator of frequency ω_0, the functions $\varepsilon''(\omega)$ and $\sigma(\omega)$ still exhibit maxima on the frequency
scale. If there is a whole spectrum of frequencies ω_{0j}, the
contributions $\sigma_j(\omega)$ might well add up to a very broad band
of $\sigma(\omega)$ as it is observed.

                Quasielastic Neutron Scattering

   Although the diffusion of atoms in liquids and of hydro-

gen in metals has been examined in various neutron scattering experiments [15], this technique was not applied to AgI-
type solid electrolytes until 1973. In the meantime, quasielastic neutron-scattering experiments on polycrystalline
samples of α-AgI have been undertaken at the "IN 5 multichopper" time-of-flight spectrometer of the Institut Laue-
Langevin, Grenoble [11], [12]. In these two experiments, the
ranges of wavevector transfer and the energy resolutions
(FWHM) were 1.4 $\text{Å}^{-1} < Q < 2.7$ Å^{-1} and 0.4 meV, and 0.56 $\text{Å}^{-1} <$
$Q < 2.16$ Å^{-1} and 0.177 meV, respectively. The temperature of
the sample was 250 $^{\circ}$C.

Fig. 6 shows one of the corrected spectra transformed
to an energy scale ($\varphi = 105^{\circ}$). There is intense low-energy
scattering, centred around energy transfer $\hbar\omega = 0$. As Q increases, its broad and almost structureless tail extending
to higher values of $\hbar\omega$ becomes more and more pronounced.
Basically, the quasielastic scattering reflects the structural disorder, because it is mostly coherent. Its shape gives
information on the diffusion of the silver ions. Its tail at

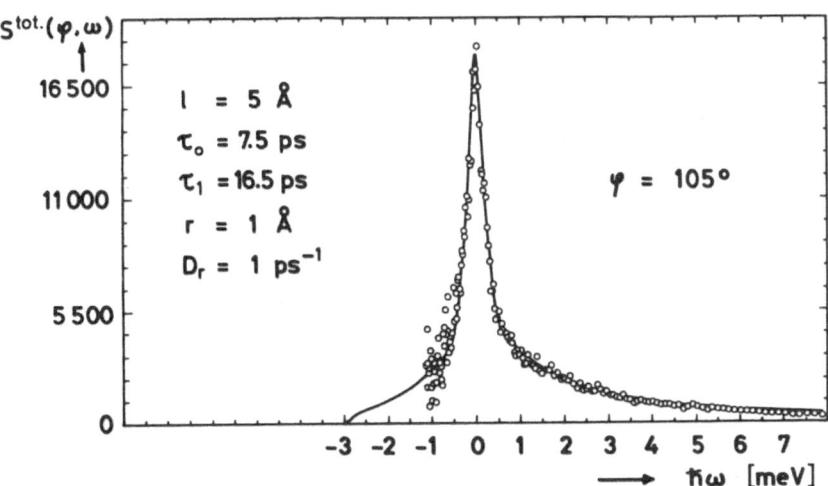

FIG. 6. Experimental scattering function $S^{tot.}(\varphi,\omega)$ of α-AgI
at 250 $^{\circ}$C as obtained with incident neutrons of $\lambda =$
5.35 Å at scattering angle $\varphi = 105^{\circ}$; not corrected for
resolution, arbitrary units. The solid line results
from our model calculation (parameter values inset).

higher energies corresponds to the broad low-frequency absorption band found in the infrared experiments.

At sufficiently large distances r, corresponding to sufficiently small Q, Fick's law of simple translational diffusion
should be fulfilled. In α-AgI, the energy widths of the quasielastic spectra are in agreement with the predictions of the
simple diffusion model in the Q-range below approx. 1 $\overset{o}{A}^{-1}$. At
larger Q, the experimental widths are found to be smaller. This
cannot be explained by the fact that most of the scattering is
coherent. Rather, it is an indication that the simple diffusion
model no longer applies.

Closer inspection of the spectra shows that they seem to
contain a narrow component - which is however broader than the
resolution function - on top of a much broader distribution.
Moreover, it is found that none of the models for translational
diffusion can explain these spectral shapes and their Q-depen-
dence. In order to derive a physical model applying to the diffusive motion of the silver ions, we make the following assumptions.

a) The spectra are free of elastic scattering [12].

b) The inelastic contributions to our spectra are negligible in comparison to the quasielastic scattering. The inelastic fraction of the scattering does not exceed ca. 10%
even at our largest scattering angles.

c) The quasielastic scattering is exclusively due to the
silver ions. In particular, its coherent part is the Fourier
transform of the silver-silver pair correlation function.

d) The silver ion is carrying out two different types of
motion on two different time scales. On the one hand it performs an overdamped or diffusive motion within a restricted
region of space, e.g. in a void of the b.c.c. iodide lattice.
Such a motion could cause a broad quasielastic distribution
plus a $\delta(\omega)$-peak on top of it. The second kind of motion is a
translational diffusion over the whole crystal volume which
might be of the jump-diffusion type. It might only slightly
broaden the spectrum which one would observe from the first
kind of motion alone; in particular it would broaden the elastic line. An example of a possible three-dimensional arrangement of the above-mentioned voids is shown in Fig. 7. These

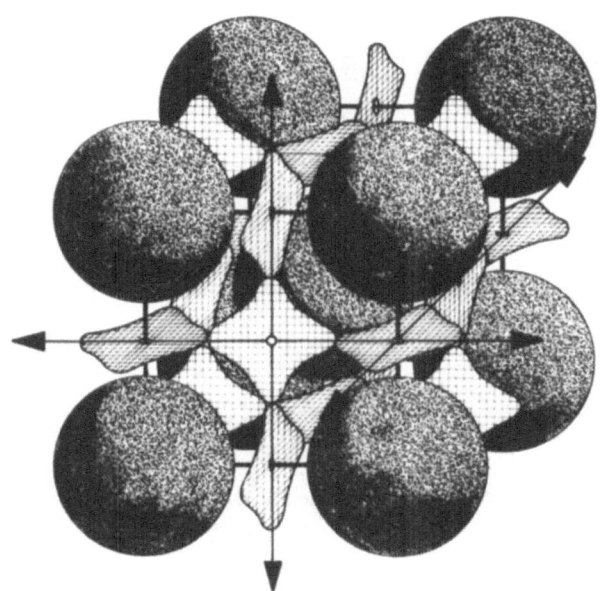

FIG. 7. Local regions of space available for the silver ions
        in α-AgI.

regions are identical with the elementary regions proposed by
Rickert [16]. The translational diffusion follows naturally from
this picture as an exchange of cations between these voids.

    e) Assuming that the silver ions perform both kinds of mo-
tion simultaneously and that these are dynamically independent
of each other, we may write the silver self-correlation func-
tion $G_s{}^{Ag}(\underline{r},t)$ and the scattering function $S_{inc}{}^{Ag}(\underline{Q},\omega)$ for the
incoherent part of the quasielastic scattering from the silver
ions as the convolutions of the individual functions. This is
illustrated in Fig. 8, where the superscripts T and L stand
for "translational" and "local" motion, respectively. $S_{inc}{}^{Ag}(\underline{Q},\omega)$
can be approximated by the use of simplifying models for both
types of motion, see assumptions g) and h).

    f) The coherent part of the quasielastic scattering is
estimated by using a phenomenological model which was proposed
a few years ago by Sköld [17]. The required structure factor for
the silver ions is taken from the experimental and theoretical
results of a recent diffuse-scattering study on α-AgI [18].

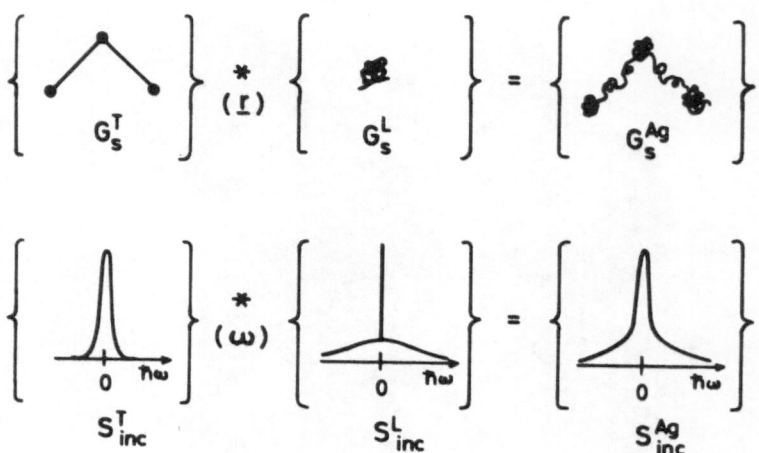

FIG. 8. Incoherent scattering function and self-correlation
function of the silver ions in α-AgI represented as
convolutions of the individual functions for translational and local motion.

g) For the translational motion of the silver ions we
apply the jump-diffusion model of Gissler and Stump [19], which
contains as parameters a mean residence time, τ_0, a jump time,
τ_1, and a fixed jump length, ℓ. The model is used in two different modifications, namely (i) isotropic, without any specific choice of jump directions, and (ii) polycrystalline,
with jumps along the <100> tunnels [20]. However, as far as the
final spectra are concerned, the numerical results obtained
from either approach are practically indistinguishable. On the
other hand, the spectra strongly depend on the choice of the
parameters τ_0, τ_1, and ℓ.

h) As for the local motion of the silver ions, the two
possibilities mentioned above, i.e. local diffusive motion
and overdamped oscillatory motion, are physically much alike.
They are both compatible not only with the neutron scattering
results but also with the infrared results if one takes into
account the different weighting factors affecting the frequency spectra when measured with the two different techniques.
For mathematical convenience, the solution of the diffusion
equation in a restricted region of space is regarded as a convenient starting point for a description of the local motion.

The model which is easiest to apply is the diffusion on the
surface of a sphere. This approximation should give results
not too different from diffusion within a full sphere. The parameters of this model are the radius of the sphere, r, and the
"rotational" diffusion coefficient, D_r.

Making the assumptions a) to h), total quasielastic scattering functions $S^{tot.}(Q,\omega)$ are calculated for various sets of
p arameter values. These functions are then transformed into
the corresponding functions which describe the scattering at
fixed scattering angles φ. Finally, convolution of these functions with the resolution function yields the calculated spectra $S^{tot.}(\varphi,\omega)$ which may be compared to the experimental ones.
Close fits were obtained at all scattering angles; an example
is shown in Fig. 6. For α-AgI at 250 °C the following parameter
values have been used:

jump distance: ℓ = 5 $\overset{o}{A}$, residence radius: r= 1 $\overset{o}{A}$,
residence time: τ_0= 7.5 ps, rotational diffu-
jump time: τ_1=16.5 ps, sion coefficient: D_r= 1 ps^{-1}.

Actually close fits to the neutron spectra can be obtained with
several different (τ_1,ℓ) combinations. If, however, we identify
τ_1 with the jump time determined by our microwave measurements
we are able to make a decision and to choose the above set of
parameter values which is not only in good agreement with the
neutron data at all Q-values, but is also consistent with the
microwave results.

Discussion

First we note that our above parameter values yield a
coefficient of self-diffusion, $D=\ell^2/(6(\tau_0+\tau_1))=1.74\cdot10^{-5}\text{cm}^2\text{s}^{-1}$,
which is in good agreement with the experimental tracer-diffu-
sion coefficient, $D \approx 2\cdot10^{-5}\text{cm}^2\text{s}^{-1}$ at 250 °C [21]. This gives
support for the treatment presented above.

The jump length, ℓ = 5 $\overset{o}{A}$, is close to the lattice constant
of the b.c.c. iodide ion lattice (a = 5.06 $\overset{o}{A}$ at 250 °C). This
is consistent with the findings from our microwave spectra
which led to the conclusion that the possible cation jump distances must have fixed values defined by the distances between
supposed cation residence regions fixed within the anion lattice.
The values of the actual jump lengths will then be limited by

the cation density. Thus a jump length of one lattice constant seems reasonable. The ions would then move along the <100> tunnels, cf. Ref. 20, and the residence regions might be those indicated in Fig. 7.

The parameter value $r = 1$ Å illustrates the spatial extension of the single particle random local motion. This motion is being carried out within a residence region during τ_o and around the ion's averaged path from one of these regions to another during the jump time τ_1. The parameter value $D_r = 1$ ps^{-1} $\hat{=}(1$ Å$)^2$ps^{-1} compares favourably with the value $D_{loc}.\approx(0.5$Å$)^2$ps^{-1} obtained from the microwave data. This is remarkable since either approach is extremely crude.

The value of the jump time, $\tau_1 \approx 16$ ps, is by a factor of ca. 10 larger than that one would expect if the ions moved with their thermal velocity directly from one position to another, one lattice constant apart. The existence of the random local motion provides a simple explanation.

Thus, at least in the case of α-AgI, our present physical model for the cation motion, cf. Figs. 1 and 8, appears to be consistent with our experimental microwave, far-infrared, and quasielastic neutron-scattering results.

References

(1) W. Jost, Ber. Bunsenges. physik. Chem. 71, 753 (1967)
(2) K. Funke, "Measurement of High-Loss Solid Dielectrics" in: "High Frequency Dielectric Measurement", J. Chamberlain and G. W. Chantry eds., IPC Science and Technology Press, Guildford 1973
(3) K. Funke and A. Jost, Nachr. Akad. Wiss. Göttingen, Nr. 15, 137 (1969); W. Jost, K. Funke, and A. Jost, Z. Naturforsch. 25a, 983 (1970); K. Funke and A. Jost, Ber. Bunsenges. physik. Chem. 75, 436 (1971)
(4) K. Funke and R. Hackenberg, Ber. Bunsenges. physik. Chem. 76, 885 (1972)
(5) C. Clemen and K. Funke, Ber. Bunsenges. physik. Chem. 79, 1119 (1975)
(6) K. Funke, Phys. Letters 53A, 215 (1975)
(7) P. Drude, Ann. Physik 1, 566 (1900)
(8) B. A. Huberman and P. N. Sen, Phys. Rev. Letters 33, 1379 (1974)

(9) P. N. Sen and B. A. Huberman, Phys. Rev. Letters 34,
     1059 (1975)
(10) R. D. Armstrong and K. Taylor, J. Electroanal. Chem. 63,
     9 (1975)
(11) K. Funke, J. Kalus, and R. E. Lechner, Solid State Comm.
     14, 1021 (1974)
(12) G. Eckold, K. Funke, J. Kalus, and R. E. Lechner, Phys.
     Letters 55A, 125 (1975); G. Eckold, K. Funke, J. Kalus,
     and R. E. Lechner, submitted to J. Phys. Chem. Solids
(13) G. Eckold and K. Funke, Z. Naturforsch. 28a, 1042 (1973)
(14) see e.g. M. Davies, G. W. F. Pardoe, J. Chamberlain, and
     H. A. Gebbie, Trans. Faraday Soc. 64, 847 (1968) and
     J. Chamberlain, Chem. Phys. Letters 2, 464 (1968)
(15) see e.g. T. Springer, Springer Tracts in Modern Physics
     64 (1972)
(16) H. Rickert, Z. physik. Chem. NF 24, 418 (1960)
(17) K. Sköld, Phys. Rev. Letters 19, 1023 (1967)
(18) H. Fuess, K. Funke, and J. Kalus, phys. stat. sol. (a)
     32, 101 (1975)
(19) W. Gissler and N. Stump, Physica 65, 109 (1973)
(20) W. H. Flygare and R. A. Huggins, J. Phys. Chem. Solids
     34, 1199 (1973)
(21) A. Kvist and R. Tärneberg, Z. Naturforsch. 25a, 257 (1970)

LATTICE DYNAMICS AND IONIC MOTION IN SUPERIONIC CONDUCTORS

H.R. Zeller, P. Brüesch, L. Pietronero and
S. Strässler
Brown Boveri Research Center
CH-5401 Baden, Switzerland

1. INTRODUCTION

Superionic conductors are characterized by an electrical conductivity comparable to that of a liquid electrolyte [1]. Classically the approach in discussing the d.c. conductivity $\sigma(0)$ has been to treat it in terms of defect concentration, defect mobility and their respective activation energies [2].

Recently several authors [3-7] have put foreward the idea that instead of aiming directly at static properties such as $\sigma(0)$ new insights into fast diffusion could be gained by investigating the full dynamics of ionic motion including the host lattice. It is one of the main purposes of this paper to explore further in this direction and to show that important information which is unaccessible otherwise can be easily and directly gained from studies of ion dynamics.

The problem of fast diffusion in solids contains three characteristic times or frequencies : i) an attempt frequency [4,8] which in a simple model [6] is the oscillation frequency ω_o of a particle in its potential well. ii) $\tau_p = \omega_p^{-1}$ is the time it takes for the lattice to relax after a particle has jumped [7,9]. iii) A jump frequency ω_j or residence time [2] $\tau_R = \frac{2}{\omega_j}$ of the mobile particle. In general $\omega_o > \omega_p \gg \omega_j$.

We will proceed as follows : first we calculate the

frequency dependent conductivity $\sigma(\omega)$ of a particle in a
periodic potential and compare it with corresponding experimental results on α-AgI and liquid AgI. Next we include effects of lattice polarizability on particle motion and show
that it accounts for the experimentally observed structure
at ~ 10 cm^{-1} in α-AgI. Finally we demonstrate that studies
of $\sigma(\omega)$ in the millimeter and microwave range yield detailed
and complete information on correlated jumps and their time
scales.

Throughout the paper we will stress simple phenomenological concepts and avoid all mathematical complexity.

2. THERMAL MOTION AND ELECTRICAL CONDUCTIVITY OF A PARTICLE IN A RIGID PERIODIC POTENTIAL

We start with a highly simplified model which neglects
polarizability of the host lattice and ion-ion interaction
(jump correlation effects). The purpose of the model is to
give a correct description at high frequencies and to serve
as a starting point for the corrections required at low
frequencies due to polarons and ion-ion interactions.

We consider the Brownian motion of a particle in the
rigid periodic potential of Fig. 1. In particular we are
interested in the case where the barrier hight is of order
kT such that strong anharmonicities and diffusion occurs.

The standard technique to treat similar problems in the
physics of liquids is due to Mori [10],[11]. In his method $\sigma(\omega)$
is written as a continued fraction expansion the coefficients

Fig. 1 Brownian motion of a particle in a periodic potential.

of which can be calculated from the hamiltonian of the system. The use of a hamiltonian for a single particle system implies that damping is excluded which in our case is a major drawback. Recently W. Schneider [12] has generalized the Mori method to equations of motion (Langevin equations) which incorporate damping.

Another more phenomenological approach is based on the memory function formalism [10,13]. For physical reasons we know that the highly nonlinear Langevin equation which describes the particle of Fig. 1

$$m\,\ddot{x} + m\,\Gamma\,\dot{x} + f(x) = K \tag{2.1}$$

(m = particle mass, Γ = damping, $f(x)$ = restoring force K = stochastic force) asymptotically describes a damped harmonic oscillator at high frequencies and diffusion at $\omega \to 0$. We now replace the nonlinear $f(x)$ by a memory function $M(t)$ which provides the correct asymptotic properties

$$m\,\ddot{x} + m\,\Gamma\,\dot{x} + m\,\omega_o^2 \int_0^t M\,(t-t')\,\dot{x}\,(t')\,dt' = K \tag{2.2}$$

It is easy to show that the simplest $M(t)$ which leads to diffusion at $\omega \to 0$ and oscillation at high ω is

$$M(t) = e^{-\gamma t} \tag{2.3}$$

which directly leads to [6]

$$\sigma(\omega) = \frac{N(Ze)^2}{m} \; \cfrac{1}{-i\omega + \Gamma + \cfrac{\omega_o^2}{-i\omega + \gamma}} \tag{2.4}$$

where N, (Ze) and m are density, charge and mass of the particles respectively and ω_o is the resonance frequency.

From a continued fraction expansion [7,12] of the exact solution of (2.1)

$$\sigma(\omega) = \frac{N(Ze)^2}{m} \; \cfrac{1}{-i\omega + \Gamma + \cfrac{\omega_o^2}{-i\omega + \gamma 1 + \cfrac{\omega 1^2}{-i\omega + \gamma 2 + \cfrac{\omega_2^2}{\ddots}}}} \tag{2.5}$$

we see that (2.4) corresponds to the first nontrivial term
of (2.5) and a computer calculation of the coefficients
of (2.5) for simple model potentials shows very rapid con-
vergence such that for all practical purposes (2.4) is
sufficiently accurate.

We have experimentlly determined $\sigma(\omega)$ of solid electrolytes [6] and molten salts [14] by a Kramers Kronig analysis
of reflectivity data. As an example Fig. 2 shows the results
obtained on α-AgI and liquid AgI together with a fit based
on (2.4). In α-AgI we clearly note the oscillatory part of
the motion in the form of a pronounced resonance at ω_o =
105 cm^{-1}. Interestingly ω_o is virtually unaffected [6] by
the β-α transition showing that a strong Ag-I interaction
persists into the α-phase.

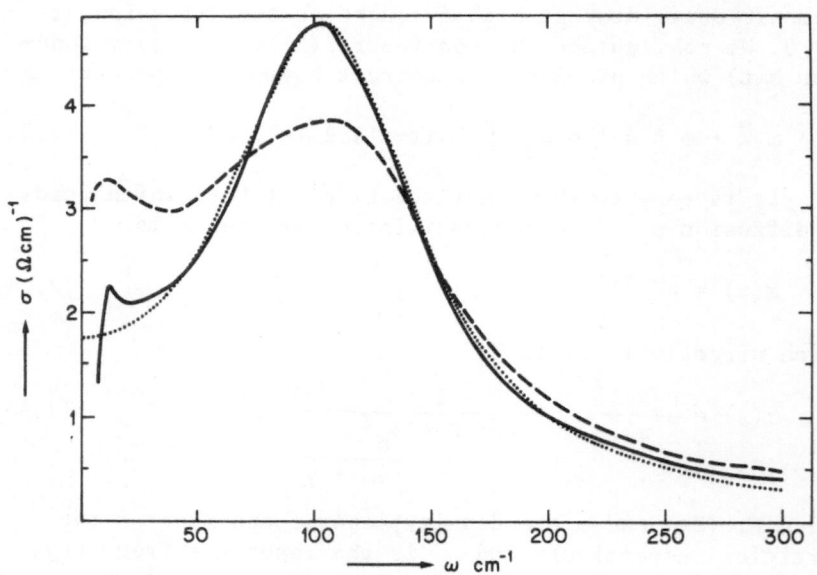

Fig. 2 Frequency dependent conductivity of α-AgI (T = 453° K,
        solid line), liquid AgI (T = 853° K, dashed line)
        and fit of equation 2.4 to α-AgI (dotted line). The
        fit parameters are : ω_o = 105 cm^{-1}, Γ = 45 cm^{-1},
        γ = 53 cm^{-1}.

There is no drastic effect upon melting except that
the resonance is almost smeared out. The fit is within experimental error except at frequencies below $\sim$ 20 cm^{-1} where
(2.4) predicts a smooth behavior whereas experimentally
structure is found. Since we are able to give an exact solution for the model which does not predict this structure,
the structure has to originate from factors outside the model such as particle correlations and lattice polarizability.

3. LATTICE POLARIZABILITY

In this section we treat complications that arise from
the fact that the potential in which the particle moves is
not rigid.

Within the model of Section 2 a sum rule holds which
is independent of the form of the potential :

$$F = \int_{o}^{\infty} \sigma(\omega) \, d\omega = \frac{\pi}{2} \frac{N(Ze)^2}{m} \tag{3.1}$$

For a polarizable lattice (3.1) breaks down for several reasons. In the oscillatory regime lattice atoms take
part in the motion and the particle moves in an effective
field and with an effective charge [15]. All these effects
combined lead for α-AgI to an enhancement of F by a factor
of 4 compared to (3.1). In Section 2 we have accounted for
this by treating F as an adjustable parameter.

However, for a classical particle all above corrections
exclusively apply to the oscillatory part of the motion. The
particle is diffusing with its true mass and charge and in
the external field. Thus we have to expect a transition
frequency at which the particle changes from its bare to its
dressed properties.

To account for this effect we again use a memory function approach. We consider a system in which mobile ions of
mass m_1 couple to lattice ions with mass m_2 and opposite
charge and make the following Ansatz for the coupled
Langevin equations :

$$m_1\ddot{x}_1 + m_1\Gamma\dot{x}_1 + m_1\omega_1^2\,\hat{M}_1\,\dot{x}_1 + m_1\omega_1^2\,\hat{M}_{12}(\dot{x}_1-\dot{x}_2) = K_1$$

$$m_2\ddot{x}_2 + m_2\Gamma\dot{x}_2 + m_2\omega_1^2\,\hat{M}_2\,\dot{x}_2 - m_1\omega_1^2\,\hat{M}_{12}(\dot{x}_1-\dot{x}_2) = K_2$$

$$(3.2)$$

where the memory function operators $\hat{M}$ are defined by :

$$\hat{M}\,u = \int_0^t M(t-t')\,u(t')\,dt'$$

The memory functions of (3.2) have to be chosen such that at high frequencies a coupled oscillation of x_1 and x_2 results whereas at low frequencies x_1 obeys the same diffusion equation as (2.2). The transition from bare to dressed motion is expected to be characterized by the same parameter γ as the transition from diffusive to oscillatory behavior. There is thus no need to introduce new parameters and we express the M_i in terms of $M(t) = e^{-\gamma t}$ of Section 2. Proper asymptotic behavior requires :

$$M_1(t) \quad + M_{12}(t) = M(t)$$

$$M_{12}(0) = 1 \quad \int_0^\infty M_{12}(t)\,dt = 0 \qquad (3.3)$$

$$M_2(t) \quad = 1 - M(t)$$

which leads in its simplest form to

$$M_1(t) \quad = \gamma t\,e^{-\gamma t}$$

$$M_{12}(t) = (1 - \gamma t)\,e^{-\gamma t} \qquad (3.4)$$

$$M_2(t) \quad = 1 - e^{-\gamma t}$$

From (3.2) and (3.4) an analytical expression for $\sigma(\omega)$ can be derived. We will report on this in a forthcoming paper [16]. Fig. 3 shows a fit to experimental results on α-AgI. In this fit we took $m_1 = m_{Ag}$ and adjusted m_2 to account for the observed oscillator strength which resulted in $m_2 = \frac{1}{3} m_{Ag}$. To expand the range to lower frequencies we have included also data of Funke and Jost [17].

Unfortunately the experiments do not extend below $\omega \sim 7$ cm^{-1} which prevents us from giving a detailed discussion of the shoulder in $\sigma(\omega)$.

Lattice polarizability does not only affect the oscillator strength but also leads to polaron effects. Polaron effects on static and dynamic properties have been discussed by Flynn and Stoneham [18], Pardee and Mahan [3,19] and by Hinkelmann and Huberman [20]. In contrast to these authors

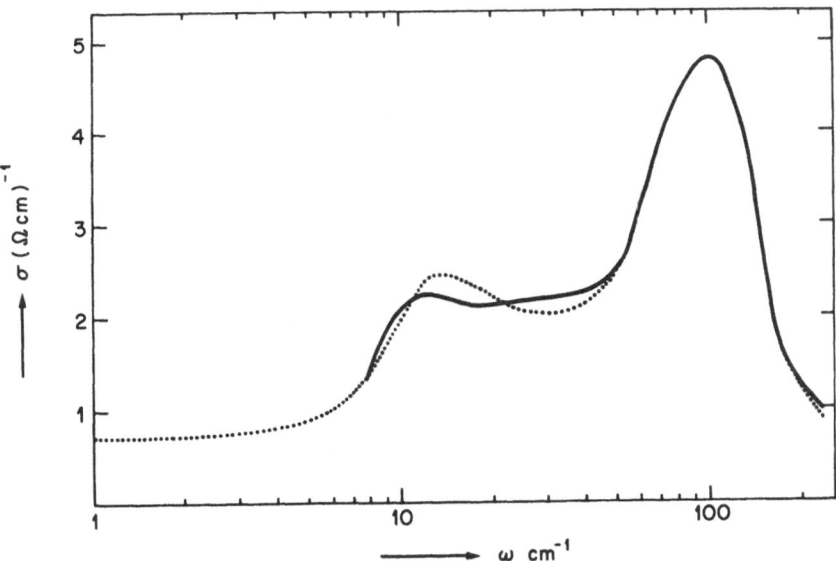

Fig. 3 Fit of $\sigma(\omega)$ calculated from (3.2) and (3.3) to the observed conductivity of α-AgI at 453° K. The fit parameters are $\omega_1 = 50$ cm^{-1}, $\Gamma = 62$ cm^{-1}, $\gamma = 24$ cm^{-1}. The structure at $\omega \sim 10$ cm^{-1} is due to the transition from bare to dressed properties.

we consider purely classical systems in which the potential barrier for diffusion is partly but not completely determined by polaron effects.

As discussed by the above authors polaron effects change the transition rate. In our model the transition rate is determined by γ which is an adjustable parameter and thus we will not discuss this effect. However, polaron dynamics may lead to structure in $\sigma(\omega)$. The relevant parameter is the polaron formation time τ_p. If $\tau_p \gtrsim \tau_R$ (residence time) the particle remembers from where it came and has a higher than random probability to jump back into its original position. Reik and Heese [9] have given an expression for τ_p which yields for α-AgI $\tau_p \sim \omega_o^{-1}$. τ_R is about two orders of magnitude longer and we expect correlation effects induced by polaron dynamics to be unimportant in α-AgI. The effect may be important in systems where the jump rate is of the same order as the lattice frequencies such as in certain metal hydrides [21]. It can be accounted for by combining (3.2) with the Ansatz of Fulde et al. [7] and results in a decrease of $\sigma(\omega)$ below the shoulder at low frequencies [16].

Thus in cases where $\tau_p \ll \tau_R$ the only structure in $\sigma(\omega)$ induced by polaron effect is at the transition from dressed to bare motion and is phenomenologically described by (3.2) by treating m_2 as a free parameter. Hinkelmann and Huberman [20] have calculated $\sigma(\omega)$ due to coherent diffusion [18] of dressed particles which results in a Drude peak at $\omega = 0$. This process is impossible in classical systems [18] such as AgI. To sum up we believe that polaron effects strongly influence the transition rate in materials built up of highly polarizable ions and that the structure observed at $\omega \sim$ 10 cm^{-1} in α-AgI is at least in part due to the transition from bare to dressed motion of the mobile particles.

4. CORRELATION EFFECTS

In most superionic conductors the number of mobile ions is not negligibly small compared to the number of available sites which leads to ion-ion interaction effects. Statically such effects show up as short range order and dynamically as correlated jumps. Experimental studies of correlation

effects have been restricted to the determination of correlation factors [22]. As opposed to diffusion of diluted defects the calculation of correlation factors in superionic
conductors is very complex [23,24]. It is the purpose of this
section to show that an analysis of $\sigma(\omega)$ provides a direct
and complete picture of correlation effects and their characteristic times.

For a homogeneous system the velocity correlation functions $S(t)$ can be written as :

$$S(t) = < v_1(0) \cdot v_1(t) > + \sum_{j>1} < v_1(0) \cdot v_j(t) >$$

$$= S_o(t) + S_1(t) \tag{4.1}$$

The brackets denote averages and the sum extends over
all particles. The autocorrelation function $S_0(t)$ measures
the correlation between successive jumps of the same particle
and the crosscorrelation function $S_1(t)$ is determined by correlation of jumps of different particles. The concentration
of radioactive atoms in tracer diffusion experiments is usually so small that the particles do not interact with each
other and hence the tracer diffusion constant D^* is completely determined by $S_0(t)$:

$$D^* = \int_o^\infty S_o(t) \, dt \tag{4.2}$$

(For simplicity we neglect the mass effect which is in general very small).

The charge diffusion coefficient D_σ on the other hand
is given by

$$D_\sigma = \frac{\sigma(0) \cdot kT}{N(Ze)^2} = \int_o^\infty (S_o(t) + S_1(t)) \, dt \tag{4.3}$$

The Haven ratio H [25] thus has the simple interpretation :

$$H = \frac{D^*}{D_\sigma} = 1 - \frac{\int_o^\infty S_1(t) \, dt}{D_\sigma} \tag{4.4}$$

According to statistical mechanics [13] $\sigma(\omega)$ can be written as :

$$\sigma(\omega) = \frac{N(Ze)^2}{kT} \int_0^\infty e^{i\omega t} (S_o(t) + S_1(t))\, dt$$

$$= \sigma_o(\omega) + \sigma_1(\omega) \qquad (4.5)$$

Furthermore $S_1(0) = 0$ which directly leads to

$$\int_0^\infty \sigma_1(\omega)\, d\omega = 0 \qquad\qquad (4.6)$$

In other words $\sigma_1(\omega)$ does not contribute to the oscillator strength and since it differs from zero in a limited frequency range it necessarily leads to structure in $\sigma(\omega)$.

By numerically performing an inverse Laplace transform of the experimentally determined $\sigma(\omega)$ $S(t)$ is obtained which contains the complete information on correlation effects. From $S(t)$ the different backward and foreward correlated events can be directly visualized and in simple cases ascribed to specific processes.

Unfortunately we do not know of a system in which $\sigma(\omega)$ is known over a sufficently wide ω range to make such an analysis possible. Instead we will make simple models for correlated events and calculate the corresponding $\sigma(\omega)$.

As a model for backward autocorrelation we consider a particle in a double well potential of the form $x^4 - x^2$. By necessity consecutive jumps are in opposite directions. $\sigma(\omega)$ can be obtained from a higher order continued fraction expansion the coefficients of which are determined by the form of the potential and temperature. Fig. 4 shows the results at different temperatures. As expected at low frequencies $\sigma(\omega)$ corresponds to Debye relaxation. Similar results but with finite dc conductivity are obtained from a periodic potential containing several inequivalent potential wells per unit cell.

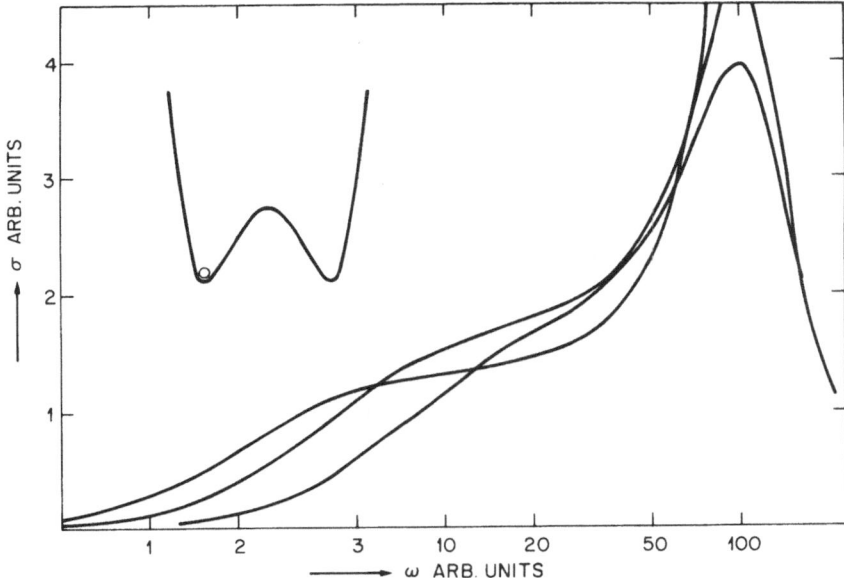

Fig. 4 Calculated conductivity at different temperatures for
a classical particle in a double well potential of
the form $V = x^4 - x^2$.

Backward correlation of the form discussed above may
occur in superionic conductors due to short range order [26,27]
or jumps between energetically inequivalent sites.

Next we turn to the discussion of foreward correlation
which enhances $\sigma(\omega)$. For instance in α-AgI $H^{-1} \sim 2$ [28] which
according to (4.4) is due to a foreward correlation in $S_1(t)$.
To demonstrate how foreward correlation leads to structure
in $\sigma(\omega)$ we make a simple model assumption for $S_1(t)$:

$$S_1(t) = A e^{-\lambda t} (1 + \cos \Omega t) - 2 A e^{-ct} \qquad (4.7)$$

The first term in (4.7) describes a foreward correlation of jumps. The second term with $c \gg \lambda$ has been added to
keep $S_1(0) = 0$ and simply leads to a constant shift in $\sigma_1(\omega)$
for $\omega \ll c$. We assume that in the frequency range of interest

also the contributions from S_0 can be incorporated into a constant background conductivity and find from a Laplace transform of (4.7) (c >> λ,ω)

$$\sigma(\omega) = a + b \left(\frac{1}{-i\omega+\lambda} + \frac{1}{-i\omega+\lambda+ \dfrac{\Omega^2}{-i\omega+\lambda}}\right) \qquad (4.8)$$

A fit of 4.8 to the structure in the microwave region of $\sigma(\omega)$ found by Funke and Jost [17] is shown in Fig. 5 $\frac{\Omega}{2\pi} < \lambda$ indicating that correlation is restricted to one induced jump. The fit should not be misunderstood as a claim that (4.7) gives a complete description of correlation in α-AgI. It merely serves to illustrate the one to one relation between particle correlation and observed structure in $\sigma(\omega)$.

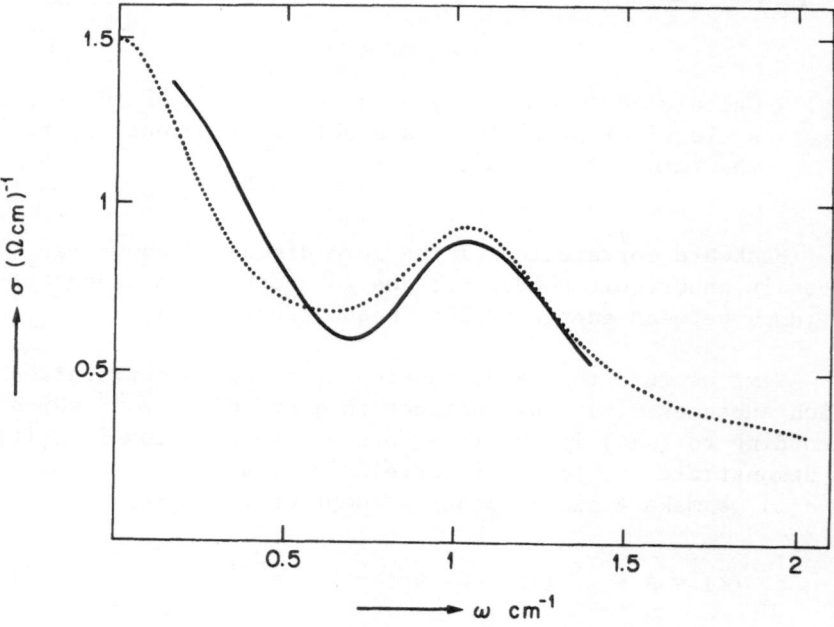

Fig. 5 Fit of (4.8) to the observed microwave structure in $\sigma(\omega)$ of α-AgI. The fit parameters are : $\Omega = 3.15 \cdot 10^{10}$ sec^{-1}, $\lambda = 1 \cdot 10^{10}$ sec^{-1}, a = 0.25 (Ωcm)$^{-1}$ b = 0.35 (Ωcm)$^{-1}$.

It is noteworthy that based on our analysis the net ef-
fect of ion-ion interaction in α-AgI is to enhance $\sigma(\omega)$ compared to the single particle conductivity of (3.2),(3.3).
This is contrary to simple expectation and indicative of a
rather interesting physics. For instance it would be tempting to introduce the soliton concept [30]. However, a closer
look shows that it leads to unphysical values for the parameters involved.

5. CONCLUSION

We have shown that studies of $\sigma(\omega)$ over a frequency
range extending from dc to optical frequencies yield important information on the dynamics of mobile ions in superionic conductors.

$\sigma(\omega)$ is dominated by a strong reststrahl resonance
which does not change its position with temperature even at
the β-α phase transition of AgI. This demonstrates that a
single particle model is valid at high frequencies and that
it provides a basis to discuss the complications at low
frequencies.

Lattice polarizability affects the oscillator strength,
the transition rate and in special cases may introduce correlation into otherwise uncorrelated jumps. Correlated jumps
originate from short range order of the mobile ions and lead
to structure in $\sigma(\omega)$ at frequencies of the order of the
transition rate ω_j. A detailed and complete picture of correlated jumps can be obtained from the velocity correlation
function $S(t)$ which is the inverse Laplace transform of $\sigma(\omega)$.

We are very much indebted to W. Schneider for his
constant support in mathematical questions.

REFERENCES

1. Fast Ion Transport in Solids, Solid State Batteries and Devices, W. van Gool ed. North Holland Publ. Co, Amsterdam (1973).

2. C.P. Flynn, "Point Defects and Diffusion" Clarendon Press, Oxford (1972).

3. W.J. Pardee and G.D. Mahan, J. Sol. State Chem. 15, 310 (1975).

4. S.J. Allen and J.P. Remeika, Phys. Rev. Letters 33, 1478 (1974).

5. B.A. Huberman and P.N. Sen, Phys. Rev. Letters 33, 1379 (1974).
   P.N. Sen and B.A. Huberman, Phys. Rev. Letters 34, 1059 (1975).

6. P. Brüesch, S. Strässler and H.R. Zeller, Phys. stat. sol. (a) 31, 217 (1975).

7. P. Fulde, L. Pietronero, W.R. Schneider and S. Strässler, Phys. Rev. Letters 26, 1776 (1975).

8. C.P. Flynn, Phys. Rev. 171, 682 (1968).

9. H.G. Reik and D. Heese, J. Phys. Chem. Sol. 28, 581 (1967).

10. H. Mori, Progr. theor. Phys. (Kyoto) 33, 423 (1965).

11. W. Götze and M. Lücke, Phys. Rev. A 11, 2173 (1975).

12. W.R. Schneider, to appear in Z. Phys. B (1976).

13. R. Kubo, Rep. Progr. Phys. 29 (Part 1), 225 (1966).

14. P. Brüesch, L. Pietronero and H.R. Zeller, to be published.

15. M. Born and K. Huang, Dynamical Theory of Crystal Lattices, Clarendon Press, Oxford (1954).

16. S. Strässler, P. Brüesch, L. Pietronero and H.R. Zeller, to be published.

17. K. Funke and A. Jost, Ber. Bunsenges. phys. Chem. 75, 436 (1971).

18. C.P. Flynn and A.M. Stoneham, Phys. Rev. B 1, 3966 (1970).

19. F.G. Mahan and W.J. Pardee, Phys. Lett. 49 A, 325 (1974).

20. H. Hinkelmann and B.A. Huberman, preprint.

21. J. Völkl and G. Alefeld in "Diffusion in Solids, Recent Developments" A.S. Nowick and J.J. Burton ed. Academic Press, New York, London (1975).

22. See chap. 6 of ref. 2 and references cited therein.

23. See paper by A.D. Leclair in ref. 1.

24. H. Sato and R. Kikucki, J. Chem. Phys. 55, 677 (1971). R. Kikucki and H. Sato, J. Chem. Phys. 55, 702 (1971).

25. See paper by M. O'Keefe in ref. 1.

26. Y. LeCars, R. Comès, L. Deschamps and J. Thery, Acta Cryst. Sect. A 30, 305 (1974). D.B. McWhan, S.J. Allen, J.P. Remeika and P.D. Dernier, Phys. Rev. 35, 953 (1975).

27. H.U. Beyeler, T. Hibma and C. Schüler, to be published.

28. A. Krist and R. Tärneberg, 3. Naturforsch. 25 a, 257 (1970).

29. A resemblence between our model and a model suggested by Clemen and Funke (Ber. Bunsenges. phys. Chem. 79, 1119 (1975)) can be found by substituting their microscopically ill defined "time of flight τ_1" by Ω^{-1} the time between correlated jumps.

30. A.C. Scott, F.Y.F. Chu and D.W. McLaughlin, Proc. IEEE 61, 1443 (1973).

Properties of the 208K Phase Transition in Single Crystals of RbAg₄I₅.* M. B. SALAMON, R. VARGAS, F. L. LEDERMAN**, AND A. SCHULTZ, University of Illinois-Urbana.-- It has been suggested that type II phase transitions in ionic conductors, in which the conductivity changes continuously, are of the order-disorder type. We present data on various properties of $RbAg_4I_5$ near the 208K phase transition which supports the view that there is partial ordering of the mobile silver ions, in a manner analogous to the cooperative Jahn-Teller transitions. The specific heat is shown to have Ising-like critical exponents and birefringence data show a linear coupling between the order parameter and the lattice strain as is characteristic of cooperative Jahn-teller systems. One of the predictions of such a model is the variation of the ionic resistivity with the degree of short range order in the system. A novel method of obtaining the temperature derivative of log R, is shown to give data strongly resembling the specific heat in the critical region which in turn demonstrates the proportionality between R and the short range order. X-ray data are presented which demonstrate that a structural component is present. Although a Landau theory analysis of this transition predicts a first order phase transition, no discontinuous behavior nor hysteresis was observed.
*Supported in part by the National Science Foundation under Grant DMR-72-03026
**Present address: General Electric Corporate Research and Development, Schenectady, N. Y.

The Distribution of Silver Ions in α-AgI. R. J. CAVA AND B. J. WUENSCH, Massachusetts Institute of Technology-- Different distributions have been proposed for the mobile Ag ions among the bcc array of I in α-AgI. Unambiguous specification of the distribution has been difficult because (a) single crystals shatter during the β→α transition at 150°C; all previous work has been powder diffraction with fewer intensities than adjustable structural parameters, and (b) the difficulty in evaluating structure factors for a "smeared" distribution of mobile ions. We have been able to reproducibly nucleate a single-crystal β→α transformation on a precession camera or four circle diffractometer through a "reverse Bridgman" technique. The silver atom distribution was completely specified, as a function of temperature and

with no adjustable parameters, through use of a probability
function derived from the potential calculations of Flygare
and Huggins. The bulk of the metal ions are concentrated
near the tetrahedral interstice. Published intensity data
obtained at 150°C provide R values of 21-23% for previously
proposed distributions, but 13.0% for the present model.
With diffractometer data obtained at 212°C, all models provide R of 17-20% but electron density and difference maps
support the proposed distribution.

Raman Spectra of the Superionic Conductors AgI and
RbAg₄I₅.* M. J. DELANEY AND S. USHIODA, Univ. of California-
Irvine.--A remarkable similarity in the Raman Spectra of the
solid electrolytes AgI and $RbAg_4I_5$ in their superionic phases
is observed. A broad peak extends from the incident laser
line out to 100 cm^{-1} in AgI and out to 70 cm^{-1} in $RbAg_4I_5$.
Both materials have a peak in their spectra at 107 cm^{-1}.
The observed features in the AgI spectrum are broader than
in $RbAg_4I_5$. In addition there is a weak shoulder at $\approx$55cm^{-1}
in the $RbAg_4I_5$ spectrum. This shoulder becomes well defined
only at low temperatures. There is a polarization selection
rule for the 107 cm^{-1} peak in AgI. The 107 cm^{-1} peak in AgI
and $RbAg_4I_5$ is observed below their superionic phases, at
room temperature in β-AgI and at nitrogen temperature in
$RbAg_4I_5$. Above 70 cm^{-1} we find a close correspondence between the frequency dependent conductivity obtained from
infrared reflectivity and that obtained from our Raman data
in both crystals. We have also measured the spectra of solid
and liquid AgBr and AgCl for comparison. The spectra of
these solids at high temperatures is very similar to those
of the melt, and liquid spectra closely resemble that of
α-AgI. The peak positions of the frequency dependent conductivity of α-AgI, AgBr and AgCl scale roughly as the
inverse square root of the reduced mass of the silver halide
pair.
*Supported in part by NSF-DMR-73-02480 A01.

Raman Measurements of AgI and RbAg₄I₅. G. BURNS,
F. H. DACOL AND M. W. SHAFER, IBM Yorktown Heights, New
York.--Using the Raman Technique, we have observed large
abrupt reversible changes in the phonon spectrum at the

transition temperature, $T_c=147°C$, of AgI. These can be
seen in the figure. The defect nature of the high temperature phase completely breaks down the selection rules and
allows the Raman measurements to give a measure of the frequency dependent conductivity, $\sigma(\omega)$, assuming a frequency
independent matrix element. Similarities as well as differences in $\sigma(\omega)$ above and below T_c can be seen.

We have also per-
formed Raman measurements
between 77°K and 296°K on
RbAg₄I₅. Careful atten-
tion was paid to the tem-
perature region of the
two phase transitions at
121°K and 208°K. We can
detect no shifts in any
of the numerous phonon
modes except one mode,
at 22.9 cm¹, abruptly and
reversibly appears in the
lowest temperature phase.
Raman results for the
isomorphic material
KAg₄I₅ are the same with
the same mode appearing
in the low temperature
phase. Thus, the results
in these systems are
markedly different from
those in AgI.

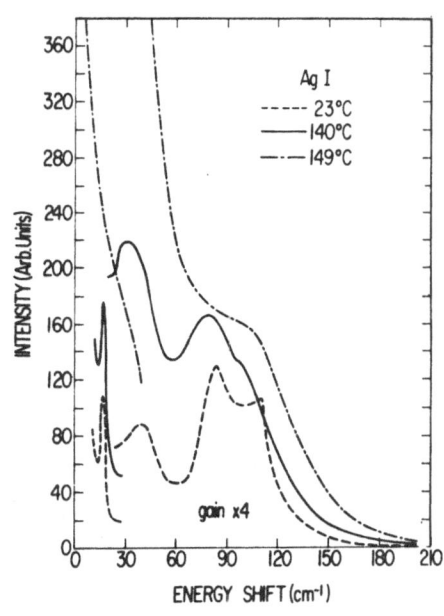

The results of
both of these studies will appear in Solid State Commun.

Raman Scattering from RbAg₄I₅.* DAVID GALLAGHER
AND MILES V. KLEIN, U. of Illinois.--Raman spectra of
oriented single crystals of RbAg₄I₅ have been measured for
all scattering geometries at temperatures up to near the
melting point and down through the second order transition
at -74C and finally down to near the first order transition
at -151C. Above -64C this crystal has a O^6 or O^7 cubic
structure and a very high ionic conductivity. It is
optically active. Below -64C the crystal deviates from
cubic symmetry and becomes trigonal and hence birefringent.

There is always a broad, structureless peak at 105 cm^{-1}
that is strongest in A_1 symmetry. This suggests that the
peak is due to the breathing modes of the iodine tetrahedra
around the silver ions. The spectra also show low frequency
scattering out to about 60 cm^{-1}. At high temperatures the
low frequency spectrum consists of a quasi-elastic peak
about the laser line plus a shoulder. It has roughly equal
T_2 and E components with little A_1 intensity. This behavior
is consistent with the assignment of the shoulder to the
attempt vibration of the Ag$^+$ ions. At room temperature
there are peaks of E-symmetry at 14 and 25 cm^{-1} superimposed
on a broad shoulder at 20 cm^{-1}, whereas in T-symmetry there
is only the shoulder. The A_1 spectrum shows a peak at 16.5
cm^{-1} and no shoulder. All three symmetries show the tail
of the quasi-elastic peak. At -63C these features sharpen
somewhat and are joined by a broad shoulder at 50 cm^{-1}.
Below the second order transition the optical polarizations
become mixed due to the presence of multiple birefringent
domains.

*Research supported by ARPA and monitored by AFOSR under
Contract No. F44620-75-C-0091.

Ion-Ion Correlations in Superionic Conductors

RbAg$_4$I$_5$. T. KANEDA, Fuji Photo Film Co., Saitama, Japan
AND T. HATTORI, Osaka University, Osaka, Japan.--Raman
scattering and infrared optical properties of superionic
conductor RbAg$_4$I$_5$ single crystals are studied. There are
two broad maxima in Raman scattering and conductivity as a
function of frequency. Using the Lyddane-Sachs-Teller
relation and a theory of motions of cations in superionic
conductors by Huberman and Sen, it is concluded that interactions between the ionic system and the host crystals have
an important role in the dynamics of RbAg$_4$I$_5$. Motions of
the silver ions may be explained in terms of a small
polaron model.

Raman Spectra of Some Ionic Conductors.* D. F.

SHRIVER, D. GREIG, G. JOY, AND M. B. LEAL, Northwestern
University.--The compounds Cu$_2$HgI$_4$, Ag$_2$HgI$_4$ and Tl$_2$HgI$_4$ all
exhibit an intense Raman feature around 123 cm^{-1}, which
corresponds to a totally symmetric HgI$_4$ stretching mode.

For the copper and silver compounds, transformation to the conducting α phase is accompanied by considerable broadening of the 123 cm^{-1} band. This feature remains narrow right up to the melting point of Tl_2HgI_4, which is not a good solid state ionic conductor. These results suggest that Raman linewidth measurements should provide a useful technique for screening potential ionic conductors. The origins of the line broadening will be discussed and vibrational assignments will be given for the above compounds as well as $MAgI_3$, MAg_2I_3, and MAg_4I_5.

*Sponsored by the National Science Foundation, through the Northwestern Materials Research Center.

Local Order in the Solid Electrolyte α-HgAg$_2$I$_4$.

T. HIBMA AND H. U. BEYELER, Brown Boveri Research Center, Baden,Switzerland.--Until now very little is known about the influence of correlations between conduction ions on the conduction mechanism in solid electrolytes. From diffuse X-ray studies the average positional pair correlations can be estimated. We have applied this technique to the classical solid electrolyte $HgAg_2I_4$, which exhibits a phase transition at 50°C from the tetragonal ordered β-phase, to the face centered cubic α-phase, in which the cations are disordered.

The diffuse X-ray patterns of α-HgAg$_2$I$_4$ consist of disk-shaped superstructure lines, centered around reciprocal lattice points, for which the average structure factors are zero. A similar diffuse intensity distribution has been observed from α-HgCu$_2$I$_4$ single crystals. The corresponding local arrangement of the cations will be discussed, as well as the consequences for the conduction mechanism in these compounds.

The Gradual Order-Disorder Transition in PyAg$_5$I$_6$.

T. HIBMA, Brown Boveri Research Center, Baden, Switzerland.--The specific heat of PyAg$_5$I$_6$ shows two features: (1) A first order phase transition at ~220°K. The low temperature structure is multiple and consists of a small monoclinic distortion of the hexagonal unit cell at high temperatures; (2) A Schottky type anomaly, associated with the disordering of the Ag$^+$ ions. The density of Ag$^+$ ions on interstitial sites and the specific heat were calculated in the Chemical

Approximation assuming two types of sites with different site energies and an interaction energy between neighboring sites. It was concluded that the interaction energy between neighboring Ag^+ ions is larger than the site energy difference.

<u>Studies of Ionic Conduction and Diffusion in Solid Silver Halides</u>. ROBERT J. FRIAUF, <u>University of Kansas</u>.-- AgCl and AgBr are simple ionic crystals displaying unusally large conductivity ($\sigma \sim 10^{-1}$ ohm $^{-1}$ cm^{-1}) and diffusion ($D \sim 10^{-5}$ cm^2 sec-1) near the melting point, values comparable to other superionic conductors. A review is given of the many experimental methods and theoretical concepts used to elucidate the nature of ionic transport processes in these materials. From conductivity in pure and doped samples it is established that Frenkel defects are dominant, with interstitial cations appreciably more mobile than vacancies. Comparison of silver tracer diffusion to theoretical correlation factors shows that interstitial silver ions move by at least two kinds of interstitialcy or knock-on processes, and this picture has been confirmed by isotope effect measurements. More elaborate computer analysis of conductivity results, taking into account short and long range coulomb interactions and using chi square statistical tests, shows a large high temperature anomaly that is only partly accounted for by the Debye-Huckel-Lidiard treatment. The suggestion of a further anomalous rise in defect concentration because of general softening of the lattice is supported by recent results for diffusion of sodium in these crystals.

Beta Alumina

STUDIES OF STABILIZATION AND TRANSPORT MECHANISMS IN BETA

AND BETA" ALUMINA BY NEUTRON DIFFRACTION[1]

W. L. Roth[2]

General Electric Research and Development Center

P. O. Box 8, Schenectady, NY   12301

F. Reidinger[3]

State University of New York

Albany, New York

S. LaPlaca[4]

Brookhaven National Laboratory

Upton, New York

INTRODUCTION

It is well known that defects are responsible for material transport in solids. In most crystalline solids their concentration is small and transport proceeds by interchanging atoms on lattice sites with vacancies or interstitial atoms. Crystal structure analysis by diffraction methods is especially valuable for determining the average atomic arrangement in a solid, but ordinarily it provides

[1]Part of this research was performed at Brookhaven National Laboratories under the auspices of the US Energy Research and Development Administration
[2]This research was supported in part by AFOSR contract F44620-72C-0007
[3]Present address, Brookhaven National Laboratory, Upton, NY
[4]Present address, IBM Watson Research Center, Yorktown Heights, New York

only limited information about defects or mechanism of
transport. The situation is different for superionic
conductors because the fraction of mobile ions is large and
scattering from them can be a significant part of the whole.
The charge carriers interact with each other and their
surroundings, and form complex configurations which are
involved in ion transport in ways that are only poorly
understood. Many experimental and theoretical approaches
are being used to determine the mechanism of conduction in
superionic conductors. In this paper, we describe advances
in identifying structural defects and disorder that are
important factors determining stability and conductivity in
sodium aluminates with the beta alumina structure.

The sodium aluminates known as beta and beta" alumina
are excellent subjects in which to study the crystal-
chemical parameters responsible for transport in superionic
conductors because their conductivity is constrained to two-
dimensional sheets in a rigid crystalline framework whose
structure is well known. The beta aluminas are exceptional
conductors of sodium ions at ambient temperature and a great
deal has been done to determine the relation of their
structure to transport. In this paper, we will briefly
review the structural features responsible for their high
conductivity, emphasizing results which give information
about the distribution of sodium on an atomic scale. We
then will describe several new defects that have been
identified in the rigid framework and the conduction plane,
and will discuss their significance to stability, conductivity, and ion transport.

STRUCTURAL BASIS OF CONDUCTIVITY IN THE BETA ALUMINAS

Beta alumina is a sodium aluminate which has nominal
composition $Na_2O \cdot 11Al_2O_3$. The general features of the
structure were solved by Bragg in 1937[1], and Beevers and
Ross in 1938.[2] They showed the compound crystallized in a
hexagonal unit cell and was composed of $Al_{11}O_{16}$ spinel-like
blocks that were separated by NaO layers. (Fig. 1) There
are three sites for sodium in the NaO layer, (2d), (2c),
and (6h) in space group $P6_3mmc$, and it was determined that
sodium occupied the (2d) site. These positions are now
called the BR (Beevers-Ross), aBR (anti Beevers-Ross), and
mO (mid oxygen) sites, respectively. Bragg noted there were

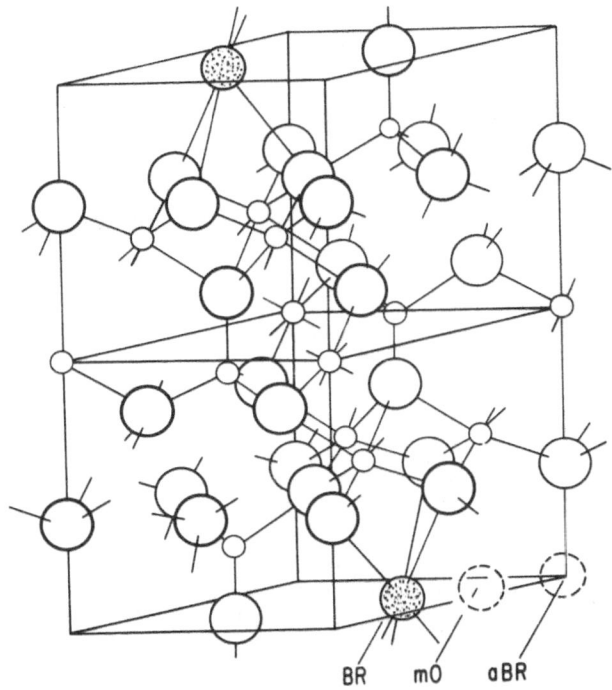

BR   mO   aBR

Fig. 1   Spinel block in beta alumina.   Large open circles
         are oxygen, small open circles are aluminum.
         Sodium is shown in the BR (Beevers-Ross) site.
         Dotted circles are the alternative mO (mid oxygen)
         and aBR (anti Beevers-Ross) sites.   The height of
         the spinel block is $1/2C = 11.26A$

difficulties reconciling the composition that was obtained
by chemical analysis with the crystallographic data.   The
discrepancies subsequently were reported to have been
resolved by improved chemical analysis, but this proved to
be illusory and the anomalies in composition are now known
to have been real.

     The name beta alumina has come into general use for
numerous beta-like phases which have related structures.
Sodium can be replaced by many monovalent cations, and
aluminum by gallium and iron, to form isomorphs with similar
unit cell dimensions.[3]   There also are phases with nearly

identical a parameters that have c parameters which are
multiples of the height of a spinel block. These correspond
to different stacking sequences of spinel blocks and their
composition may differ somewhat from beta alumina. With
the exception of beta and beta" alumina, little is known
of the structures or properties of these pseudo-polytypes.
The subject matter of this paper is restricted to the sodium
forms of beta and beta" alumina, although many of the results
should be applicable to the other forms.

The exceptional transport properties of beta alumina
were recognized in 1968 when Weber and Kummer disclosed its
use as a solid electrolyte for Na/S batteries.[4] The reason
for fast sodium diffusion was established by Peters et al.[5]
in 1972 by a three-dimensional refinement of the structure
from single crystal x-ray data. They confirmed the structure
found by Beevers and Ross, but found that sodium did not
simply occupy BR sites but was smeared out in a complex
disordered pattern in the basal plane. About 0.75 sodium
was near each BR position, spread out in a triangular
density pattern. The remaining sodium was in the mO
position, and the aBR position was empty. This disordered
structure accounts for rapid diffusion in the basal plane
since sodium can easily exchange its position between the
occupied and vacant sites. The activation energy for diffusion is determined largely by the dimensions of the channels
in the conduction plane, and increases with size of the
mobile ion (Table 1). An exception is Li^+ which may be too
small and becomes attached to the channel wall.
In a series of measurements starting in 1971,

TABLE 1   Activation Energy and Ion Size for Diffusion in
Beta Alumina

Ion	Radius, A	$E(kcal\ mole^{-1})$
Li^+	0.60	8.71
Na^+	0.95	3.81
Ag^+	1.26	4.05
K^+	1.33	5.36
Tl^+	1.44	8.22
Rb^+	1.48	7.18

Whittingham and Huggins[6] determined that beta alumina
probably conducted by an intersticialcy mechanism. By this
time, it was known the compound contained a greater concentration of sodium than the theoretical formula, and they
suggested the excess sodium charge was balanced by Al^{3+}
vacancies. This was supported by the x-ray evidence of
Peters et al.[5] who proposed the excess sodium was stabilized
by counterion defects of aluminum vacancies distributed in a
12-fold set of aluminum atoms. Models were proposed for the
distribution of sodium in BR and mO sites that consisted of
an aluminum vacancy, three BR vacancies, and six mO interstitials. A somewhat different distribution of the conduction ions was found in the isomorph Ag-beta alumina.[7] The
silver was distributed in a similar triangular pattern about
the BR sites, but the balance of the silver occupied aBR
and mO sites. The aluminum vacancy in the spinel block was
not confirmed, but comparison of the crystal and x-ray
densities of the sodium and silver isomorphs supported an
alternative model that excess sodium is compensated by
interstitial oxygen.

The second sodium aluminate for which detailed information is available is beta" alumina. It is a rhombohedral
variant of beta alumina and has three spinel blocks related
by a three-fold screw axis in the triply primitive hexagonal
cell. The principal features of the atomic arrangement were
solved by Yamaguchi and Suzuki[8] from powder diffraction
patterns, and by Bettman and Peters[9] from single crystal
data. The spinel blocks in beta" and beta alumina are
virtually identical, and both are spaced apart by oxygen
bridges which form interconnected open channels for sodium
diffusion. The sodium in beta" alumina was found to reside
in incompletely filled sites in the conduction plane, but
the chemical composition was uncertain and a detailed determination of the distribution of sodium in the plane was not
made.

The study of beta" alumina has been hampered by lack of
single crystals of the binary compound; compositions of Na_2O-
Al_2O_3, when cooled to room temperature, are mixtures of beta
alumina, beta" alumina, and other sodium aluminates. The
rhombohedral structure is "stabilized" with additives of
lithium, magnesium, and nickel. The crystal studied by
Bettman and Peters was stabilized by magnesium. Its composition was assumed to be $Na_2O \cdot MgO \cdot 5Al_2O_3$, which is an

integral formula close to the approximately known stoichiometry of the ternary compound. The similarity of the x-ray
form factors of magnesium and aluminum precluded determining
the location of magnesium in the structure.

An unusual feature of the structures of beta and beta"
alumina is their thermal parameters. The thermal vibrations
of atoms in the spinel blocks are well behaved and typical
of tightly bound ions in a crystal with high melting point.
The thermal parameters in the conduction plane are extremely
anisotropic, and the vibration amplitude of sodium in the
conduction plane at room temperature is 0.3A, or more. The
thermal effects contribute to uncertainty about the significance of the spread out density and occupation parameters
in the conduction plane, since it is difficult to differen-
tiate the contributions of defects, static displacements,
and anharmonic vibrations from the intensity data, which
essentially determine only the average structure.

NEUTRON DIFFRACTION

The well established atomic arrangements of beta and
beta" alumina explain their conductivity, but leave many
questions unanswered about the charge compensation, stoichiometry, conduction mechanism, and other crystal-chemical
factors which limit their stability and conductivity. These
properties depend on the composition and the average structure,
and also on defects and disorder. The structures of beta
alumina and magnesium stabilized beta" alumina were re-
investigated by neutron diffraction in order to identify the
defects and their relation to transport and stabilization.
Neutrons were used because their scattering and absorption
properties were especially favorable for this particular
study. The small absorption of neutrons by many materials
used to construct cryostats and furnaces makes it feasible to
study beta alumina from -190° to +600°C and determine the
effect of temperature on the sodium disorder in the conduction
plane. Neutrons were essential for studying the stabilization of beta" alumina because the nuclear scattering by
magnesium is much greater than that of aluminum which made
it possible to distinguish the ions in the structure.

The analysis of diffraction data to determine the
relatively small perturbations of structure caused by defects

and disorder requires accurate measurement of intensity
data from single crystals. The intensities were integrated
and corrected for the effects of absorption and extinction.
Multiple sets of symmetry equivalent reflections were
collected and averaged to yield a final set with intensities
that were reliable to about one percent. These data were
interpreted by least square refinement in which site occupation probabilities were treated as variables, in addition
to the usual structure parameters. Special difference
fouriers were calculated and found particularly useful for
" seeing" magnesium atoms in the spinel block, and sodium
atoms in the conduction plane.

BETA ALUMINA[10]

The excess sodium in beta alumina is 18 percent of the
total and the number of compensating defects that are
required for electrical neutrality is sufficiently large to
observe by diffraction methods. Reidinger's refinement of
neutron diffraction data confirmed the general features of
the structure which had been determined with x-rays. In
addition, an unique defect was discovered in the spinel
block framework, and this defect is an essential component
determining sodium disorder and transport in the conduction
plane.

The defect has the formula

$$V_{Al}Al_i \text{ ------ } O_i \text{ ------ } Al_iV_{Al}$$

where $V_{Al}Al_i$ is an aluminum vacancy-interstitial pair, i.e.
a Frenkel defect that is formed by the displacement of an
Al(1) atom into an adjacent interstitial site. By symmetry,
the Frenkel defect is attached through an oxygen atom, O_i,
in the conduction to its mirror image in the next spinel
block. The c-axis is horizontal in the formula above.

This compound defect resolves the controversy whether
the charge of excess sodium in beta alumina is compensated
by aluminum vacancies[5] or extra oxygen.[7] The Frenkel
defects confirm the existence of aluminum vacancies in the
spinel block. However, there is an equal number of interstitial aluminum atoms in adjacent tetrahedral sites and
the vacancies, therefore, do not neutralize the positive

charge of excess sodium. The sodium excess is compensated by the O_i ions which are located in the mO sites of the conduction plane.

The defect in the spinel block plays an important part determining the sodium disorder. A disorder model for the conduction plane follows from associating two sodium ions with O_i to preserve electrical neutrality:

$$V_{Al}Al_i \text{---} \overset{\textstyle Na}{\underset{\textstyle Na}{\overset{|}{\underset{|}{O_i}}}} \text{---} Al_iV_{Al}$$

The population parameters of the deficit aluminum in Al(1) and the interstitial aluminum which comprise the Frenkel defect correspond to the formula $1.22\ Na_2O \cdot 11Al_2O_3$.

Figure 2 is a model for disorder in the conduction plane. The defect is formed from two sodium atoms which have been displaced by O_i, and are located in or near mO sites. The sodium stoichiometry and occupation probabilities are consistent with one defect for every 9-10 unit cells.

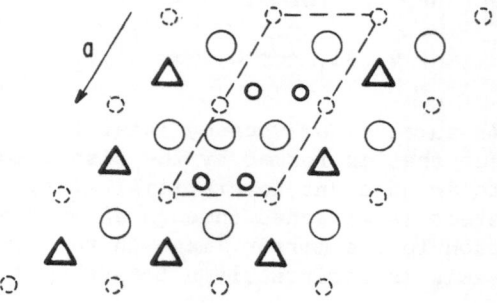

Fig. 2   Defect in beta alumina conduction plane at 80°K. The outlined region contains an interstitial oxygen atom O_i, two vacant BR sites, and four sodium atoms in mO positions. The triangles are sodium in BR sites, small circles sodium in mO sites, large circles oxygen. The small dashed circles are aBR sites which are not occupied at low temperature.

This disorder model refers to the non-conducting state
in which sodium is distributed in BR and mO sites, and the
aBR sites are empty. This is the state of affairs at -190°C.
Figure 3 displays the gradual development of sodium disorder
with temperature. At elevated temperature, aBR sites fill
with sodium, the mO site loses the characteristic of an
equilibrium position, and at 600°C is merely a broad diffuse
shoulder on the main BR maximum.

An estimate of the distribution of sodium among the
BR, mO, and aBR sites from -190° to $+600^{\circ}$C is given in
Table 2. Although the occupation probabilities can be determined only approximately, it is clear that with increasing
temperature sodium is transferred from BR and mO sites to
aBR sites. At the highest temperature, 600°C, the distribution of mobile ions is still highly non-uniform and the
most probable location for sodium is near BR sites.

BETA" ALUMINA[11]

Compared to beta alumina, there are relatively few
crystal imperfections in magnesium stabilized beta" alumina.
The spinel block is essentially free of vacancies and interstitial aluminum atoms. The only defect found was magnesium
substituted for aluminum in regular lattice positions. The
Frenkel defects in beta alumina were concentrated in a
particular part of the unit cell, and there is a similar
specificity in beta" alumina. There are four crystallographically independent aluminum sites in the spinel block
and magnesium replaces aluminum preferentially in one of
these, the tetrahedrally coordinated Al(2) sites. Slight
substitution is observed in Al(3), which is the other

TABLE 2   Site Occupation Probabilities for Sodium in
Beta Alumina from 80° to 873°K

$T^{\circ}K$	BR	mO	aBR
80	.63	.36	.01
300	.66	.30	.04
623	.59	.36	.05
873	.61	.31	.08

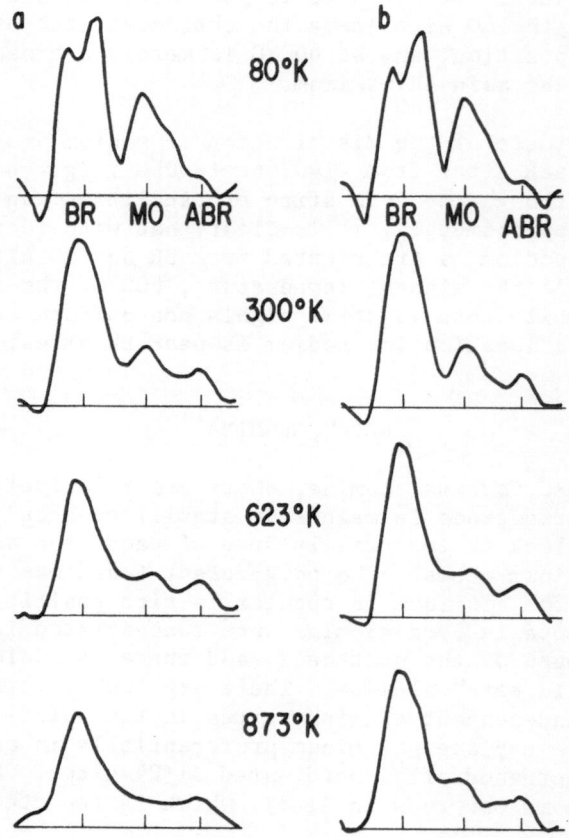

Fig. 3   One dimensional Fourier synthesis of sodium density
          at different temperatures; a, one dimensional
          section through BR, mO, and aBR sites; b, partial
          projection to clarify the density near the BR site
          at low temperature and near the aBR site at elevated
          temperature

tetrahedral site, and no substitution at all in the two
remaining octahedrally coordinated sites. The cation
distribution summarized in Table 3 shows that 38 percent of
the aluminum in Al(2) sites was replaced by magnesium,
compared to four percent of the Al(3).

The driving force for substitution of Mg^{2+} for Al^{3+}
ions in Al(2) sites is believed to be release of local
strain in the immediate environment about the site. Spinel,
$Mg[Al_2]O_4$, is the reference state for judging strain because
it is thermodynamically stable, all of the oxygen ions are
close packed, and the cations are distributed in the same
kind of site as in the beta aluminas. The diameter of a
spherical ion that can just be inserted in a tetrahedral
interstice of spinel is 1.32A, which is approximately equal
to the diameter of Mg^{2+} (1.44A) which occupy tetrahedral
sites in the low temperature (normal) state. Replacing
Mg^{2+} with Al^{3+} ions which have a diameter of only 1.00A,
and in addition are more highly charged, polarizes the
surrounding oxygen structure. Since Al(2) is in the middle
of the spinel block the strain cannot easily be relieved by
displacement of oxygen because their movements are constrained
by repulsion from neighboring atoms in the close packed
structure. Magnesium stabilized beta" alumina crystals are
grown from a high temperature melt and diffusion of cations
is sufficiently rapid for the magnesium and aluminum ions
to be distributed in sites which minimize the total energy
of the system. Support for this mechanism is supplied by
the Al-O bond distances (Table 4). An estimate of the strain
based on average Al-O bond distances in the coordination
polyhedra shows that the partial replacement of aluminum by
magnesium in Al(2) sites increased the bond distance 2.4
percent, equal to 38 percent relaxation relative to spinel.

TABLE 3   Cation Distribution in Stabilized Beta" Alumina

Atom	Coordination	Atoms per spinel block
Al(1)	6	6.03(5) Al
Al(2)	4	1.23(2) Al + .77(2) Mg
Al(3)	4	1.92(2) Al + .08(2) Mg
Al(4)	6	.98(1) Al

TABLE 4   Average Interatomic Bond Lengths in Spinel, Beta
          Alumina, and Beta" Alumina

M-O Bond Length, A

Site	Coordination	Spinel	Beta	Beta"	$\Delta \%$
Al(1)	6	1.929	1.918	1.917	-0.05
Al(2)	4	1.919	1.805	1.849	+2.44
Al(3)	4	1.919	1.745	1.744	-0.01
Al(4)	6	1.929	1.895	1.896	+0.05

The availability of an alternative mechanism for
relaxing strain about Al(3) is probably the reason only
four percent of the aluminum is replaced by magnesium.
Al(3) is coordinated to three oxygen atoms in the spinel
block and O(5) in the conduction plane. Strain is reduced
by displacement of O(5), which can easily move in directions
perpendicular to the c axis. This mode of relaxation is
supported by the anistropic thermal parameters of O(5):
$(\bar{u}_{11}^{2})^{1/2}$ = 0.218A, compared to $(\bar{u}_{33}^{2})$ = 0.068A. The partial
substitution of magnesium for aluminum is in agreement with
the suggestion by Bettman and Peters[9] that electrostatic
considerations favor Mg^{2+} in sites near the conduction plane,
but the much greater substitution in Al(2) suggests that ion
size probably is a more important consideration for determining the cation distribution.

The incorporation of magnesium in beta" alumina
increases the number of sodium atoms in the conduction plane
and in this way accomplishes a result analogous to that of
interstitial oxygen in beta alumina. The sodium distribution
in beta" alumina is simpler than in beta alumina because interstitial oxygen is not a significant factor. A further
simplification is that the principal sites which correspond
to the BR and aBR positions in beta alumina have the same
energy in beta" alumina because they are crystallographically
equivalent. Figure 4 displays the density of sodium in the
conduction slab at room temperature. There is no conduction
plane in beta" alumina because the sodium diffuses in an
undulating path through a thin slab; to display the total
sodium, it is necessary to project the density in the slab
on the basal plane. The density on the three-fold axis is

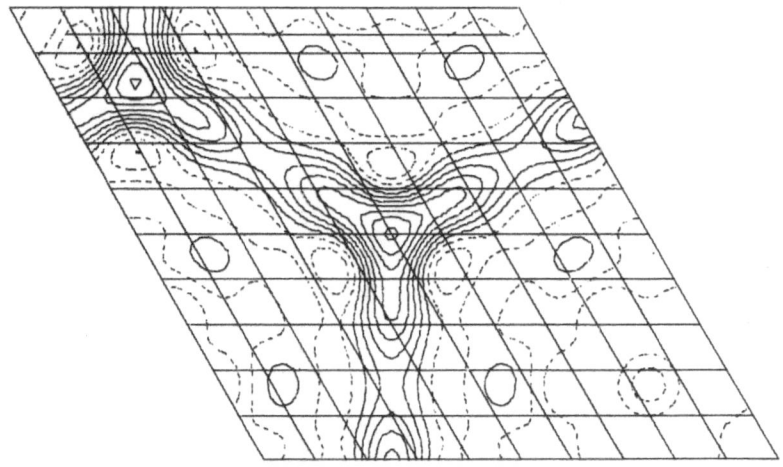

Fig. 4   Sodium density in conduction slab of beta'' alumina
         projected on the basal plane. The difference
         fourier was calculated to eliminate the oxygen
         nucleus which is located at x = -1/3, y = 1/3

equal to 1/3 sodium atom; the density in the smeared out
triangular region about the axis is equal to 1/2 sodium
atom, corresponding to 1/6 sodium in each of the subsidiary
plateaus.

    A defect model which accounts for the principal features
of the sodium disorder is shown in Figure 5. The model
assumes some of the BR-type sites are vacant. Sodium atoms
which do not have vacancies for neighbors lie on the three-
fold axis, sodium atoms with a vacancy neighbor are displaced
about 0.4A toward the vacancy. When averaged over the
symmetrically related positions, there is an equal number
of displaced sodium atoms in the three directions which
correspond to a random distribution of vacancies in principal
sites. This model gives the formula $Na_{1.67}Mg_{0.67}Al_{10.33}O_{17}$
for a composition in which there is a 2 to 1 ratio of cells
with both sites occupied to cells with one site occupied.
This stoichiometry is equal to two magnesium atoms distributed at random in the Al(2) sites of three spinel blocks
in the hexagonal cell.

    This simple model probably requires some qualification
since our best refinement of site occupation parameters

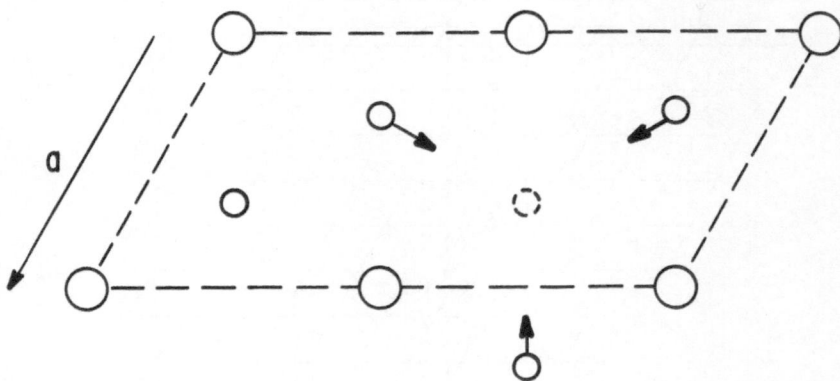

Fig. 5   Defect in conduction plane of beta" alumina. Sodium
atoms adjacent to a vacant site are displaced 0.4A
toward the unoccupied site.

indicates the magnesium content of our crystal was somewhat
greater than 2/3 magnesium per spinel block. There were 5/3
sodium per block, as predicted, but error limits for sodium
are large and the possibility cannot be eliminated that
magnesium in Al(3) is compensated by additional sodium,
probably in mO positions. This interstitial sodium would
account for the small density plateau at mO positions.

STABILITY AND CONDUCTIVITY

Relative to the nearly ideally close packed arrangement
of normal spinel in which magnesium occupies tetrahedral
sites, the spinel blocks in beta and beta" alumina are
stressed by the small highly charged aluminum ions in tetrahedral sites. Partial relaxation of these stresses is
accomplished by the Frenkel defects in beta alumina, and by
substitution of larger and less positive cations for some
Al^{3+} ions in stabilized beta" alumina. With the exception
of Al(3), which is next to the conduction plane, the average
bond distances in the coordination polyhedra correlate with
the electrostatic bond strengths[12] calculated from Pauling's
valence rule: $p = z/CN$ where z is the formal charge on the
cation and CN is the coordination number. Fig. 6 compares

the average interatomic distance in beta alumina, beta"
alumina, and spinel with bond strengths computed from the
formal charges on the ions and the site probabilities.   The
average m-O bond length decreases with increasing electrostatic bond strength in proportion to the aluminum content.
The residual strain in tetrahedral sites containing Al^{3+}
is clearly evident.   The anomalous short bond supports the
interpretation we have offered that the strain in the
Al(3) site is relaxed by displacement of O(5) in the conduction
plane.   Although it will not be treated further in this
paper this mechanism for stabilization of beta" alumina by
substitution of Mg^{2+} for Al^{3+} in Al(2) sites is also valid
for interpreting the effect of magnesium additives on the
properties of beta alumina.

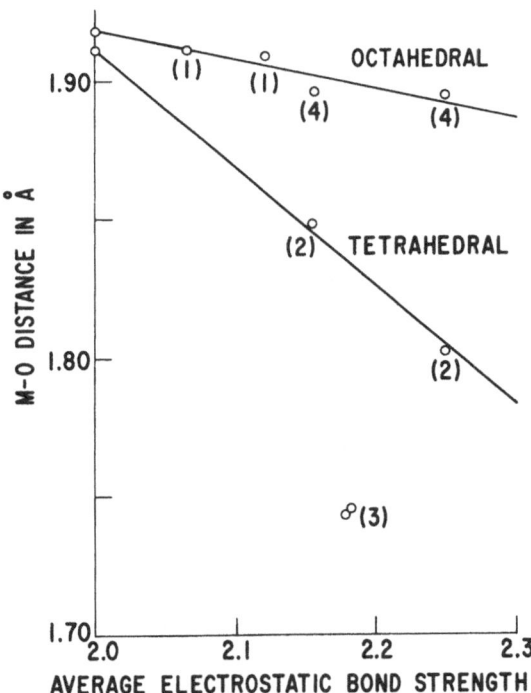

Fig. 6   Average bond length in beta alumina, magnesium
         stabilized beta" alumina, spinel, and bond strength

The structure of a solid is the average distribution
of atomic density and consequently usually does not provide
much information abvout time dependent properties, such as
material transport. Inferences nevertheless can be made
about the mechanism of sodium conduction if it is post-
ulated that the defects and disorder in the conduction
planes of beta and beta" alumina are part of the conduction
process. Bettman and Peters[9] discussed the importance of
the channel dimensions on sodium transport. In beta alumina,
the minimum channel width is about 2.4A, which is large
compared to the diameter of a sodium ion, and appreciably
smaller than the minimum channel width in beta" alumina,
which is 2.8A. The defect and disorder structures that
have been found suggest the mechanism of sodium transport is
less similar than previously assumed and we believe there
are likely to be more profound differences between the two
phases.

The principal barrier to sodium diffusion in beta
alumina we believe is thermally activated hopping from mO
to aBR sites. Evidence for this is the temperature
dependence of the sodium occupation probabilities.
The conductivity undoubtedly is influenced by interstitial
oxygen in the mO position since some of the channels are
blocked and unavailable for sodium transport. The major
effect of O_i is probably the complete or partial immobiliza-
tion of some sodium in defects such as shown in Figure 2.
It is not possible to predict the effect of O_i on conductivity without detailed information about the binding energy
of sodium in the various sites, but this model suggests
there are alternative interpretations of the correlation
coefficients determined from the Nernst-Einstein relation.
In the limit, only a fraction of the sodium ions may
actively participate in the conduction process, particularly
at low temperature.

The transport mechanism in magnesium stabilized beta"
alumina is expected to be less complicated than in beta
alumina. The interpretation of correlation coefficients
and activation energy in terms of microscopic processes
should be considerably more straightforward. Unfortunately,
few transport measurements have been reported for this
system.

SUMMARY

The disorder structures we have found in the conduction planes of beta and beta" alumina should provide a start for developing quantitative atomistic mechanisms of sodium transport. The difficult problem which remains is a quantitative treatment of activation energy which may eventually point the way to developing superior superionic conductors by control of crystal-chemical parameters and defects. The temperature dependence of the sodium concentration in beta alumina sites should help to select appropriate potential energy functions for describing the energy barriers in the BR, mO, and aBR sites.

The discovery that point defects in the spinel block interact strongly with mobile ions in the conduction plane is an important result of this work. It suggests that purposeful or accidental manipulation of defects in the rigid framework of superionic conductors may result in considerable modification of the transport properties.

There appear to be two principal driving forces for producing the defects that have been observed: polarization of the spinel block by Al^{3+} ions in tetrahedral sites; electrostatic forces due to low density NaO conduction planes each fifth layer in an otherwise closely packed ionic structure. The aluminum vacancy-interstitial pair in beta alumina and the substitutional defects in magnesium stabilized beta" alumina can both be considered responses that minimize strains which result from these forces.

The extent of the long range order that is associated with these defects remains to be determined. There is about 2/3 Mg in the average rhombohedral cell of stabilized beta" alumina, which is equivalent to two Mg in the three spinel blocks of each hexagonal cell. Models with either random defects or domains can be constructed that are compatible with this average; they must account for some spinel blocks free of magnesium, and others with either one or more magnesium atoms in Al(2).

The defects we have observed are both intrinsic and extrinsic, and little is known of the mechanism of their formation, or their relation to mass transport. It is probable that the Frenkel defects in beta alumina depend on thermal history, the temperature at which the crystals are grown, and the rate at which they are cooled to temperatures where diffusion in the spinel block is negligible. The defect concentration can also be varied by impurities, particularly by incorporating monovalent and divalent cations.

ACKNOWLEDGEMENT

One of us (WLR) wishes to acknowledge the generous contribution of experimental and computation facilities by the Brookhaven National Laboratory.

REFERENCES

1.   W. L. Bragg, C. Gottfried and J. West, Z. Krist., 77, 255 (1931)

2.   C. A. Beevers and M. A. S. Ross, Z. Krist., 97, 59 (1937)

3.   J. T. Kummer, Progr. Solid State Chem., 7, 141 (1972)

4.   J. T. Kummer and Neill Weber, US Patent No. 3,404,035 and 3,404,036, October 1, 1968

5.   C. R. Peters, M. Bettman, J. W. Moore and M. P. Glick, Acta Cryst., B27, 1826 (1971)

6.   M. S. Whittingham and R. A. Huggins, J. Electrochem. Soc., 118, 1 (1971)

7.   W. L. Roth, J.S.S.C., 4, 60 (1972)

8.   G. Yamaguchi and K. Suzuki, Bull. Chem. Soc., Japan, 41, 93 (1968)

9.   M. Bettman and C. R. Peters, J. of Phys. Chem., 73, 1774 (1969)

10. The refinement of the structure of beta alumina by neutron diffraction was done by F. Reidinger at Brookhaven National Laboratory and is to be submitted in a thesis to the State University of New York at Albany

11.   A preliminary account of this work is given by W. L.
      Roth, W. C. Hamilton, and S. J. LaPlaca, <u>Am. Cryst.</u>
      <u>Assoc. Abstr.</u>, Ser 2, <u>1</u>, 169 (1973)

12.   L. Pauling, <u>The Nature of the Chemical Bond</u>, 3rd ed.,
      Ithaca, Cornell Univ. Press (1960)

1.

2.

X-RAY DIFFUSE SCATTERING FROM β ALUMINA

J.P. Boilot[a+b], G. Collin[b], R. Comès[b],

J. Théry[a], R. Collongues[a], A. Guinier[b]

(a) Chimie Appliquée de l'Etat Solide
     Rue P. & M. Curie - 75005 PARIS (France)

(b) Physique du Solide - Université Paris-Sud
     91405 ORSAY (France)

INTRODUCTION

The well known superionic conductor β alumina belongs to a class of mixed oxydes with a general formula :

$$n\ A_2O_3 - B_2O$$

with $A = Al^{3+}$, Ga^{3+}, Fe^{3+}

and $B = Na^+$, Ag^+, K^+, Rb^+, Tl^+.

The original material is sodium β alumina, the composition of which has been found to range from $n = 5,33$ to $n = 8,5$ (1). The exceptionnally high ionic conductivity of these compounds (2, 3) is due to two-dimensional diffusion of cations in the low density layers which separate the successive spinelle blocks which form the construction elements of the bulk crystal structure (4, 8). Temperature dependent planar structural disorder (figure 1) without real phase transition, as observed both from conventional structure determinations and from diffuse scattering (9, 12) is clearly related to the temperature dependent ionic conductivity. While the structure determinations as deduced from Bragg diffraction give an average picture of the disorder, diffuse scattering on the opposite will provide insight on the real local organization of the cations at a scale of a few unit cells.

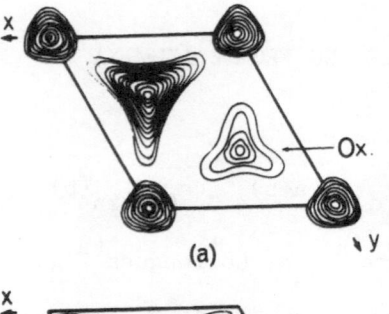

(a)

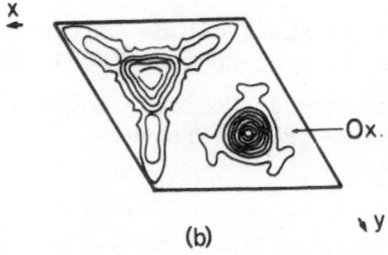

(b)

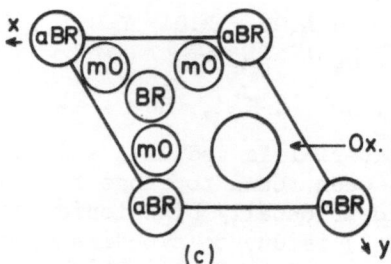

(c)

Fig 1
------

a) and b)   Electronic density maps in the
            low density layers of Ag and Na
                Alumina (5, 7).

c)   Simplified site model used for the
            determination of the local order
            with the conventional site denomi-
            nation B.V. = Beevers-Ross,
            a B.V. = anti-Beevers,
            mO = mid-Oxygen

X-RAY DIFFUSE SCATTERING

Any deviation from the perfect periodic structure
which gives rise to the Bragg diffraction, produces an
extra scattering at general positions in reciprocal
space. In the case of chemical disorder as for example in
alloys, this scattering is proportional to the square of
the deviation of the local electronic density from its
mean value in the crystal. In the case of displacive disorder or phonons, the scattering is proportional to the mean
square displacement of the atoms out of their average position. Generally, the scattered intensity is proportional
to I_d given by :

$$I_d = <F - <F>>^2 \exp - 2M$$

where F is the local structure factor calculated over
one ordered area, $<F>$ the average structure factor calculated over the same area, and M the Debye Waller factor.
Random chemical disorder, or ordinary phonons, give only
rise to a slowly varying background which is difficult
to analyse, but if there is some kind of local order or
correlated motion of several atoms, the scattering will
concentrate in particular directions and localized weak
maxima of intensity will be observable in reciprocal
space. From this localization, it is then possible, in a
similar way as for ordinary structures, to determine the
nature of such a local order, or of correlations in the
case of dynamical effects corresponding to larger amplitudes than those of ordinary phonons. A minimum in a
dispersion spectrum for instance yields an intensity I :

$$I \propto (\text{amplitude})^2 \propto \frac{1}{(\text{frequency})^2}$$

As a first approach, because one ignores a priori
how such an intensity will be localized, the fixed crystal fixed film technique with the schematic set up shown
in figure 2a is best suited. The intensity collected on
the photographic film corresponds then, point by point,
to all general positions in reciprocal space which simultaneously satisfy the relations :

$$\vec{Q} = ha^x + kb^x + lc^x = 2\pi (\vec{S} - \vec{S_0})/\lambda$$

where a^x, b^x, c^x are the reciprocal lattice parameters,
$\vec{S}$ is the unit vector of each scattered beam, $\vec{S_0}$ the

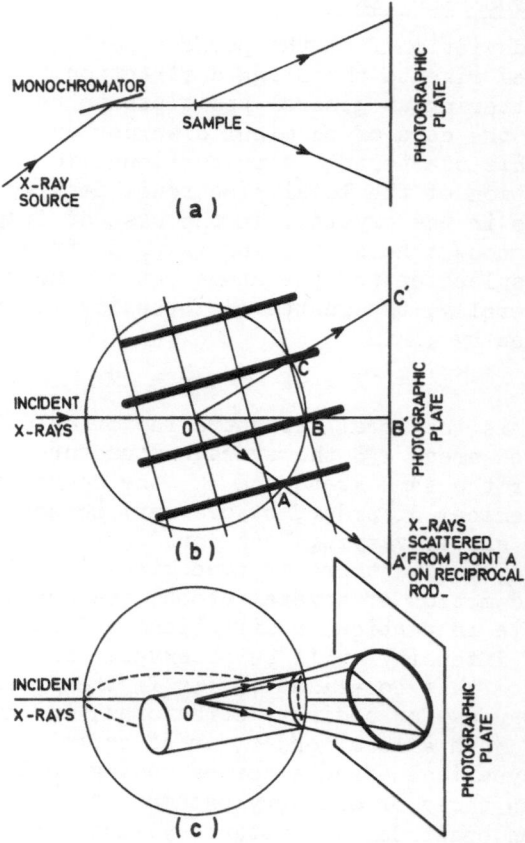

<u>Fig 2</u>   - a) Schematic set up of the fixed crystal fixed
               film X-ray diffuse scattering technique.

        b) Corresponding view of a vertical section of
            reciprocal space, showing the intersection
            of a lattice displaying intensity localized
            on rods,with the Ewald sphere and the resul-
            ting scattered beams.

        c) Corresponding view in reciprocal space in the
            case of a two-dimensional liquid giving rise
            to intensity localized on a cylinder. When the
            2-d liquid is not perpendicular to the inci-
            dent beam, the resulting halo is elongated.

unit vector of the incident beam and λ is the wavelength.
h, k, l are here not restricted to integer values as for
usual Bragg diffraction. Figure 2b shows a vertical section containing the incident beam, the locus of the
points $2\pi(\vec{S} - \vec{S}_0)/\lambda$ (Ewald sphere), a reciprocal space
corresponding to intensity located along reciprocal rods
as expected from two dimensional local order. Figure 2c
shows a similar construction in the case of a two dimensional liquid which corresponds in reciprocal space to
an intensity maximum localized on a cylinder.

SAMPLE PREPARATION

Several β Alumina crystals with M = Na, Ag, K, Tl
were studied with this method. The initial sodium β
Alumina crystals have been prepared from a melt of sodium
carbonate and β Alumina ; their composition as determined
from neutron activation analyses yielded a formula of
8.3 Al_2O_3 - Na_2O. The other compounds were obtained from
sodium β Alumina by molten salt total ion exchange (13).
Most the detailed analyses of the local order was
obtained on silver β Alumina because of the higher scattering factor of the Ag cations ($I_{\beta Ag} = 2 0 I_{\beta Na}$ for instance), and only a qualitative comparison will be made
with the other crystals of the series. All patterns shown
were obtained using MoKα radiation (λ = 0.709 Å).

ORGANIZATION OF THE CONDUCTING IONS IN SILVER β ALUMINA

Figure 3 summarizes the qualitative temperature
dependence of the organization of the silver ions (9).

At the highest temperature of 750 K, the pattern of
figure 3a, besides a few Bragg spots from the crystal
matrix, displays a diffuse halo. Such a halo, despite the
fact that it is not completely isotropic, recalls the
scattering characteristic of a liquid or an amorphous
substance. It bears for instance much resemblance with
the diffuse halo obtained from neutron irradiated quartz
before this crystal has become completely amorphous (14);
there the stirring up of atoms due to the "thermal spike"
also produces a highly disordered quasi liquid region
(15), but the surrounding ordered crystal still imposes
some preferred position giving some structure to the

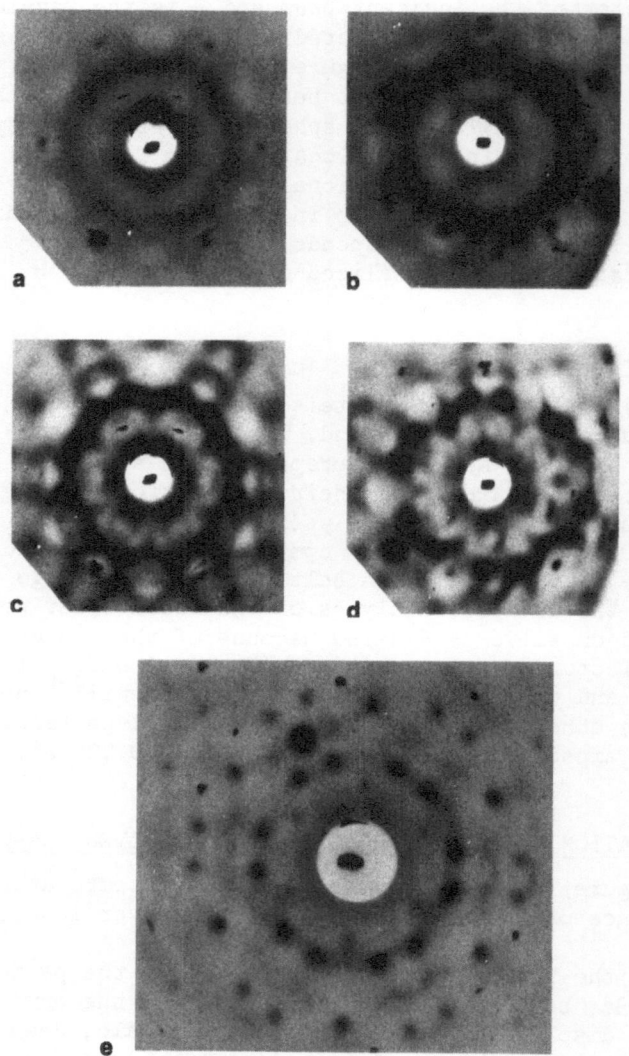

Fig 3 :   Diffuse scattering from Ag β Alumina

          a) 700 K   c axis perpendicular to incident beam.
          b) 700 K   c axis tilted away from incident beam.
  c and d) same as a)and b) at room temperature
          e) 77 K   c axis perpendicular to incident beam.

diffuse halo. In β Alumina, the same type of order can be
expected for the conducting ions, here however the quasi-
liquid should be two-dimensional. An ideal two-dimensional
liquid, as already mentioned above (figure 2c), gives rise
in reciprocal space to a scattered intensity which has a
maximum located on a cylinder, and not on a sphere as is
the case for a usual three-dimensional liquid. That this
is the case in β Alumina can be seen on the patterns of
figure 3. If the c axis of the silver β Alumina crystal is
oriented parallel to the incident beam, the diffuse halo
is nearly circular (figure 3a) ; if on the contrary the c
axis is tilted relative to the incident beam the halo becomes elongated (figures 3b and 2c). This clearly demonstrates that the silver ions, at high temperature, are in a
two-dimensional quasi-liquid state, which gives a qualitative explanation of the high ionic conductivity in the
same temperature range.

When the temperature is lowered, the intensity which
was relatively uniform in the high temperature halo,
begins to concentrate on broad spots within the halo
first, and around room temperature other broad spots outside the halo become also observable (figure 3c). This
evolution is characteristic of a progressive built up of
a periodic small range order between the conducting ions.
Here again, tilting the c axis away from the incident beam
(figure 3d), establishes that the diffuse spots originate
from the intersection with the Ewald sphere of rods in
reciprocal space, showing the two-dimensional character
of the local order.

At lower temperatures, at 77 K (figure 3c), there is
only little left from the diffuse halo, but a regular pattern of diffuse spots can be easily observed. The corresponding periodic local order is a hexagonal superlattice
with a superlattice constant of $A = a\sqrt{3} = 9.68$ Å, oriented
at 30° from the average cell of β Alumina with $a = 5.59$ Å.
Qualitatively, the progressive built up of a quasi-crystalline arrangement of the conducting ions, account for the
progressive decrease with temperature of the ionic conductivity.

From this evolution towards a low temperature
superstructure, it is tempting to predict the occurrence
of a real phase transition giving rise to a long range

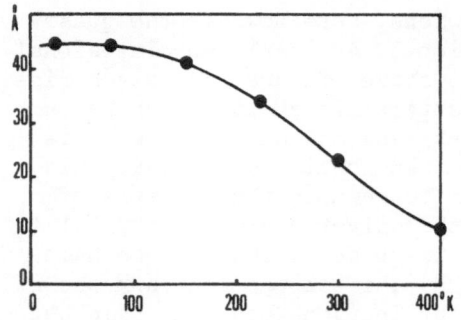

Fig 4 : Size of the low
temperature or-
dered regions
as a function
of temperature
in Ag β Alumina

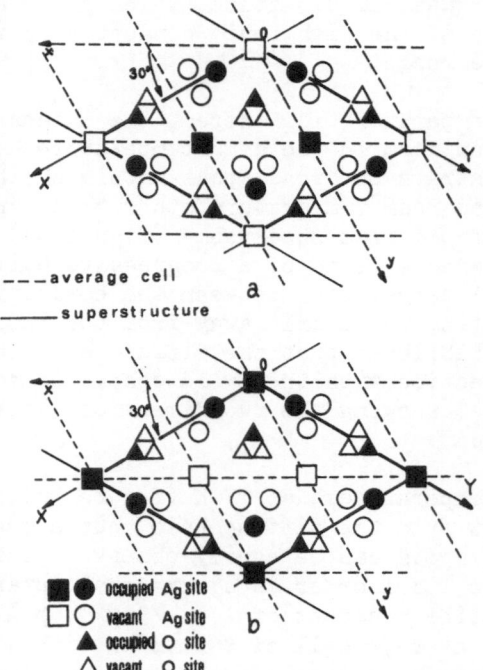

Fig 5 : Two models which can account for the a√3
superlattice of Ag β Alumina. The super cell
is three times bigger than the average cell,
and arises in both cases because of the occupation of the 3 possible mid-oxygen sites.

ordered arrangement of the silver ions below 77 K. This is
however not the case, as can be seen from the temperature
dependence of the size of the ordered regions shown in
figure 4. This size is easily obtained from the extension
in reciprocal space of the diffuse spots as a function
of temperature. After a rapid evolution just below room
temperature, the size of the ordered regions clearly saturates around 100 K with no further change down to 20 K.

A quantitative analyse of the local order at high
temperature is a very difficult task, but in the low
temperature quasi-crystalline state, it was possible
to get some insight on the type of local order (10).
Using the schematic average disordered structure model
shown in figure 1c, with the following simplified occupation factors deduced from the electronic density distribution determined by Roth (5) and shown in figure 1a for
silver β Alumina ;occupation of each Beevers Ross site=0,
each anti-Beevers-Ross site = 0.5 and each mid-oxygen
site = 0.8/3, two simple models shown in figure 5 can
produce a a√3 superlattice. These two models are unique
if only one of each mid-oxygen site can be occupied in
the low density layer of a given unit cell, assumption
which is justified by the size of the silver ions and
the value of the positional parameters.

In fact, the analysis of the intensity of 12 superlattice reflexions at 77 K showed that neither of these
ideal models could give a satisfactory agreement, but
that partial order with a mixture of both models yielding
enriched occupation of the 3(c) and 1(a) sites of model
b, and non zero values for the 6(d) and 2(b) positions
of model a, resulted in R factors as low as 0.07, which
is very good for such an analysis (the site denomination
used here is that of the P 31m 2-d space group Wyckoff
positions). Figure 6 shows the temperature dependence
obtained for these occupation factors. It is remarkable
that the occupation factors tend towards the average
occupation around room temperature, where superlattice
diffuse spots are hardly visible.

One surprising aspect of the partially ordered low
temperature micro domains of about 45 Å is that the atoms
in the mid-oxygen site tend to relax towards an occupied
antiBeevers-Ross site, which results in clusters of four

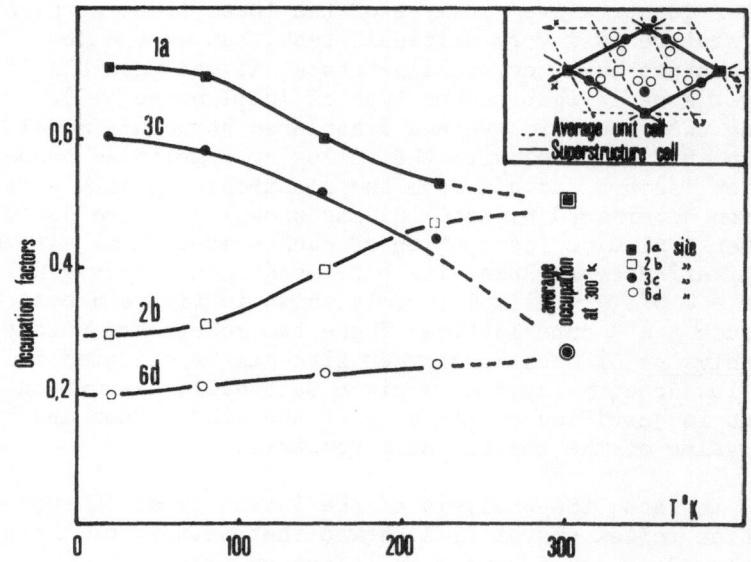

Fig 6   :   Occupation factors of the sites of the
              mixed model which best accounts for the
              observed diffuse scattering, versus
              temperature.
              Note that the occupations tend towards
              the average value around room temperature.

Ag ions at the origin of the superlattice cell.

How reliable are such models of local order as
deduced from a very small number of reflections ? Disa-
greement between the present description (10) and an
independent investigation by Mc Whan et al (11) initially
casted some doubt, but since most major discrepancies
could be attributed to different sample preparation, melt
grown crystals are different from flux grown ones (16).
Another example of sample dependent differences in the
X-ray scattering will be provided by Tl β Alumina.

OTHER β ALUMINA CRYSTALS (Na, K, Tl)

Silver β Alumina is of course an ideal case to be
investigated by X-ray diffuse scattering because of the
high scattering factor of silver, compared to all other
atoms of the crystal.

One of the hardest case is unfortunately sodium β
Alumina, which happens to be the most interesting material
of the series, because of its higher ionic conductivity.
Figure 7 shows 2 patterns obtained at 20 K and 800 K.
There is a clear modification of the scattering demonstrating increased order when going towards lower temperatures. At high temperature, there is no sign of a
diffuse halo as observed with silver β alumina, but it is
difficult to state whether this absence is only due to
the smaller scattering power of sodium (sodium represents
only 5 % of the scattering power of the material) or if
it reflects an organization of the cations which is even
more disordered than a liquid. At lower temperature, a
few diffuse spots become observable and seem to correspond to a 2a√3 superlattice (which would be twice as
big as the Ag β Alumina superlattice). Attempts to construct models to account for the low temperature scattering of Na β Alumina have so far only given poor agreement, it seems that the weak effects (on the diffuse
scattering) which arise from the charge compensation
mechanism cannot be neglected here (16). A clear result
however concerning Na β Alumina is that this compound
is more disordered, in agreement again with its higher
ionic conductivity.

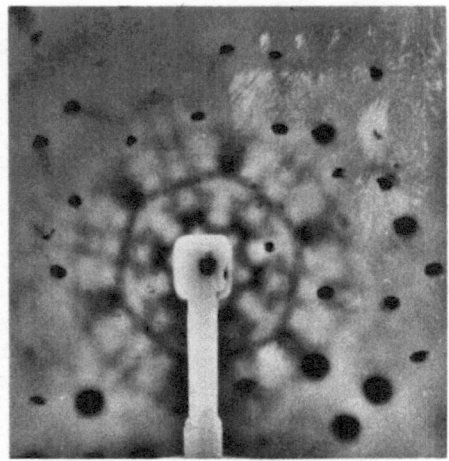

T = 20 K

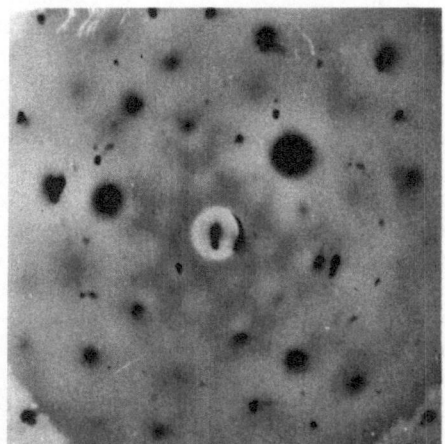

T = 800 K

Fig 7 : X-Ray patterns from Na β Alumina at 20 K
         and 800 K.
         Note the absence of diffuse halo at high
         temperature.

Potassium β Alumina is a somewhat more favorable case ; figure 8 shows two patterns obtained at 78 K and 770 K. These patterns clearly reveal diffuse spots at all temperatures, showing a higher degree of order in this compound, which one more is in agreement with the lower ionic conductivity. There is however some ambiguity about the superlattice arising from the organisation of the potassium ions. From the smaller angle diffuse reflections one finds a superlattice of a√3 (as for Ag β Alumina) in agreement with the superlattice reported by Mc Whan et al (11), but some weak reflections only obwerved at higher scattering angles seem to indicate a real superlattice of 2a√3, as suggested here for sodium β alumina. A detailed analysis of the diffuse scattering was here delayed because the average structure was not known, which is an absolute perequisite. This structure has now been determined (11, 17) and is found to be closer to the Na β Alumina structure, than to the Ag β Alumina one. Most atoms have been found on Beevers-Ross and mid-oxygen sites.

For Potassium β Alumina, such a structure and the size of the conducting ions only allows a simultaneous occupation of several mid-oxygen sites in a given unit cell. The simplest superstructure one can think of, therefore leads to a 2a√3 superlattice as observed. The doubling of the superlattice (relative to Ag β Alumina) arises from an alternate occupation of Beevers-Roos sites and mid-oxygen sites in successive average size unit cells, and from the three fold symmetry. Preliminary attempts to account for the intensity of the diffuse reflections with models of this type gave encouraging agreement.

The case of Tl β Alumina raised another difficulty. The pattern shown in figure 9 displays only broad diffuse spots surrounding the regular Bragg reflections, that is to say no superlattice, implying an occupation of one Tallium atom per average unit cell. Other patterns however, from different samples, on contrary showed the same type of pattern as Potassium β Alumina. This marked difference in the diffuse scattering is found to be due to incomplete ion exchange in the case of figure 9. This result illustrates, together with the controversial observations already mentionned for different Ag β Alumina samples (10, 11), the importance of sample prepara-

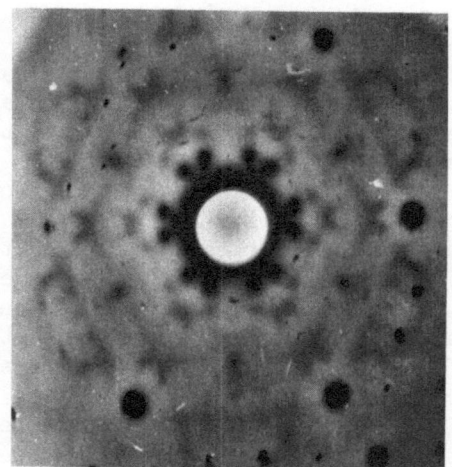

T = 78 K

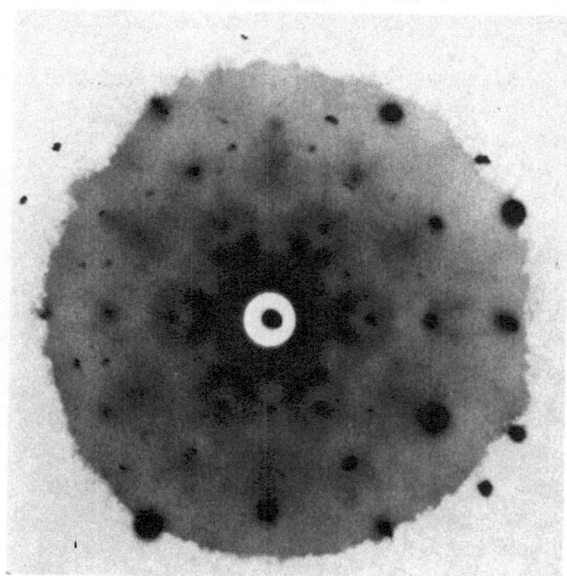

T = 770 K

Fig 8 : X-ray patterns from K β Alumina at 78 K
           and 770 K.

           Note the presence at all temperatures of
           superlattice diffuse spots.

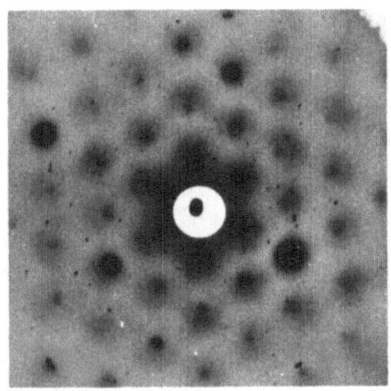

Fig 9

Pattern obtained with
Tl β Alumina at 300K
in the case of incom-
plete ion exchange.
Diffuse spots are
observed at all tem-
peratures but only
surrounding the Bragg
reflections from the
main lattice. Complete
ion exchange results
in patterns very simi-
lar with those of
K β Alumina (fig 8).

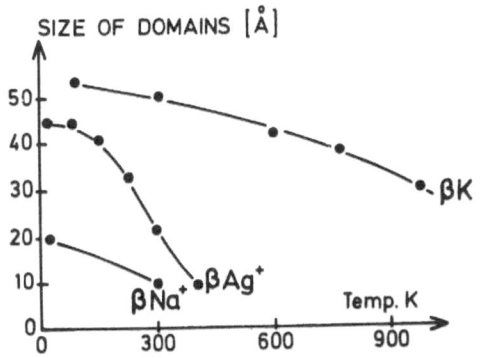

Fig 10

Size of ordered regions
versus temperature and
as a function of the
conducting ion.

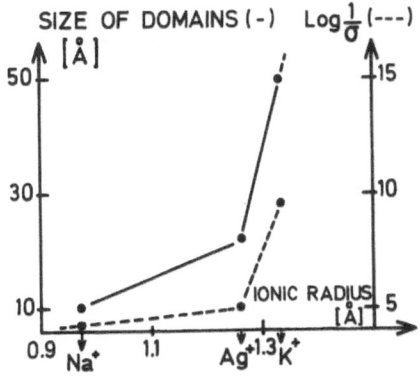

Fig 11

Correlation at a given
temperature between the
size of the ordered
regions, the ionic radius
of the conduction ion and
the inverse conductivity.

and characterization, but also the sensitivity of the diffuse
scattering technique.

C O N C L U S I O N

        The different efforts which have been made in order to
clarify the organization of the conducting ions in β Alumina
using X-ray diffuse scattering, as well with photographic
techniques (9→11) as with more precise counter detection (16)
have up to now only given preliminary results. Two important points seem however well established :

i)   as shown above the changes in the ionic conductivity as
     a function of temperature, or as a function of the con-
     ducting ions are correlated with changes in the degree
     of the 2-D order among the conducting ions. Higher the
     degree of order, lower the conductivity. Figure 10 il-
     lustrates the temperature dependence of the "size" of
     the ordered region as a function of temperature, and as
     a function of the conducting ions. Figure 11 shows the
     size of the ordered regions and the conductivity as
     function of the ionic radius of the conducting ions ;

ii) as shown by Mc WAHN et al (16), the analysis of the in-
     tensity of the diffuse scattering in c direction (along
     a diffuse rod), confirms the existence of interstitial
     oxygen in mid-oxygen sites, with a shift along the c axis
     of the nearby aluminium atoms from their octahedral
     sites in the spinelle blocks to the tetrahedral sites
     closer to the conducting planes, as was earlier sug-
     gested by ROTH et al (8).

        In all this work, we have analyzed spatial order ; as
is well known, X-rays cannot distinguish a static ordered
cluster or domain, from a dynamic correlated motion. Diffuse spots observed by X-rays can arise from inelastic effects, some indication in that direction, at least at sufficiently high temperature can be found in the neutron scattering study of SHAPIRO et al (12).

        This work has benefited from several discussions with
B.A. HUBERMAN, D.B. Mc WHAN and S.M. SHAPIRO.

R E F E R E N C E S

(1)   LE CARS Y., THERY J. and COLLONGUES R., Rev. Int.
       Haute Temp. et Refract., 9, 153, (1972).

(2)   RICE M.J. and ROTH W.L., J. Sol. State Chem., 4,
       294, (1972).

(3)   WHITTINGHAM M.S. and HUGGINS R.A., J. Chem. Phys.,
       54, 414, (1971) and J. Electrochem. Soc., 1, 118,
       (1971).

(4)   BRAGG W.L., GOTTFRIED C. and WEST J., Z. Kristallogr.
       77, 255, (1931).

(5)   ROTH W.L., J. Sol. State Chem., 4, 60, (1972).

(6)   BEEVERS C.A. and ROSS M.A.S., Z. Kristallogr., 97,
       59, (1937).

(7)   PETERS C.R., BETTMAN M., MOORE J.W. and GLICK M.D.,
       Acta Cryst., B 27, 1826, (1971).

(8)   ROTH W.L., REIDINGER F. LAPLACA S., - to be
       published ibid.

(9)   LE CARS Y., COMES R., DESCHAMPS L., THERY J., Acta
       Cryst., A 30, 305, (1974).

(10)   BOILOT J.P., THERY J. , COLLONGUES R., COMES R. and
       GUINIER A., Acta Cryst., A 32, 250, (1976).

(11)   Mc WHAN D.B., ALLEN S.J., REMEIKA J.P., DERNIER P.D.,
       Phys. Rev. Lett., 35, 953, (1975).and 36, 344(E),(1976)

(12)   SHAPIRO S.M. et al - to be published ibid.

(13)   YU YAO Y.F. and KUMMER J.T. - J. Inorg. Nucl. Chem.,
       29, 2455, (1967).

(14)   COMES R., LAMBERT M., GUINIER A., in Interaction of
       Radiation with Solids, p. 319 (Plenum Press, 1967).

(15)   BRINKMANN J.A., Am. J. of Phys., <u>24</u>, 246, (1956).

(16)   Mc WHAN D.B. - private communication.

(17)   R. COLLONGUES et al - to be published.

NEUTRON SCATTERING STUDIES OF SOLID ELECTROLYTES *

S. M. Shapiro

Brookhaven National Laboratory

Upton, New York 11973

INTRODUCTION

Neutron scattering is now a well-established technique for probing atomic positions and motions in a solid or liquid.[1] To a first approximation the neutron impinging on a sample is scattered by the nuclei and the energy and momentum of the scattered neutron is determined by the collective nature of the sample's nuclei (coherent scattering process) or by the individual atom's motion (incoherent scattering process).

The conservation laws of momentum and energy in a scattering process are

$$\vec{Q} = \vec{k}_i - \vec{k}_f$$
$$\hbar\omega = \frac{\hbar^2}{2m_N}\left(k_i^2 - k_f^2\right) \tag{1}$$

$\vec{Q}$ and $\hbar\omega$ are the crystal momentum and energy, k_i and k_f and the initial and final momentum of the neutron, and m_N is the mass of the neutron. Neutron scattering offers important advantages over optical and x-ray techniques. Because k_i is comparable to Q_{ZB}, the size of the Brillouin zone ($1-5\text{Å}^{-1}$), studies of excitations throughout the Brillouin zone are possible. In light scattering $k_i \ll Q_{ZB}$ and

* Work performed under the auspices of the U. S. Energy Research and Development Administration.

only excitations very near the zone center are studied.
Since the neutron energy is also comparable to the lattice
vibrational frequencies (1-50 meV) studies of the lattice
modes can be performed. In x-ray scattering the x-ray
energy is much greater than the lattice frequencies and
energy analysis with sufficiently high resolution is presently impossible. The scattering amplitude for neutrons
has a limited variation over the periodic table whereas
the scattering amplitude for x-rays is proportional to Z^2,
the charge in the nucleus. Thus neutrons allow visibility
of light elements when heavier atoms are present.

In this talk I shall review some of the general formalism for elastic and inelastic neutron scattering and
demonstrate what quantities can be measured. Next I shall
show how neutron scattering has been applied to the studies
of three different problems: 1) the anion disorder in the
fluorite systems; 2) the dynamical behavior in the beta-
alumina; and 3) the cation diffusion in αAgI.

The scattering cross section for neutrons is the sum
of a coherent and an incoherent part. Dealing first with
the coherent scattering, the partial differential cross
section for a solid is:[1,2]

$$\frac{d^2\sigma}{d\Omega d\omega} = \frac{k_f}{k_i} \mid F(\vec{Q},\omega) \mid^2 \tag{2}$$

where $F(\vec{Q},\omega)$ is the generalized structure factor. It is
composed of an elastic part which measures the average
structure and an inelastic part which probes the time
dependent fluctuations in the average structure. In the
harmonic approximation

$$\mid F(\vec{Q},\omega) \mid^2 = (2\pi)^3 \frac{N}{V} \mid F_0(\vec{Q}) \mid^2 \delta(\vec{Q}-\vec{G}) \delta(\hbar\omega)$$

$$+ \sum_s \mid g_s(Q) \mid^2 (n(\omega)+1) \delta(\vec{Q}\pm\vec{q}-\vec{G}) \delta(\hbar\omega\pm\hbar\omega_s) \tag{3}$$

Here N is the number of unit cells in a crystal, V is the
volume of the unit cell , $\vec{G}$ is a reciprocal lattice vector
and $\vec{q}$ is the wave vector reduced to the first Brillouin
zone. s denotes the mode of wave vector $\vec{q}$ and branch j.
$n(\omega)+1 = \{ (\exp(\hbar\omega/kT)-1 \}^{-1} + 1$ is the population factor

for creation of a phonon. The first term in Eq. (2) is the structure factor for elastic Bragg scattering

$$F_o(\vec{Q}) = \sum_k b_k \, e^{i\vec{Q}\cdot\vec{R}_k} \, e^{-W_k(\vec{Q})} \tag{4}$$

where R_k is the position of the kth atom in the unit cell and $W_k(\vec{Q})$ represents the Debye-Waller factor. This is valid for point particles whose position is smeared out by their thermal motions. Eq. (4) can be written in a more general way by introducing an average density:

$$\wp(\vec{r})> = b_k \, e^{-W_k(Q)} \, \delta(\vec{r}-\vec{R}_k)$$

and $F_o(Q)$ becomes

$$F_o(\vec{Q}) = \int_{cell} \wp(\vec{r})> \, e^{i\vec{Q}\cdot\vec{r}} \, d\vec{r} \tag{4'}$$

For solids with long range order elastic peaks appear at well-defined positions in reciprocal space which satisfy the δ function, namely $\vec{Q}=\vec{G}$. If the long range order is interrupted due to some defects, the Bragg scattering now represents the scattering from an average cell. By examining the intensities of these peaks one can obtain an effective value of the scattering strength $b_k' = m_k b_k$ where m_k is the fractional occupancy of the kth atom at site R_k. In a disordered system additional scattering appears which is not as well localized in reciprocal space. Its intensity and q width is related to the size of the disordered regions and the interaction among the defects. It is measurements of this diffuse scattering that provides detailed information of the disorder occurring in many solid electrolytes and has been amply discussed for the x-ray case by R. Comès.[3]

The second term in Eq. (2) is the dynamical structure factor for coherent inelastic scattering:

$$g_s(Q) = \sum_k \frac{b_k(\vec{Q}\cdot\vec{\xi}_{s,k})}{(2NM_k \, \omega_s/\hbar)^{\frac{1}{2}}} \, e^{i\vec{Q}\cdot\vec{R}_k} \, e^{-W(\vec{Q})} \tag{5}$$

By satisfying the δ functions in Eq. (2) with a judicious choice of incident and scattered neutron momentum the phonon dispersion curves can be measured. Additionally, the eigenvector of the atomic displacements, ξ_k, determines the intensities of the scattered neutrons. If there is disorder in the solid or there are important anharmonic effects the normal modes become less well defined and the observed spectra contain a measurable energy width. Below I shall show how inelastic neutron scattering has provided some key information into the lattice dynamical behavior of the beta alumina electrolytes.

The coherent scattering cross section discussed above probes the correlations of pairs of particles distributed in space. The incoherent scattering, on the other hand, probes correlations of individual particles with themselves. The time dependence of these correlations is related to the motion of the isolated atoms which can usually be described by a diffusion process. If the atom diffuses freely over the entire sample the neutrons scattered by it will have a slightly different energy. The resulting spectra will show a quasielastic peak whose width is proportional to the diffusion constant of the material. Neutron scattering offers the possibility of simultaneously measuring the time scale of the diffusion via the energy width and the spatial correlations via the Q dependence of the scattering. Hydrogen has an exceptionally large incoherent scattering strength and the studies of the quasielastic scattering from the hydrogen atoms has provided valuable data about the diffusive processes of hydrogen dissolved in metals[4]. Similar techniques have been applied to the superionic conductors. Axe et al[5] have measured the rotational diffusion of NH_4^+ in beta alumina and Eckold et al[6] have used the incoherent scattering from silver ions to study the diffusion processes in $\alpha-AgI$.

TECHNIQUES

There are two conventional types of inelastic neutron scattering spectrometers: a Triple-Axis Spectrometer(TAS) or a Time-of-Flight Spectrometer(TOFS). I shall not go into the details of these instruments since they are well described in the literature[1,2,7], but I would like to discuss where each spectrometer is most applicable. The energy and momentum transfer ranges covered by these instruments with existing apparati are: $0.005 < \hbar\omega (meV) < 300$ and $0.02 < Q(A^{-1}) < 20$. The TAS is most useful for single crystal spectroscopy where one has flexibility to scan along a chosen direction in $\hbar\omega$-q space. The conventional types of scans are either constant Q scans (where the energy is varied) or constant E scans (where the momentum is varied). These methods are most useful in measuring phonon dispersion curves in solids.

In polycrystalline materials where the momentum direction does not play an important role, or in systems where the scattering is nearly isotropic as for incoherent scattering, the TOFS is a more efficient instrument. In these instruments the scattering angle is kept constant and the scan is along an oblique direction in ω-q space as energy is transferred to the sample. The added efficiency of this instrument is the ability to measure several scattering angles simultaneously by use of multiple detectors. Since the time for the neutron to travel a known path is being measured the entire energy spectrum at each angle is obtained for a given counting period. Below I shall discuss the results of Eckold et al[6] who have measured the diffusion processes in a polycrystalline sample of α-AgI using a TOFS.

A most interesting property to measure in the superionic conductors is the diffusion constant, D, of the conducting species. To set the scale of what energy resolution is needed consider a diffusion constant $D = 10^{-5}$ cm^2/sec. The observed quasielastic line width (Γ) at a $Q=0.5$ A^{-1} (assuming simple diffusion) is $\Gamma = 2DQ^2 \approx 0.2$ meV a value obtainable with conventional instruments. However, for solid electrolytes neutron scattering methods have been utilized only in α-AgI[6] and NH_4^+-beta alumina[5] to measure the diffusive motion. Na^+ and Ag^+-beta alumina possess diffusion constants of $\sim 10^{-5} cm^2/sec$ near $T=400°C$ and should be measurable.[8]

DEFECT STRUCTURE IN FLUORITES

Fig. 1 shows the face centered cubic structure of the fluorite class of ionic conductors. The cation (Ba or Pb

in our study) is located at the cube center of Fig. 1 and
surrounded by 8 fluorines which occupy tetrahedral interstices (FI sites) of the cation sublattice. As the material is heated it is the anion that disorders and contri-
butes to the high ionic conductivity.[9] Some of the fluorine ions move from their regular (FI) positions and occupy the empty octahedral positions (FII) denoted by "x"
in Fig. 1. The redistribution of the fluorine ions with
rising temperature causes an anomaly in the heat capacity
at temperatures near the melting temperature.[10] For BaF_2[11]
and PbF_2[12] the heat capacity and ionic conductivity indicate that there is a smooth or "diffuse" phase transition
which can be associated with fluorine disordering.

Elastic neutron scattering experiments were performed
on polycrystalline samples of BaF_2 and PbF_2 in order to
measure directly the anion disorder. The details of the
experiment will be published elsewhere[13] but we essentially
performed accurate intensity measurements of as many reflections as possible (about 20 at room temperature and
~15 near the melting point). These intensities are pro-
portional to $|F_o(Q)|^2$ of Eq.(4). One common way of obtain-
ing the disordered structure would be to assume a model of
the disorder and perform a least-squares fit to the observed intensities allowing the R_k's and $W_k(Q)$'s, and the
occupancies to be adjustable parameters. An alternative
method that we adopted is to calculate directly the nuclear
density in real space by taking the Fourier inversion of
the observed structure factors. The average nuclear density in the unit cell obtained from Eq. (4')is:

$$< \rho(\vec{r})> = \frac{1}{V_c} \sum_{G=-\infty}^{\infty} F_o(\vec{G})\ e^{-i\vec{G}\cdot\vec{r}} \tag{6}$$

This Fourier method is usually quite difficult in complicated structures because it requires the knowledge of
the phase of the structure factor for each reflection. However, for the simple fcc fluorite lattice the phase can be
obtained by calculating from the known atomic arrangement
of the ordered structure. An additional problem arises due
to the limited number of reflections measured such that the
sum in Eq. (6) is not complete. This results in termina-
tion errors which can yield nonsensical negative scattering densities. This is solved by multiplying the structure

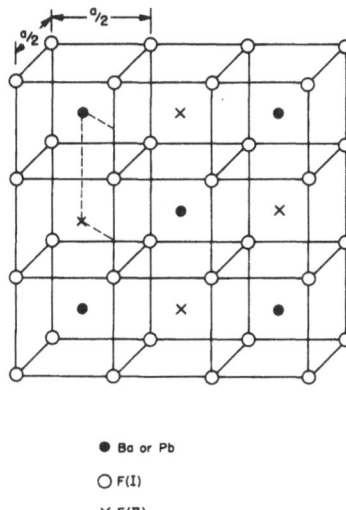

● Ba or Pb
○ F(I)
✕ F(II)

Fig. 1.   Fluorite structure

by an apodizing function (well known in optical interfero-
metry[14]) which gives strong weight to the small G reflections and forces the large G reflections to be negligible.
The Fourier inversion of the apodized structure factor
smears out the nuclear density over a region determined by
the size of the apodizing function. This sets a limit to
the resolution of the Fourier method. We used as an apodizing function a cube in real space. The size of the cube
was chosen such that the densities of the ordered fluorite
structure were reproduced at room temperature.

Fig. 2 shows a portion of our BaF_2 results for the
nuclear densities in the region of the fluorite cell encom-
passed by the dashed lines of Fig. 1. This is a difference
plot in the nuclear density between 1100°C and 25°C:
$\rho(1100°C)-\rho(25°C)$. The top of the figure shows the resolution of this method where δ represents the size of the apodizing box. It can be seen that the barium density has
spread out compared to room temperature. This is expected
because of the large thermal motions generated by the high
temperature, which appears as an increased Debye–Waller factor in Eq. (4). A similar effect appears for the normal
fluorine sites, FI. The new feature is the appearance of
additional fluorine density centered at FII. There is no

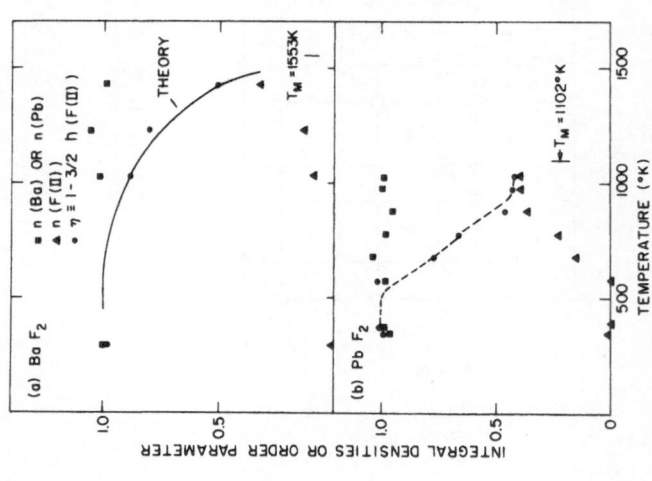

Fig. 3. (a) n(Ba), n(FII) and η for BaF₂.
The curve is the result of simple alloy
theory described in the text. (b) n(Pb),
n(FII) and η for PbF₂. The curve is a
guide to the eye. After Axe et al.[13]

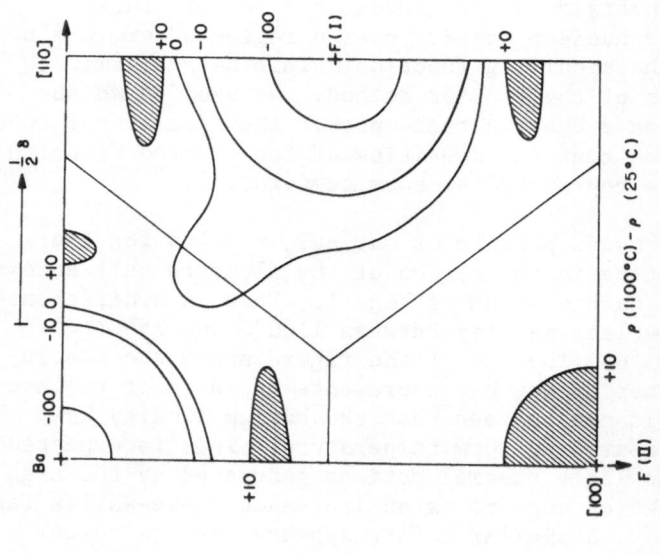

Fig. 2. Difference in the nuclear densities between 1100°C and 25°C for BaF₂.
After Axe et al.[13]

evidence that the fluorine atoms occupy off-centered positions as obtained from substitutionally induced disorder in other fluorite systems.[15]

In order to obtain quantitative information we can calculate the total nuclear density near the Ba,FI, and FII sites. This is obtained by dividing the fcc cell into Wigner-Seitz cells centered at the Ba, FI, and FII positions. In this way all the space is filled. The integrated densities for the kth atomic species, n(k), normalized to the ordered densities is shown in Fig. 3a for BaF_2 and Fig. 3b for PbF_2. For both systems the metal densities remain constant over the entire temperature range as expected. The FII densities initially are zero at room temperature but begin to increase as the temperature is raised. They reach a value of $\sim 40\%$ for BaF_2 and PbF_2 just below the melting points. (n(FI) is not shown but can be obtained from the obvious fact the n(FI) + n(FII) = 1.)

We now have a quantitative measure of the amount of disorder. An order parameter for this phase transition can be defined:

$$\eta = 3n(FI) - 2$$
$$\text{or } \eta = 1 - 3/2 \, n(FII)$$

The order parameter satisfies the conditions that $\eta=1$ in the ordered state (n(FII)=0 and n(FI)=1) and $\eta=0$ in the disordered state (n(FI)=n(FII)=2/3). The transition can be analyzed in the mean field or Bragg-Williams approximation by using conventional alloy theory.[16] The configurational energy E can be calculated and one obtains:

$$E = \beta\eta + \gamma\eta^2 \tag{7}$$

where β is proportional to the single particle energy and γ represents the interaction between the FI and FII fluorines. Once the configurational entropy, W, is calculated the free energy becomes

$$F = E - kT \, \ell nW \tag{8}$$

By minimizing this expression with respect to η the

temperature dependence of the order parameter is obtained
by solving the following expression:

$$\frac{2(1-\eta)^2}{(\eta+2)(2\eta+1)} = \exp\left(\frac{9}{2kT}(\beta+2\gamma\eta)\right) \qquad (9)$$

The temperature dependence for η with $\beta=0$ is shown in
Fig. 3a as the solid line. It agrees reasonably well with
our experimental data for BaF_2. This solution $\gamma\neq0$ suggests
that strong collective effects between the fluorines play
an important role in the disordering and thus the conductivity. For PbF_2 the situation is not as clear and no simple approximation of Eq. (9) reflects the behavior of the
order parameter.

For BaF_2 the order parameter shows a featureless smooth
change as the melting temperature approaches. This is true
of the conductivity also; there is no evidence of an anomalous change as the sample is heated to the melting point.[9]
For PbF_2, however, there is a rapid change in the order
parameter centered about $T \sim 750$ K. This corresponds closely with the behavior of the ionic conductivity[12] which also
shows a rapid variation in the same temperature region.

BETA ALUMINA

The most widely studied solid electrolytes are the
beta aluminas.[8] We have measured the dispersion of the low
frequency lattice modes in an effort at understanding the
lattice dynamics of these materials.[17,18]

Recent optical measurements have revealed lattice vibrations whose frequencies are very sensitive to the type of
cation.[19,20] These modes were interpreted as the attempt
frequency for the diffusing ion in a random-hopping model
of diffusion. They were treated as Einstein-like oscillators since the cation is moving independently of the host
lattice. Fig. 4 shows the dispersion of the low frequency
modes along the two high symmetry directions in the mirror
plane for Na-beta alumina. The normal longitudinal and
acoustic modes are observed in addition to several optical
branches. A striking feature is a branch at ~ 6 meV which
shows very little dispersion. It is this mode that is

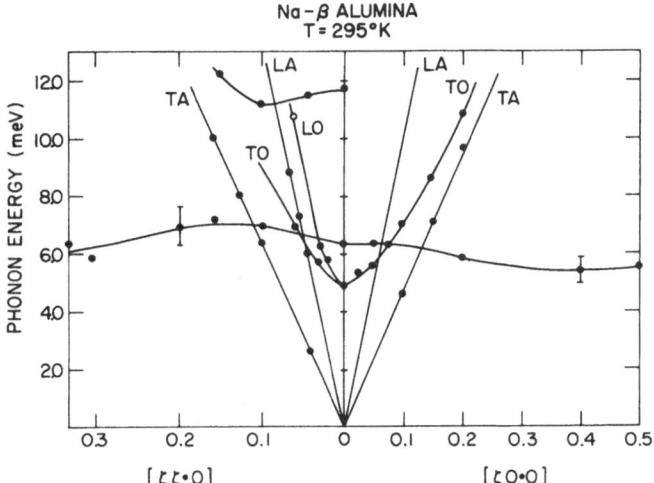

Fig. 4. Phonon dispersion curve for Na-beta alumina. After Shapiro et al.[18]

assigned to the cation vibration. The weak dispersion is evidence that the mode is truly Einstein-like and moves independently of the rest of the lattice. This mode is only seen when the mirror plane coincides with the scattering plane, which proves that the displacements are confined to the mirror plane. Fig. 5 shows the spectra of this mode at two different Q values. The intensity is weak compared to the other modes in this crystal. This is expected for the cation vibrations since only 4% of the atoms are contributing to the scattering. The energy width is ∿2.5 meV which is much larger than the instrumental resolution (0.8 meV). This suggests that there is a finite lifetime of this mode of ∿3 x 10^{11} sec^{-1}. Another possibility to explain the broadening is that there are more than one excitation in this frequency region and our resolution is insufficient to resolve them. Recent infrared transmission studies do observe several overlapping modes in this frequency region.[21]

Substitution of Ag$^+$ for Na$^+$ in the beta alumina host changes the low frequency spectra drastically as shown in Fig. 6 and observed in the optical measurements.[19,20] Instead of a well-defined propagating excitation there now is a broad feature centered about ℏω=0 meV. As the temperature

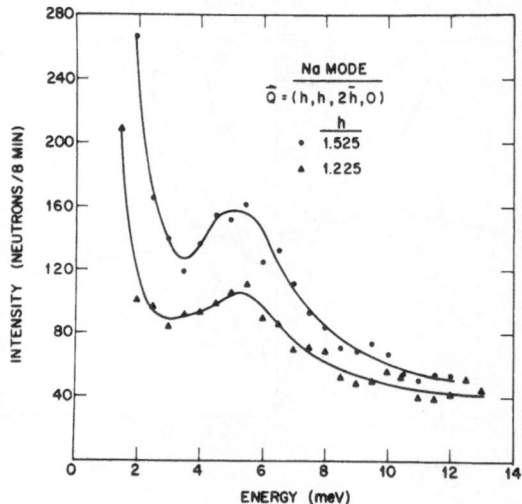

Fig. 5.   The cation mode in Na-beta alumina.
After McWhan et al.[17]

is decreased the line width remains constant but the intensity drops due to the temperature factor in the dynamical
structure factor, Eq. (5). There also is no change in the
line width as the sample is heated to 425°C. It is not
understood why no well-defined excitations are observed
since the optical measurements show features in the 3-4 meV
range.[19,20] It is likely that there is enough dispersion
in these low frequency optical branches that the neutron
scattering sees some average of several overlapping, damped
excitations.

The energy line width of the E=0 component in Fig. 6
is resolution limited. We attempted to measure the line
width at temperatures up to T=425°C in an effort to observe
any diffusion broadening. Even with a resolution of 0.24
meV no broadening was observed.

The intensity of the central E=0 component of Fig. 6
shows a strong temperature dependence. This is shown more
clearly in Fig. 7 where E=0 scans are performed along the
[ξξ·0] direction. At room temperature there is a broad
asymmetric peak centered about ξ∿1.65 which is most likely
related to the general disorder in the solid. As the temperature is cooled a sharp peak develops at ξ=1.35 which

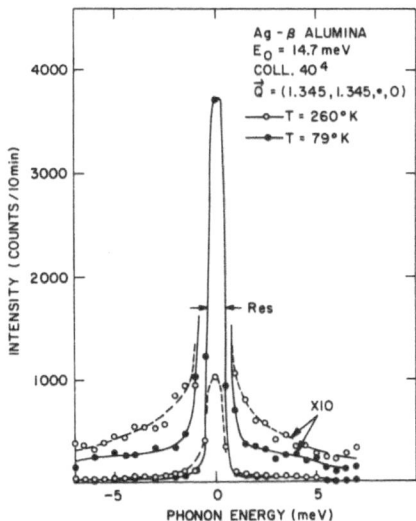

Fig. 6.   Inelastic spectra of Ag-beta alumina.
After Shapiro et al.[18]

becomes very intense at 26 K. This is similar to the x-ray
results of LeCars et al[22] who interpreted this scattering
as an ordering of the Ag^+ sublattice. There are, however,
some important differences between the neutron and the x-
ray results. At room temperature the x-ray studies observe
a spot at $\xi=4/3$. Since the x-ray technique automatically
integrates over energy it cannot distinguish between dynamical or static phenomena. In Fig. 7 the spectrometer is
set for zero energy transfer and only dynamical behavior
with frequencies within the instrumental resolution (0.8
meV) is observed. Since at room temperature no spot-like
scattering is observed we conclude that the feature in the
x-ray experiment must be due to some dynamical cation motion
with relatively large frequency.

QUASIELASTIC SCATTERING IN α-AgI

Quasielastic neutron scattering experiments were performed by Eckold et al[6] on α-AgI at T=250°C where they successfully measured the diffusion motion of the Ag^+ ion.
The experiments were performed with the multichopper TOFS,
IN5 at the Institut Laue-Langevin. Their results are
shown in Fig. 8 for two different scattering angles. The

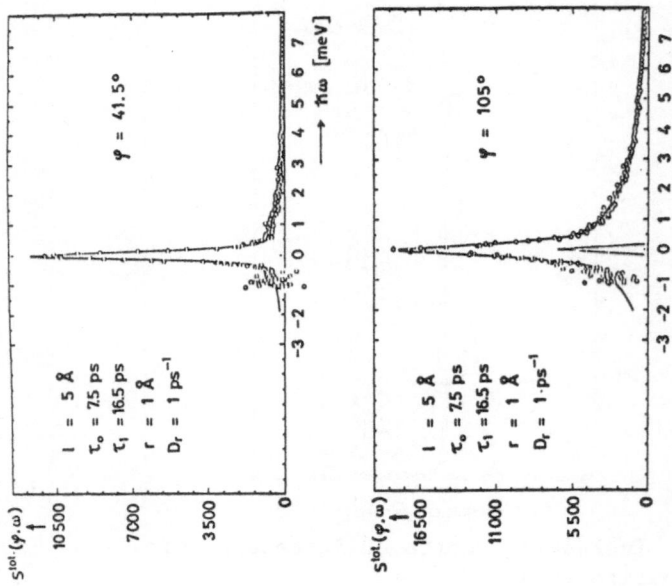

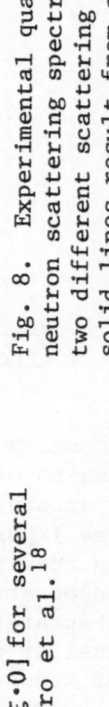

Fig. 8. Experimental quasielectric neutron scattering spectra of α-AgI at two different scattering angles. The solid lines result from a model calculation. After Eckold et al.6

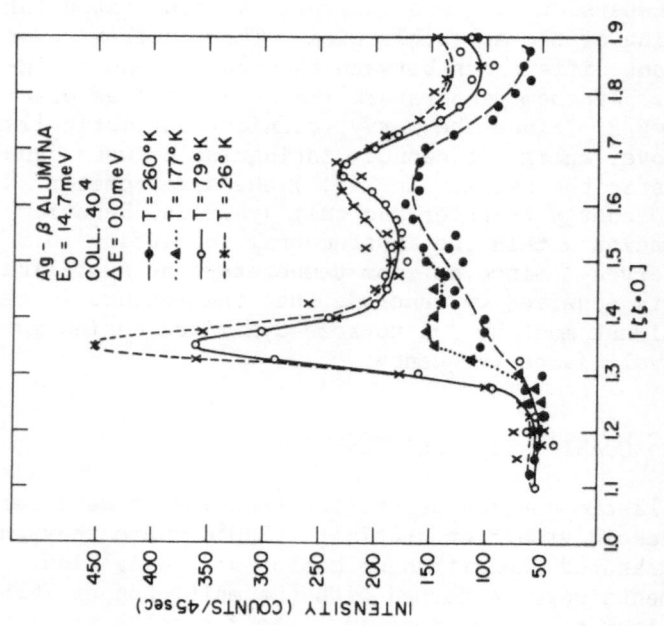

Fig. 7. E=0 scan along [ξξ·0] for several temperatures. After Shapiro et al.18

observed scattering was assumed to come from only the Ag
ions. These spectra are composed of two distinct features:
a narrow peak centered about $\hbar\omega=0$ which has an energy width
larger than the instrumental width; this is resting on top
of a much broader peak which has a larger width and a
strongly Q dependent intensity. No simple model of translational diffusion can explain these spectral shapes.
Eckold et al assumed that there were two types of motion of
the Ag^+ ion on two different time scales. The broad peak
arises due to a diffusive "local" motion within a restricted region of space given by the structure of the iodide
ion lattice. The narrow peak is due to a translational
diffusion over the entire crystal volume. This latter component corresponds, at small Q, to the diffusion measured
by the tracer method.

The model proposed for the translational jump diffusion
process consists of a mean residence time τ_o, a jump time
τ_1, and a jump length ℓ. In most discussions of jump diffusion the jump time is instantaneous: $\tau_1 \rightarrow 0$. However, in
this case a non-zero value of the jump time is necessary
in order to fit the data. For the local motion it was assumed that the Ag^+ ion diffuses within a limited spherical
region of radius r. It was further assumed that the Ag^+
ions move along the surface of the sphere with a rotational
diffusion constant D_r. The results of the fit of their
model with the data is shown in Fig. 8 and the agreement is
quite good. The values of the parameters obtained from
this model are also given for the two angles measured. It
is seen that the jump time is roughly one-half of the residence time.

It now appears that the diffusion of Ag^+ ions in AgI
is composed of a large amplitude local random motion super-
posed with a translational jump diffusion with a finite
jump time.

CONCLUSION

I have demonstrated how neutron scattering can play an
important role in determining the nature of the disorder and
the conducting mechanism in the solid electrolytes. There
remains much to understand, especially in the studies of
the β-alumina systems. It is expected that as new electro-

lytes are developed neutron scattering techniques will continue to play a fundamental role in understanding their
behavior.

ACKNOWLEDGMENTS

I gratefully acknowledge my collaborators J. D. Axe,
D. B. McWhan, J. P. Remeika, G. Shirane, and N.Wakabayashi
whose cooperative efforts contributed to the above results.
I thank Dr. Eckold for allowing me to discuss his excellent
work on α-AgI. I also thank F. Reidinger for continued
discussions on the structure of the beta aluminas.

REFERENCES

1.   See Thermal Neutron Scattering,Ed. P. A. Egelstaff
      (Academic Press, New York, 1965).

2.   B. N. Brockhouse, in Phonons and Phonon Interactions,
      Ed. T. A. Bak (W. A. Benjamin, Inc. New York, 1964)
      p. 221.

3.   R. Comès, these Proceedings.

4.   T. Springer in Springer Tracts in Modern Physics,
      (Springer, Berlin, 1972) Vol. 64.

5.   J. D. Axe, L. M. Corliss, J. M. Hastings, and W. L.
      Roth, these Proceedings and to be published.

6.   G. Eckold, K. Funke, R. E. Lechner, and J. Kalus,
      Phys. Letters 55A, 125 (1975); J. Phys. Chem. Solids,
      to be published.

7.   R. Pynn in Methods in Computational Physics: Advances
      in Research and Applications, Vol. 15 (Academic Press,
      New York, 1976).

8.   J. T. Kummer, Progr. Solid State Chem. 7, 141 (1972).

9.   M. O'Keefe in Fast Ion Transport in Solids, W. Van
      Gool(North-Holland, Amsterdam, 1973).

10. A. S. Dworkin and M. A. Bredig, J. Phys. Chem. 72, 1277 (1968).

11. V. R. Belosluov, R. I. Efremova, and E. V. Natizen, Soviet Phys. Solid State 16, 847 (1974).

12. C. E. Derrington, A. Navrotsky, and M. O'Keefe, Solid State Commun. 18, 47 (1976).

13. J. D. Axe, S. M. Shapiro, and N. Wakabayashi (to be published).

14. See J. Strong, Concepts of Classical Optics (W. H. Freeman & Co., San Francisco, 1958) p. 206.

15. A. K. Cheetam, B. E. F. Fender, and M. J. Cooper, J. Phys. C: Solid State Phys. 4, 3107 (1971); D. Steele, P. E. Childs, and B. E. F. Fender, J. Phys. C: Solid State Phys. 5, 2677 (1972).

16. M. A. Krivoglaz and A. A. Smirnov, The Theory of Order-Disorder in Alloys (MacDonald, London, 1964).

17. D. B. McWhan, S. M. Shapiro, J. P. Remeika, and G. Shirane, J. Phys. C: Solid State Phys. 8, L487(1975).

18. S. M. Shapiro, D. B. McWhan, G. Shirane, and J. P. Remeika, to be published.

19. S. J. Allen, Jr. and J. P. Remeika, Phys. Rev. Lett. 33, 1478 (1974).

20. L. L. Chase, C. H. Hao, and G. D. Mahan, Solid State Commun. 18, 401 (1976); C. H. Hao, L. L. Chase, and G. D. Mahan, to be published in Phys. Rev. B.

21. S. J. Allen, Jr., F. DeRosa, and J. P. Remeika, Bull. Amer. Phys. Soc. 21, 284 (1976); and these Proceedings.

22. Y. LeCars, R. Comès, L. Deschamps and J. Thery, Acta Cryst. A30, 305 (1974).

EXPERIMENTAL PROBES OF MICROSCOPIC INTERACTIONS CONTROLLING DIFFUSION IN CATION SUBSTITUTED β-ALUMINA

S. J. Allen, Jr., L. C. Feldman, D. B. McWhan,

J. P. Remeika and R. E. Walstedt

Bell Laboratories

Murray Hill, N. J. 07974

The large concentration of diffusing species poses two interesting problems for our understanding of mass diffusion and ionic conductivity in solid electrolytes or superionic conductors. The first concerns the stability or structural chemistry of the solid electrolyte phase. What causes the transition to the electrolyte phase in AgI? What causes β-alumina to grow with large deviations from stoichiometry? The second concerns the diffusion kinetics. Given a solid electrolyte can we microscopically understand the factors contributing to the activation energy and prefactor? What is the role of mobile ion-ion interaction and mobile ion-defect interactions? The former is clearly a difficult problem but crucial to the synthesis of new solid electrolytes or the improvement of existing systems. The experiments described herein make little or no contribution to this particular area. We are concerned with the latter, elucidating the microscopic interactions that play a role in controlling the diffusion in a particular system, the cation substituted β-aluminas.

In normal ionic conductors, diffusion can be understood in terms of isolated defects, their formation and motion through an otherwise ideal lattice. This gives rise to Arrhenius expressions, over a suitable temperature range, for diffusion and ionic conductivity

$$D = D_o e^{-U/kT} \tag{1}$$

$$\sigma T = (\sigma_o T_o) e^{-U/kT} \tag{2}$$

The prefactors for diffusion and conductivity are proportional to the concentration of diffusing species which are quite low in normal solids, and the microscopic interactions that determine the prefactor and activation energy are controlled by local interaction of the defect with its environment.

Solid electrolytes can be described phenomenologically by Arrhenius expressions for D and σ. However, we note two important distinctions: first, the activation energies are usually small, of the order of .1 - .2 eV compared to ≈ 1 eV for normal solids, second, the density of defects or mobile ions is sufficiently large that defect interactions very likely influence the activation energy and prefactor.

In the following, wherever possible, we attempt to relate the results to the observed diffusion coefficient and ionic conductivity, which we write here exposing some of the microscopic parameters controlling diffusion.

$$\sigma = N_d \frac{e^2 a^2 \nu}{kT} e^{-U/kT} \tag{3}$$

$$D = N_d a^2 \nu e^{-U/kT} f \tag{4}$$

N_d is the density of defects participating in the diffusion, a, a jump distance, ν the attempt frequency, and U the barrier height. f is a correlation factor and we have ignored the entropy factor. Although it is conceivable that (3) and (4) are totally inappropriate for β-alumina or that we cannot identify these parameters with single ion motions, it does provide a useful framework in which to describe some of these results. Furthermore, the picture that begins to emerge, however incomplete, does suggest that certain aspects of the problem can be related to local mobile defect motion.

The experiments described below were performed on β-alumina from two sources. The first were grown in our laboratory from a flux at temperatures not exceeding 1300°C; the second were purchased from Union Carbide Company and grown from the melt at temperatures near 2000°C. In the following, it will become apparent that there are quantitative and qualitative differences between these two samples and we can not consider our understanding satisfactory unless these differences are understood.

The experiments that have been performed on these samples at Bell Laboratories concern both the structure and the dynamics. Structural information has been gained from the following:

> X-ray scattering, diffuse and Bragg
> Neutron diffuse scattering
> Ion back scattering
> Mossbauer effect

Dynamic information has been extracted from:

> Infrared absorption and reflection
> Microwave absorption and reflection
> Nuclear magnetic relaxation
> Specific heat at low temperatures
> Inelastic neutron scattering

The first section describes some experiments that extend our knowledge of the structure of the average unit cell of the β-aluminas. The wealth of information contained in various dynamical probes is discussed in the second section while we return to structural information obtained from diffuse X-ray scattering in the third. Our understanding, which is by no means complete, is summarized in the last section.

STRUCTURE

Diffusion in β-alumina occurs in planes separated by spinel-like blocks of aluminum and oxygen.[1-4] The diffusion plane, shown in Fig. 1, contains a loose packed arrangement of oxygens as well as Na ions distributed over three possible sites, the Beevers-Ross (BR), the anti-Beevers-Ross (ABR) and the mid-oxygen (M-O) sites. It is presumed that stoichiometric Na β-alumina, $Na_2O \cdot 11 Al_2O_3$, would give one

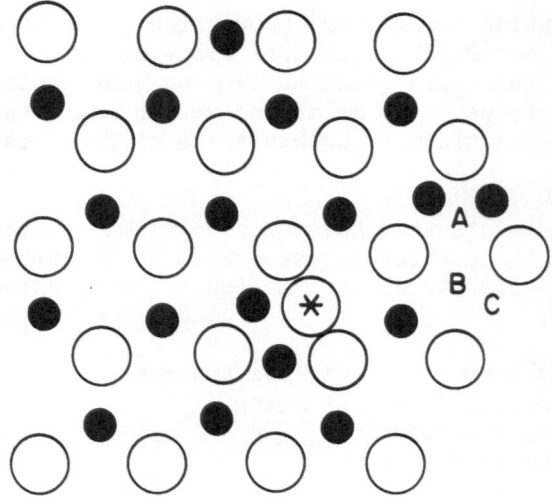

Fig. 1. The diffusion plane in β-alumina. Open circles, $O^=$. Closed circles, Na^+. A, BR site. B, ABR site. C, M-O site. The starred $O^=$ is an interstitial compensating defect.

Na per oxygen in the plane and the Na would reside on the BR site. Approximately 25% deviations from stoichiometry are invariably found for the mobile cation. This excess Na is accommodated by replacing a Na at the BR site with two ions in the M-O positions. Ag appears to be unlike the alkalis; the extra Ag goes into the ABR.[2] At the same time the excess charge is compensated by some other charged defect. In the past, this has been thought to be Al^{3+} vacancies[5] in the spinel block or O^{2-} interstitials[2] in the diffusing plane. Most recently, Roth[6] and Roth, Reidinger and LaPlaca[4] have suggested that interstitial oxygen in the plane, bound by two Al^{3+} Frenkel defects is the most likely mechanism for charge compensation. Such a possibility is shown by the starred oxygen in Fig. 1.

Crystallographic Refinement of K β-Alumina

The X-ray crystal structure of K β-alumina has been refined by Dernier and Remeika[7] so that detailed structural information on K as well as Na and Ag β-alumina is available. K occupies the same sites as Na.

Ion Back-Scattering

Ion back-scattering experiments have been performed by Feldman, Remeika, Silverman and Komaki[8] on Ag and K β-alumina. In this experiment 1.6 MeV He$^+$ ions are channeled down the c-axis of the β-alumina. The He$^+$ ion is back scattered from an ion with an energy loss determined by the mass of the ion and its distance from the surface. As the beam angle with respect to the channeling direction increases, the yield of back scattered He$^+$ from a particular ion increases in a manner which indicates how far the ion is displaced from a high symmetry position; e.g., how far an average Ag is displaced from the BR or ABR position.

Figure 2 shows back scattering intensity versus beam angle for He$^+$ scattered off Al and Ag. The solid line is the calculated yield assuming Ag displacements from BR and ABR determined by Roth.[2] The agreement is good. Similar

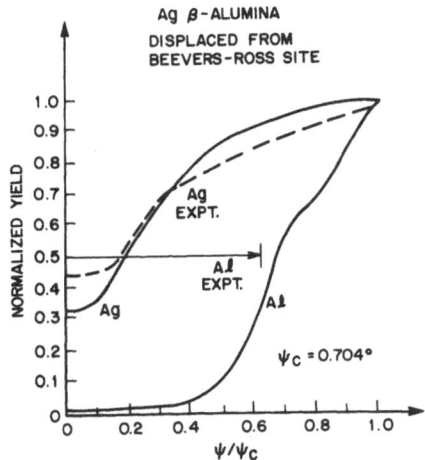

Fig. 2. Back scattering yield for Al and Ag. ψ is the angle of incidence measured from the channeling direction. The solid curve for Ag is based on Roth's X-ray refinement, Ref. 2.

experiments on K β-alumina require that the K ion suffer
$\approx 1/2$ the displacement from high symmetry positions required
by the X-ray refinement.[7] This difference is not of over-
riding importance but does give a measure of the uncertainty
in the determination of the cation position.

Mössbauer Effect

Cohen, Remeika and West[9] have studied the isomer shift,
hyperfine splitting and temperature dependence of the Debye-
Waller factor in Eu β-aluminas by using Mössbauer spectroscopy of the Eu^{151} species. Although all the europium is
divalent, two types of sites are detected. The first has
an isomer shift appropriate to an oxygen environment and a
Debye temperature, θ_D, similar to Eu O. The second has the
lowest isomer shift ever reported for Eu^{2+} corresponding to
unusually low electron density at the nucleus and a θ_D
nearly one half the first site. Presumably the second site
is "roomy" with the Eu^{2+} somewhat loosely bound. Presumably
the second kind of site lies in the diffusing plane.

DYNAMICS

A direct measure of the local vibrational frequency of
the mobile cations is revealed by the far infrared conductivity and Raman scattering. Allen, DeRosa and Remeika[10]
have measured the far infrared conductivity of Na, Ag, K,
Rb and Eu β-alumina, both flux and melt grown material.
Infrared absorption by Ag and Na has also been observed by
Armstrong, Sherwood and Wiggins,[11] by Strom, Taylor,
Bishop, Reinecke and Ngai[12] and by Barker and Remeika.[13]
Chase, Hao and Mahan[14,15] have observed Raman scattering
from Na, Ag, K and Rb β-alumina melt grown. In both measurements, a low frequency mode is seen whose intensity and
resonant frequency is sensitive to cation substitution.
(Barker and Remeika[13] have also studied the multitude of
infrared active modes associated with the Al^{3+} and O^{2-} in
the structure and find them insensitive to mobile cation
substitution. The mobile cation mode for Na β-alumina is
also seen by inelastic neutron scattering.[16,17]

We show in Figs. 3A and 3B the infrared conductivity
of flux and melt grown material and the Raman scattering[15]

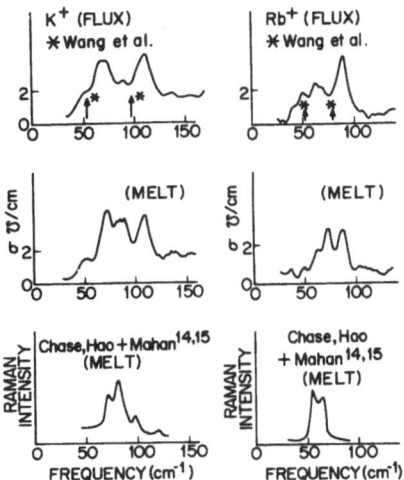

Fig. 3A and B.   Infrared conductivity of K and Rb β-alumina at room temperature, flux and melt grown.   Raman scattering taken from Ref. 15.   Calculated defect and single ion vibrations from 18.

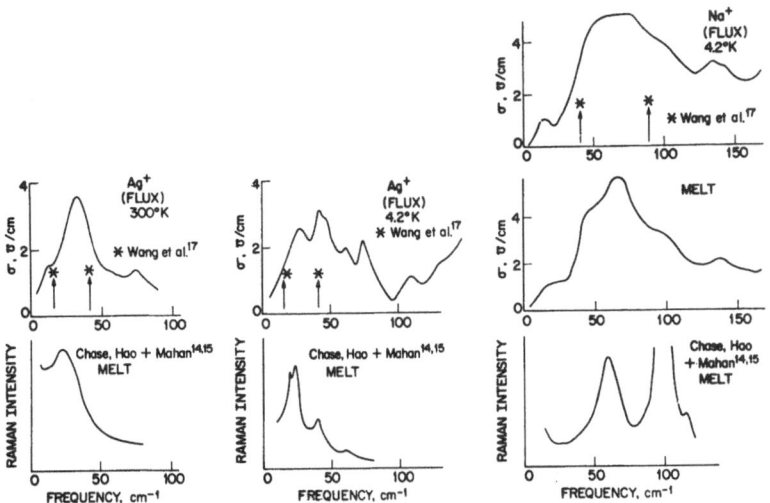

Fig. 3C, D and E.   Infrared conductivity of Ag β-alumina at 300 and 4.2°K and Na β-alumina at 300°K.   Raman scattering taken from Ref. 15.   Calculated defect and single ion vibrations from 18.

for Rb and K β-alumina. (The Na^+ and Ag^+ results at 300°K have been published elsewhere.[10,14] The Ag^+ results at 4.2°K show a striking temperature dependence[14] which is not seen in the other systems.) It seems clear that the resonance is inhomogeneously broadened, i.e. a composite of vibrations coming from cations in different environments - singly occupied cells, doubly occupied cells or doubly occupied cells near a compensating defect.

We note that the Raman scattering is more selective in that it sees only the low frequency features or part of the much broader infrared conductivity. This is also the case in Na and Ag. The low frequency edge is also that part sensitive to growth conditions of the crystal. Two important points may be inferred. 1. The low frequency part of the infrared absorption is caused by cation defect vibrations. (These are the vibrations of the doubly occupied cells calculated by Wang, Gaffari and Choi[18] and the mode involved in the intersticialcy mechanism of Whittingham and Huggins.[5]) 2. The Raman scattering appears to see only these modes. Why the latter should be so is not clear.

This interpretation relies heavily upon the calculation of Wang, Gaffari and Choi[18] which predicts vibrational frequencies for singly occupied cells and doubly occupied cells indicated by the arrows in Fig. 3. The agreement between the calculation of Wang et al.[18] and the overall breadth and position of the far infrared absorption is also obtained for Na and Ag. In Fig. 4 we plot the calculated vibrational frequencies of Wang et al.[18], the Raman results[14,15] and the far infrared results.[10] These results lend considerable support to the intersticialcy mechanism[5] and the model calculation of Wang et al.[18] for the potential well in which the mobile cation sits.

McWhan, Hsu, Remeika and Varma[19] have measured the low temperature specific heat of Li, Ag, Na, K and Rb β-alumina. In Fig. 5 we show the excess specific heat caused by the mobile cation. For a given statistical weight and for kT less than the vibrational energy $\hbar\omega$, low frequency modes may be expected to contribute more specific heat.

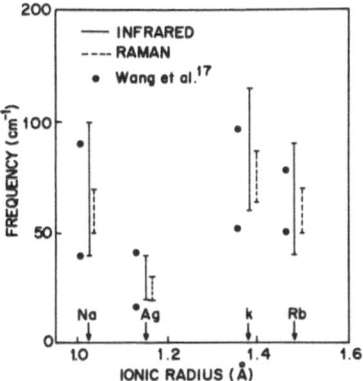

Fig. 4. Vibrational frequency from infrared absorption,[10]
Raman scattering[14],[15] and theory of Ref. 18.

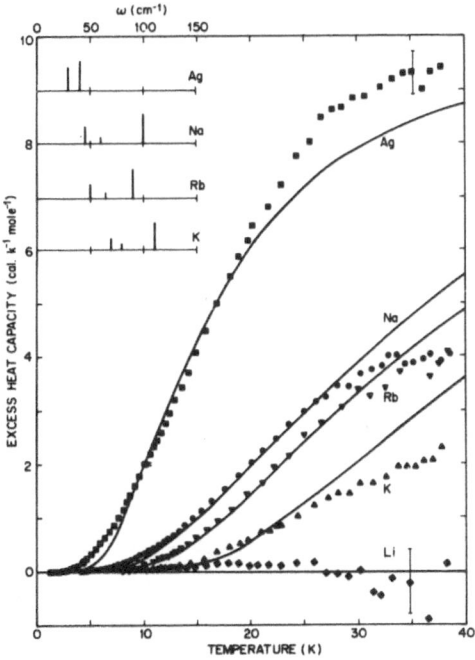

Fig. 5. Excess specific heat versus temperature.[19] Upper
left hand corner, the distribution of modes that reproduces
the observed data.

We note that the ordering of the vibrational frequencies $\nu(Ag) < \nu(Na) < \nu(Rb) < \nu(K)$ is observed in the specific heat, as it was in the Raman scattering,[14,15] infrared absorption[9] and predictions of Wang et al.[18] Indeed, McWhan, Hsu, Remeika and Varma[19] are able to fit the specific heat shown in Fig. 5 by the distribution of modes shown in the top left hand portion of the figure which is consistent with the infrared result.[10]

The local vibration is related to the activation barrier. In Fig. 6 we plot activation energy versus ion radius for Ag, Na, K and Rb. We also show the calculation by Wang et al.[18] If the vibrational frequency is determined by the same potential that determines the activation energy, U, we expect U to scale like $\nu^2 M$ where M is the ion mass. $\nu^2 M$ for the Raman scattering and $\nu^2 M$ for the infrared are shown in Fig. 6. The lower edge of the infrared, the Raman result, and the calculation of Wang et al.[18] follow the observed activation energy. This implies that the barrier height seen in diffusion is controlled by the potential determining the vibrational frequency of the mobile cation, i.e. the barrier is determined by ion interactions with its local environment.

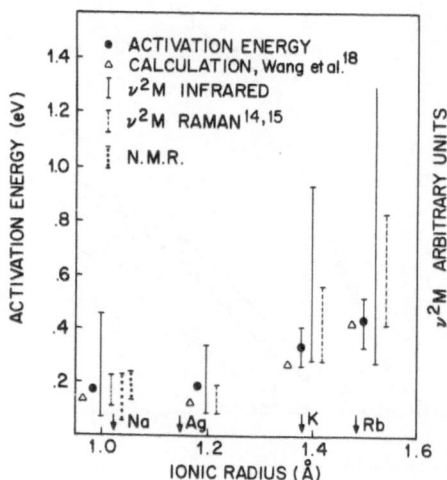

Fig. 6. Activation energy versus ion radius. Calculation by Wang et al.[18] $\nu^2 M$ from the infrared. $\nu^2 M$ from Raman scattering. NMR distribution of barrier height from Ref. 20.

Barker and Remeika[13] and Strom, Taylor, Bishop, Ngai and Reinecke[12] at the Naval Research Laboratory have measured the dielectric response of Na and Ag β-alumina in the microwave regime. There one finds a power law dependence of the conductivity on frequency which is characteristic of hopping motion in disordered systems. This implies a distribution of hopping times and barrier heights.

The specific heat measurements of McWhan, Hsu, Remeika and Varma[19], at sufficiently low temperatures, give a linear dependence of C_p on T which is also characteristic of disordered systems with a distribution of barrier heights.

A distribution of barrier heights is most evident in the nuclear magnetic relaxation measurements on Na β-alumina performed by Walstedt, Dupree and Remeika.[20] In Fig. 7 we show the nuclear spin-lattice relaxation time, T_1, of Na β-alumina, melt grown, measured at two different frequencies, as a function of 1/T. Walstedt et al.[20] find that a detail fit to the observed dependence of T_1 on temperature can be made only by assuming a distribution of activation energies

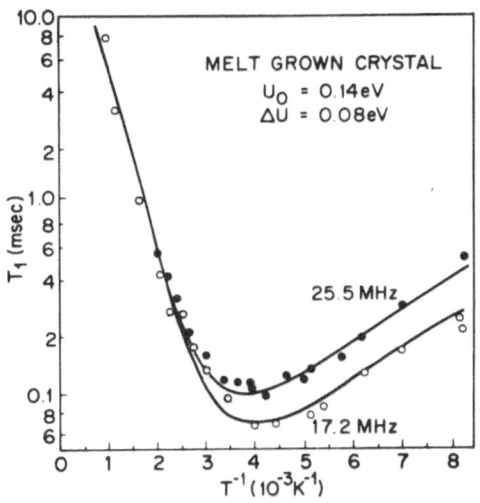

Fig. 7. Nuclear spin relaxation time, T_1, versus 1/T at two different frequencies.[20] The data can be fit to the distribution of relaxation times shown in the figure.

for the hopping time τ. In essence one assumes that the
total relaxation rate, $1/T_1$, is given by

$$1/T_1 = \int_0^\infty dU e^{-(U-U_o)^2/\Delta U^2} \frac{\Omega^2 \tau(U)}{1+\omega^2 \tau(U)^2} \qquad (5)$$

where $\tau(U)$ is given by

$$\frac{1}{\tau(U)} = \nu_o e^{-U/kT} \qquad (6)$$

The data in Fig. 7 are fit with $U_o = .14$, $\Delta U = .08$ and
$\nu_o \approx 6$ cm^{-1}.

There are marked changes in Fig. 7 when one uses flux
grown material. Remarkably, they can be accounted for by
using a different average activation energy, U_o, and
width, ΔU. In Fig. 5 we have also included the activation
energy from NMR and we see that the distribution of U's
includes the activation energy obtained from d.c. conductivity.

The prefactor in the conductivity or diffusion should
be proportional to the vibrational frequency observed in
the far infrared or Raman scattering. In Fig. 8 we plot
the observed diffusion coefficient prefactor versus ion
radius as well as $1/4\, a^2 \nu$, derived from the Raman scattering
or infrared absorption if every ion could participate in
the diffusion process. Na and Ag have prefactors that
approach those given by the vibrational frequency. The
departures may be easily accounted for by site availability
factors in the Sato and Kikuchi[21] theory or by the density
of interstitials following Wittingham and Huggins[5] and
Wang et al.[18] Rb and K, however, depart substantially from
these values. At the same time, Barker and Remeika[13] tell
us that the d.c. conductivity of the β-alumina is four times
smaller in flux grown material than in the melt material.
This is indicated by the arrow in Fig. 8. It appears that
the prefactor in the conductivity or diffusion coefficient
is far more sensitive to growth conditions and mobile
cation than is the activation energy.

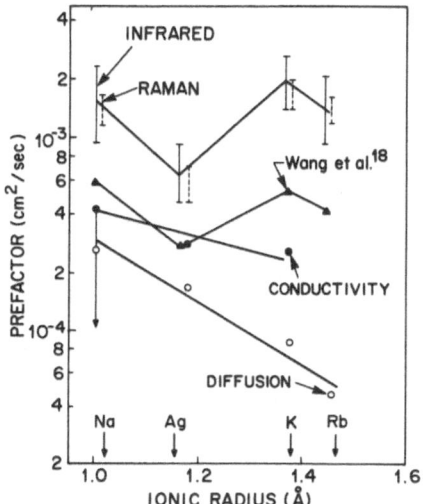

Fig. 8. 1/4 a²ν, infrared and Raman, top of figure. ▲, Wang et al.[18], ●, conductivity, o, diffusion coefficient. The arrow for Na indicates variation seen by Barker and Remeika.[13]

The attempt frequency extracted from NMR is nearly an order of magnitude smaller than the vibrational frequency seen in Raman scattering and far infrared absorption. This is somewhat disconcerting for NMR is sensitive to both diffusive and nondiffusive hopping motions. At the same time it seems clear that most of the ions are executing vibrations at the observed infrared vibrational frequency. A successful theory of diffusion in β-alumina will have to reconcile this apparent contradiction.

These results are surprising. If the vibrational frequencies correlate with anything, they should with the attempt frequency. On the contrary, they correlate with the barrier height but poorly with prefactor. We speculate that the attempt frequency for diffusion is determined by the vibrational frequency but that the density of ions or defects that can participate in the diffusion process is controlled by growth conditions and mobile ion.

DIFFUSE X-RAY SCATTERING

Inhibition of defect or ion diffusion is related to local order which can be deduced from diffuse X-ray scattering. Diffuse X-ray scattering has been observed by McWhan, Dernier and Remeika[22] on Na, Ag, K, Rb and Eu β-alumina. (R. Comes discusses diffuse X-ray scattering elsewhere in these proceedings.)[23] The diffuse scattering is produced by the disorder in the diffusing plane caused by the excess mobile cation, doubly occupied cell and the compensating defect. The upper corner of Fig. 9 shows a schematic of the X-ray diffuse scattering. The open circles are peaks in the diffuse scattering caused by the short range order in the diffusing plane. The peaks in the diffuse scattering are a measure of the order amongst the doubly occupied cells

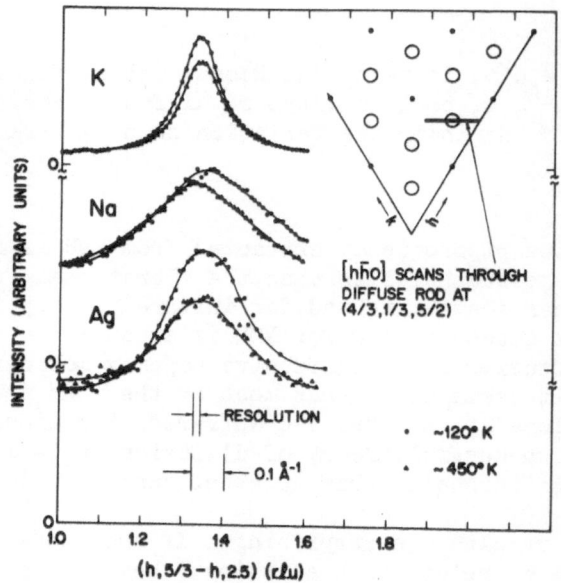

Fig. 9. Measurement of the correlation length for different ions in β-alumina, flux grown, at 120 K (circles) and 450 K (triangles). The direction of the scan through the diffuse rod in reciprocal space is shown in the inset. The correlation length increases with increasing size of the diffusing ion but is almost independent of temperature.

and compensating defects and in K β-alumina may be indexed
on a $\sqrt{3}a \times \sqrt{3}a$ super lattice. Although long range order is
not observed, such a super lattice of doubly occupied cells
would give nearly the right deviation from stoichiometry
while keeping the doubly occupied cells as far removed from
one another as possible.

In Fig. 10 we show the width of the diffuse peaks,
which is inversely proportional to an average correlation
length, versus ion radius for Ag, Na, K and Rb, for flux
grown and melt grown material. Here we note that the flux
grown material shows a marked increase in correlation
length[24] with ion radius, whereas the melt grown does not.
In Ag, Na, K, β-alumina, flux grown, and for Ag and K
β-alumina, melt grown, the correlation length is essentially
<u>temperature independent</u>. The largest changes occur for Ag
β-alumina, melt grown, where it increases by 30% when the
temperature is dropped from 298 to 120°K. There are,
however, some marked changes in intensity with temperature
which may be caused by redistribution of the ions amongst
the possible sites. (As we noted earlier, at room temperature the Ag ion favors the ABR site over the M-0 while the
opposite is true for the alkali ions.)

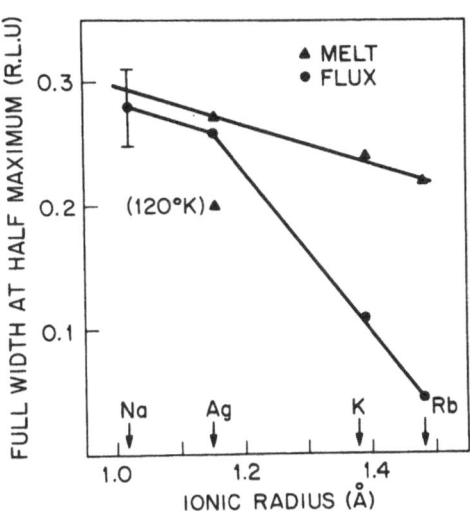

Fig. 10. Width of the diffuse peak versus ion radius for
melt and flux grown material. 300°K except where noted.

Diffuse elastic neutron scattering[17] has been observed
on Ag β-alumina, melt grown. Although the intensity varies
with temperature in an interesting way, the widths and,
hence, correlation lengths, agree with the X-ray diffuse
scattering.

Current speculation is that the melt grown material
grown at high temperatures has a greater degree of disorder
amongst the compensating defects than does the flux grown
material. This disorder prevents the K and Rb from estab-
lishing the degree of short range order that is seen in the
flux grown material. Since the correlation length is
sample dependent, it is clear that the interesting correlation between correlation length and prefactor pointed
out by McWhan, Allen, Dernier and Remeika[24] may not be well
founded and requires more investigation. The greater disorder in the melt grown material is consistent with the
larger distribution of barrier heights seen by Walstedt
et al. in the NMR measurements.[20]

The structure factor for the diffuse scattering tells
us real space structure of the defects responsible for the
disorder. Whereas the structure factor in the plane relates
to both the mobile cation and compensating defect, the
structure factor along the inverse c axis is specific to
defect structure above or below the diffusing plane. Since
the contribution to the overall disorder from the mobile
cation is restricted to in-plane displacements, this part
of the structure factor is specific to the compensating
defect. The modulation of the diffuse rods seen in K
β-alumina, flux grown, Fig. 11, has been fit by McWhan,
Dernier, Vettier and Remeika[25] with the model of the compensating defect proposed by Roth, Reidinger and
LaPlaca.[2,4]

The density of diffusing ions or defects, N_d in (3)
and (4) is probably intimately related to the local order
and hence the diffuse scattering. If the intersticialcy
mechanism is operative in the β-alumina as suggested by
Whittingham and Huggins[5] and modeled quantitatively by
Wang, Gaffari and Choi,[18] then the mutual interaction
between doubly occupied cells and between double occupied
cell and compensating defect may play a role in determining N_d. These interactions manifest themselves in the
diffuse X-ray scattering. It is clear that a complete

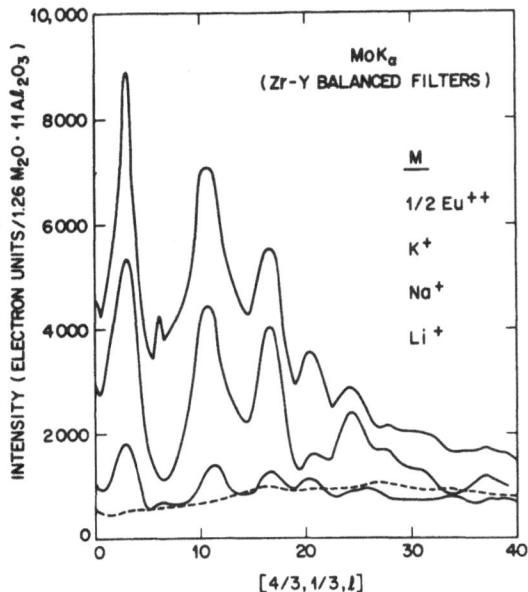

Fig. 11. Modulation of the diffuse intensity along the reciprocal c axis for β-alumina, flux grown. This structure is due to the compensating defect and can be fit to the interstitial oxygen model, Ref. 2,4.

understanding of the diffuse scattering will make a strong contribution to our understanding of the role of defect-defect interactions in controlling diffusion in the β-alumina.

    Presumably the sensitivity of the diffusion coefficient and ionic conductivity prefactors, to mobile cation and growth condition of β-alumina, reflects the fact that the number of ions that can participate in the diffusion processes depends on these factors. We speculate that the ability of ion to participate in the diffusion process depends on the configuration of its environment, i.e. long range order or short range order. Hence, we hope that the detailed structural analysis of local order around a defect will tell us why only a fraction of the defects diffuse in certain of the β-aluminas.

SUMMARY

The potential that determines the barrier height and activation energy for diffusion is determined by local interactions of the ion or defects with the host lattice. As a result, the vibrational frequencies correlate with the activation energy. The success of Wang, Gaffari and Choi[18] in predicting both vibrational frequencies and activation energies indicates that the intersticialcy mechanism of Whittingham and Huggins[5] is probably operative.

At the same time, there is evidence for a large distribution of barrier heights in the system, presumably caused by the inherent disorder in the diffusing plane. This is evidenced by low temperature specific heat as well as microwave conductivity, but most directly by measurements of the temperature dependent nuclear relaxation time in Na β-alumina. The melt grown material is more disordered than the flux grown material.

Diffuse X-ray scattering has given us information on the degree of local disorder and the structure of the cation defect and its position relative to the compensating defect. The melt grown material is more disordered which correlates with the NMR result and the observation of greater conductivity in the melt grown material. Presumably a detailed understanding of the local order around the defects will explain the apparent sensitivity of the prefactors to mobile cation and growth conditions.

It is apparent that β-alumina is a complex system and no single measurement will successfully tell all we need to know to understand its high conductivity. Only a synthesis of many different experiments will enable us to draw a coherent picture of the conduction process in this system.

Acknowledgment. We would like to thank our colleagues at Bell Laboratories who have shared their results and ideas with us, in particular C. M. Varma, A. S. Barker, P. D. Dernier, and R. L. Cohen.

REFERENCES

1. C. R. Peters, M. Bettman, J. W. Moore and M. D. Glick, Acta, Cryst. B27, 1826 (1971).

2. W. L. Roth, J. Solid State Chem. 4, 60 (1972).

3. J. T. Kummer, Prog. Sol. State Chem. 7, 141 (1972).

4. See also W. L. Roth, F. Reidinger and S. LaPlaca, this Conference. F. Reidinger, thesis, unpublished.

5. M. S. Whittingham and R. A. Huggins, J. Electrochem. Soc. 118, 1 (1971).

6. W. L. Roth, Trans. Am. Cryst. Assoc. 11, 51 (1975).

7. P. D. Dernier, to be published, J. Sol. St. Chem.

8. L. C. Feldman, J. P. Remeika, P. J. Silverman and K. Komaki, this Conference and to be published.

9. R. L. Cohen, J. P. Remeika and K. W. West, J. Phys. 35 Supp. #12, 34 (1974).

10. S. J. Allen, Jr. and J. P. Remeika, Phys. Rev. Lett. 33, 1478 (1974); S. J. Allen, Jr., F. DeRosa and J. P. Remeika, to be published.

11. R. D. Armstrong, P. M. A. Sherwood and R. A. Wiggins, Spectrochimica Acta 30A, 1213 (1974).

12. U. Strom, P. C. Taylor, S. G. Bishop, T. L. Reinecke and K. L. Ngai, to be published Phys. Rev.

13. A. S. Barker and J. P. Remeika, to be published.

14. L. L. Chase, C. H. Hao and G. D. Mahan, Solid State Comm., to be published.

15. C. H. Hao, L. L. Chase and G. D. Mahan, to be published Phys. Rev.

16. D. B. McWhan, S. Shapiro, J. P. Remeika and G. Shirane, J. Physics C 8, L487 (1975).

17. S. Shapiro, this Conference.

18. J. C. Wang, M. Gaffari and Sang-il Choi, J. Chem.
     Phys. 63, 772 (1975).

19. D. B. McWhan, J. P. Remeika, F. L. S. Hsu and
     C. M. Varma, to be published.

20. R. E. Walstedt, R. Dupree and J. P. Remeika, to be
     published, and this Conference.

21. H. Sato and R. Kikuchi, J. Chem. Phys. 55, 677 (1971).

22. D. B. McWhan, P. D. Dernier and J. P. Remeika, to be
     published.

23. R. Comes, this Conference.

24. D. B. McWhan, S. J. Allen, Jr., J. P. Remeika and
     P. D. Dernier, Phys. Rev. Letters 35, 953 (1975); 36,
     344(E) (1976).

25. D. B. McWhan, P. D. Dernier, C. Vettier and
     J. P. Remeika, to be published, and this Conference.

LIGHT SCATTERING MEASUREMENTS IN SOLID IONIC CONDUCTORS

L. L. Chase

Clarendon Laboratory[+], Oxford OX1 3PU, U.K. and

Indiana University[x], Bloomington, Ind. 47401

I.   INTRODUCTION

The application of light scattering techniques to investigate factors affecting ionic conduction in solids is a
very recent development. As a consequence, many of the
experimental results, most of which have been reported in
the past year, are not well understood. Also, some interesting phenomena which have been predicted theoretically have
not yet been observed experimentally. The aim of this
article is to review the current situation in both of these
categories. Wherever possible, examples of experimental
results are presented to illustrate the potential of light
scattering measurements for examining the mechanisms
responsible for ionic conduction in solid electrolytes.

Light scattering has become established as an effective
probe of vibrational excitations in ordered solids, and considerable progress has been made in applying it to disordered
solids. The many advantages offered by this technique
include experimental simplicity, high resolution, and applicability to small samples and at high temperatures. The
polarization dependence of the spectra can also be valuable
in interpreting the observed excitations, despite the massive
disorder inherent in ionic conductors.

The excitations of importance to ionic conduction range
in frequency from the very low frequency diffusive motions
up to the highest vibrational frequencies < 1000 cm^{-1}, of
the mobile ions. The three available light scattering
techniques, Raman scattering, Brillouin scattering and

photon-correlation spectroscopy span this entire frequency
range. Each of these experimental methods is discussed in
turn in the remaining sections.

II. RAMAN SCATTERING

Two types of motion of the mobile ions can lead to
fluctuations of the polarizability with frequency components
greater than a few cm^{-1} ($\sim 10^{11}$ Hz) which is the accessible
region of Raman scattering. These are the vibrations of the
ions about their equilibrium positions and their transitions
between available sites in the lattice. If the potential
barrier between available sites is considerably larger than
both the zero point vibrational energy of the ion and the
mean thermal energy per ion, these processes may be experimentally distinguishable by means of the frequency distribution and temperature dependence of the scattered light.
If, in addition, the vibrations of the mobile ions are
largely decoupled from those of the rest of the lattice,
there will be a reasonably well defined frequency for the
vibrational motion. This frequency has a particular significance, in the context of hopping models, as the attempt
frequency, ω_0. In terms of the attempt frequency, the dwell
time, or the mean time between hops is given by $\tau =$
$\exp(U/k_BT)/\omega_0$, where U, the activation energy, is the height
of the potential barrier between sites. The measurement of
the frequency and linewidth of this vibration as a function
of temperature can be used to test the validity of this
hopping model and estimate corrections to it arising from
correlation effects   and   site occupancy factors, as has
been done by Allen and Remeika[1]. The vibrational levels
of an ion in a multiple well potential become broader and
more closely spaced as their energy approaches U[2]. It is
therefore to be expected that the Raman scattering spectrum
for such a system will become asymmetric and shifted to
lower frequency for $k_BT \gtrsim U/2$.

In ionic conductors there may be several sites in each
cell for each mobile ion occupying it. Since the polarizability of each cell will depend on the configuration of
these ions, light is scattered due to fluctuations in the
polarizability resulting from the hopping of ions between
sites. Such scattering processes have been discussed
recently by Huberman and Martin[3] and, in more detail by

Klein[4]. Klein's analysis is based on the simplifying
assumptions that: (1) there is only one mobile ion per unit
cell; (2) the polarizability, $\tilde{\alpha}_i$, of each inequivalent site
is independent of the configuration in neighbouring cells;
(3) the dwell time at a site is much longer than the time
spent in jumping between sites. If N_i is the occupation
number of the ith site in a unit cell, the second assumption
allows the cell polarizability tensor to be written as

$$\tilde{\alpha}(t) = \sum_i \tilde{\alpha}_i N_i(t) \tag{1},$$

where the sum is taken over all the inequivalent sites of a
primitive unit cell.

The limitations of this approach are apparent from the
fact that no DC conductivity exists without fluctuations in
the numbers of mobile ions in the unit cells. Also, the
polarizability of each cell will generally depend on the
configurations of neighbouring cells, especially where sites
exist at the boundaries between cells. These limitations
are particularly evident in the fluorite lattices where
anion sites are nearest neighbours of anion interstitial
sites of adjacent cells and in β-alumina where there is an
excess cation concentration.

If the third assumption about the relative magnitude
of dwell and jump times is used, the light scattering cross
section is a sum of lorentzians[4],

$$\frac{d^2\sigma}{d\omega d\Omega} = \left(\frac{\omega_i}{c}\right)^4 \frac{1}{\pi} \sum_\gamma \frac{\Gamma(\gamma)\alpha^2(\gamma)}{\Gamma(\gamma)^2 + \omega^2} \tag{2},$$

where $\Gamma(\gamma)$ and $\alpha(\gamma)$ are the inverse relaxation time and
polarizability strength of the γth "hopping mode". Klein
has shown how the $\Gamma(\gamma)$ and $\alpha(\gamma)$ can be obtained as linear
combinations of the intersite hopping rates and site polarizabilities, $\tilde{\alpha}_i$[4]. These hopping modes are specified by
a set of eigenvectors with components proportional to
linear combinations of the differences, $(N_i - \langle N_i \rangle)$, of the
site occupations from their mean values at thermal equilibrium. The eigenvectors transform as irreducible representations of the crystallographic space group of the lattice for a stoichiometric crystal. Their number and
symmetry can be determined by applying group character
analysis to the representation of all equivalent ion sites,

as Klein has done for CuI and AgI[4]. The integrated
scattering cross section of equation (2) should have the
temperature dependence of a Debye-Waller factor, quite different from that of a vibrational excitation. Assuming that
the differences in polarizabilities of the various sites are
comparable to the polarizabilities themselves, the integrated scattered intensity from these hopping modes will be
comparable to the Rayleigh scattering from a random array of
the same number of fixed ions. Materials in which both the
mobile ions and their nearest neighbours are highly polarizable, such as AgI, are obviously most promising for
studies of hopping modes. In principle, much very useful
and detailed information could be obtained from the spectral
shape and polarization dependence of the scattered light.
However, extracting this information is much more difficult
than it is for vibrational modes because the spectra of the
hopping modes overlap each other as well as the Rayleigh
scattering from static defects and surfaces.

The dwell time can be comparable to, or shorter than
the jump time if either $kT \gtrsim U$ or $\hbar \omega_0 \gtrsim U$. Then the spectral distribution of the scattered light becomes more like
that of an anharmonic, and perhaps heavily damped, vibrational motion encompassing the inequivalent sites. In
that case, the intensity distribution may resemble that of
a set of Lorentzian oscillators,

$$I(\omega) \propto \sum_{\gamma} \frac{\left[n(\omega,T) + 1\right]\omega}{(\omega^2 - \omega_{\gamma}^2)^2 + \Gamma(\gamma)^2\omega^2} \qquad (3),$$

where $n(\omega,T)$ is the Bose-Einstein distribution function.
However, if $kT \sim U$, ω_{γ} and $\Gamma(\gamma)$ might still be rapidly
varying functions of T. The vibrational motion is no longer
separable from all of the hopping modes, although hopping
between and among some sets of sites may still be represented by equation (2).

Several solid ionic conductors have been investigated
by Raman spectroscopy. The observed spectra have been of
two general types, depending upon the extent to which the
motions of the mobile ions are coupled to the rest of the
degrees of freedom. In the most common situation, found in
materials of "simple" chemical and crystal structure, the
mobile ions are a major constituent of the lattice and are
involved to a large degree in all the lattice modes. The

spectra reported in the high conductivity phases of such
materials as AgI[5,6,7], PbF$_2$, and similar fluorites[5], are
relatively broad, and have only one or two distinguishable
features, possibly reflecting a large portion of the
vibrational density of states of the disordered lattice.

The spectra reported by Harley et al.[5] for the
alkaline-earth halides of the fluorite structure are shown
in Figure (1). Ionic conduction in the fluorites involves
the migration of interstitial halogen ions and the corresponding vacancies. At low temperatures where the interstitial concentration is negligible, a single sharp Raman-
active vibration, due to motions of only the halide ions,
is observed. As the temperature is increased, this line
broadens considerably and a featureless low frequency wing
appears, which has an intensity decreasing monotonically
with increasing frequency shift. This alteration of the
spectrum accelerates as the temperature increases through
the region of a gradual order-disorder transition, which
has been observed in a number of fluorite crystals[5]. This
transition is associated with a large increase in the concentration of interstitial halide ions.

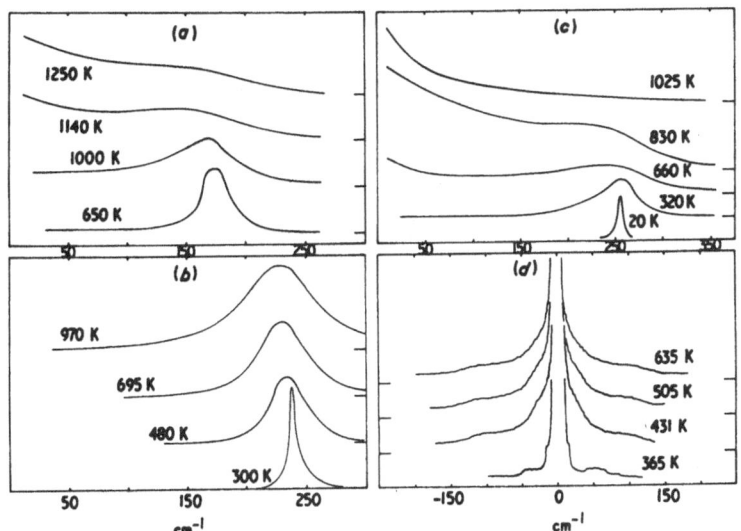

FIG. 1.   Raman spectra as a function of temperature for
(a) SrCl$_2$ (A$_{1g}$ + E$_g$ + T$_{2g}$), (b) BaF$_2$ (T$_{2g}$), (c) PbF$_2$ (T$_{2g}$),
(d) AgI (unpolarized).

In silver iodide, as well as several other Ag and Cu
binary and ternary compounds, ionic conduction is associated
with disorder on the Ag (or Cu) sublattice. At atmospheric
pressure, AgI is stable in a β-wurtzite phase below $146^{\circ}C$,
and at higher temperatures, the iodine ions form a body
centered cubic lattice in which the Ag ions are distributed
essentially at random over 21 inequivalent sites in the
unit cell[8]. Raman spectra have been reported for the low
temperature phase by Bottger and Damsgaard[6], for the high
temperature phase by Harley et al.[5], and for all five
phases which are stable up to pressures of 5 Kbar by Hanson,
et al.[7]. The spectra of the low and high temperature
phases[7] are shown in figure (2). The dominant features
are a 17 cm^{-1} E_2-symmetry phonon for T $<$ 146°C and a low
frequency wing for T $>$ 146°C, which is similar to that
observed at high temperatures in the fluorites. The peaks
at 37 cm^{-1} and 85 cm^{-1} in the low temperature phase are
assigned to two-phonon scattering from a pair of nearly
degenerate modes. The intense 17 cm^{-1} line is derived from
a very flat optic branch, of which the 37 cm^{-1} band is an
overtone. The high degree of anharmonicity of this mode is
evident from the extremely rapid decrease of the frequencies
of the 17 cm^{-1} mode and the 37 cm^{-1} band with increasing
hydrostatic pressure[7]. Bührer and Brüesch[9] have pointed
out that the eigenvector of this E_2 mode promotes Ag ionic
displacements in the direction of interstitial Ag sites,
and also involves iodine displacements twoard their positions in the high temperature BCC sublattice. They argue
that the displacements in this mode account for a large
contribution to the mean ionic displacement of 0.38 Å for
the Ag ions at 115°C. In the high temperature phase, a

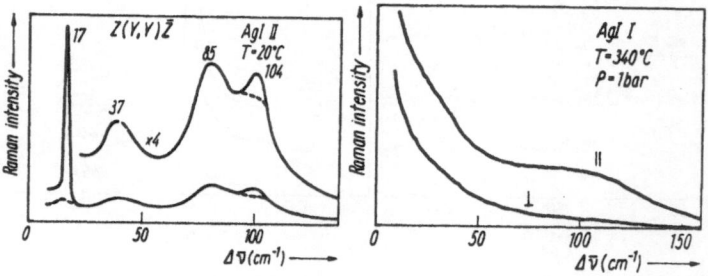

FIG. 2. Raman scattering from the Wurtzite phase (left)
and the disordered phase (right) of AgI. Polarized and
depolarized spectra are denoted by $\parallel$ and $\perp$, respectively.

broad band at about 100 cm^{-1} and a hint of a shoulder at 30-40 cm^{-1} may be broadened versions of the 37 cm^{-1} band and the bands at 85 cm^{-1} and 104 cm^{-1} of the β-wurtzite phase.

Spectra have also been reported recently for RbAg$_4$I$_5$ (10,11). Delaney and Ushioda[10] note a remarkable similarity of the spectra to those of AgI, including a band extending out to about 70 cm^{-1} from the laser line and a peak at 107 cm^{-1} in the high temperature phase.

The origin of spectra such as those in Figures 1 and 2(b) is in doubt at present. A detailed treatment of the dynamics of the disordered lattice may be required if such spectra are to yield quantitative information about factors affecting the conduction process. Since no such treatment is available, only qualitative, phenomenological interpretations have been attempted. Harley, et al.[5], have noted that if the intensity distributions of figures 1 and 2(b) are divided by the Bose-Einstein factor, $n(\omega,T)+1$, the low frequency tails produce peaks at 100 cm^{-1}, 60 cm^{-1} and 30 cm^{-1} for SrCl$_2$, PbF$_2$ and AgI, respectively. This procedure should be useful if the scattering is due to a distribution of damped oscillators as in equation (3), but not for the hopping mode distribution of equation (2). Harley, et al. suggest that such a peak could arise from either disorder-induced Raman scattering reflecting the one-phonon density of states of the crystal, or, alternatively, from the hopping of the anions in the fluorites among normal and interstitial sites. A peak of this type is also predicted from several recently proposed models such as those of Huberman and Sen[12], Fulde et al.[13], or Hubbard and Beeby[14].

A notable aspect of the experimental data for the fluorites and AgI is that no clearly defined hopping modes appear in addition to the broad wings discussed above, even for frequency shifts within the range of a Fabry Perot interferometer[7]. This suggests that the broad wings contain the scattering from intersite motions of the mobile ions. The possible significance of such an interpretation is apparent if the widths of these wings are compared with hopping rates, Γ, for the ions estimated from the measured DC conductivity and diffusion coefficients. For AgI at 300°C, and for the fluorites above the order-disorder tran-

sition, $\Gamma \sim 1$ cm^{-1}. The small size of these rates relative
to the widths of the light scattering distributions may be
due to ion-ion interactions. For example, in a stoichiometric material where interactions between mobile ions are
strong, the intracell relaxation rate may be much larger
than the rate for intercell hopping, in which an ion must
hop between, say, singly occupied cells to leave a configuration of an unoccupied cell and a doubly occupied cell.
If the increase in energy of this new cell configuration is
large enough, the intracell and intercell hopping rates may
differ by orders of magnitude. In fact, Hoshino[15] has
pointed out, on the basis of ion sizes, that an Ag ion in
AgI excludes the presence of other ions as far away as
fourth neighbour sites. In the fluorites, intercell relaxation could be impeded by a binding of an interstitial to
its associated vacancy. Of course, in a massively disordered crystal, the ion correlation problem is not that
simple. However, Pardee and Mahan[16] have emphasized the
possibility that these correlation effects may account for
most of the activation energy for migration and have
developed a model on that basis which successfully explains
the changes in activation energy at the order-disorder
transitions of several materials. The possibility of
measuring the intracell relaxation rate by Raman scattering
complements measurements of DC conductivity and diffusion
in determining the importance of these correlation effects.

The only examples of materials examined by Raman
scattering in which the motions of the mobile ions are
largely decoupled from the rest of the lattice are the β-
alumina isomorphs[17,18]. The stoichiometric chemical
structure of these materials is $M_2Al_{22}O_{34}$ where M is one of
the monovalent cations, Li, Na, K, Rb, Ag, Tl, etc.
Conduction occurs by migration of the M ions in planes of
mirror symmetry separated by 11 Å thick blocks of spinel-
like structure composed of only the Al^{3+} and O^{2-} ions[19].
A conducting plane is shown in Figure 3. The other plane
in the hexagonal cell has the oxygens and cations inter-
changed, and a center of inversion is midway between planes.
The structure in this plane is complicated by a cation concentration which is 25% in excess of stoichiometry in most
samples[19], so that the chemical structure is close to
$M_{2.5}Al_{22}O_{34}$. These excess cations occupy either the anti
Beevers-Ross (ABR) or mid-oxygen (MO) sites which are
energetically unfavourable compared with the predominantly
occupied Beevers-Ross (BR) sites.

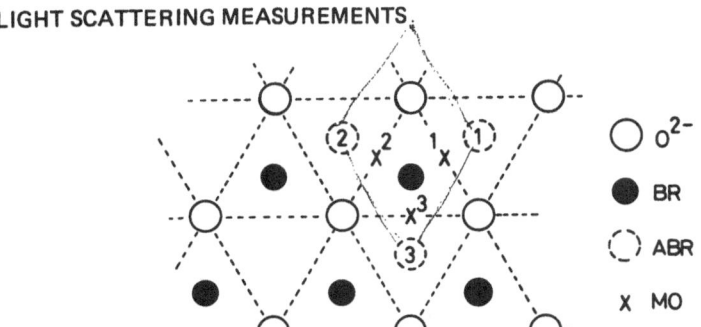

FIG. 3.   Structure of conducting plane of β-Al$_2$O$_3$.

Neglecting the disorder due to these interstitials, the
motions parallel to the conducting plane of the BR-site
cations and oxygen ions lead to two E_{2g} symmetry Raman-
active vibrations in the D_{6h} space group of β-alumina.   In
those vibrations, the cations in the two mirror planes
vibrate out of phase, as do the oxygens in those planes.
The vertical motions are not Raman active[18].   The low-
frequency E_{2g} spectra of Li, Na, Ag, K and Rb β-alumina are
summarized in Figure 4.   The peaks, or sets of peaks,
assigned to the cation vibrations are labelled B.   At least
two lines attributable to these cation vibrations are
observed in the Rb and K isomorphs and in the Ag isomorph
for T < 77 K.   The corresponding feature in Naβ-Al$_2$O$_3$ shows
no resolvable structure, even at 4.2 K.   The lack of a B
line for Liβ-Al$_2$O$_3$ is explained by the fact that the intensity of this mode scales roughly as the square of the cation
polarizability, so the Li line would be a factor of ten
weaker than that of Na.

There is evidence that the line labelled D is the other
E_{2g} vibration, due mainly to oxygen motions in the conducting plane.   Although there is a large variation in the
intensity of line D for the various cation isomorphs, its
frequency is nearly independent of cation substitution.   It
can be inferred from this that the force constant coupling
the cations to the mirror-plane oxygens is much smaller
than their coupling to the spinel block.   These Raman
spectra and the conclusions drawn from them are in agreement
with a one-dimensional model of β-Al$_2$O$_3$ used by Barker[20]
to fit IR data, with the exception of a ~115 cm^{-1} spinel
block shear labelled silent by Barker, but which Hao et al.,

[18] find to be an E_{1g} Raman mode, in agreement with the polarization data.

In β-alumina, therefore, the motions of the mobile ions are largely isolated from the lattice due to their weak coupling to the massive spinel blocks with which the cation vibrations are far from resonance. For this reason Allen and Remeika[1] have assumed that the B line frequencies in Figure 4 can be taken as attempt frequencies for the diffusion process. However, two other factors must be considered in judging the validity of this assumption. First, as was noted by Allen and Remeika, the measured frequencies must be corrected for Coulomb interactions among the cations since the B vibration, as well as its infrared active counterpart, corresponds to an in-phase motion of all the cations in the plane. The attempt frequency is that of a single cation vibrating about its equilibrium position. A calculation of these corrections reveals that the coupling between ions on different planes is negligible, which accounts for the identical IR and Raman frequences[18].

The measured Raman and IR frequencies and the attempt frequencies calculated from them by correcting for the intraplane Coulomb interactions are given in the first three columns of Table I. The Coulomb corrections amount to about 50% of the measured frequencies for the highly conducting Na and Ag isomorphs. Wang et al.[21] have calculated attempt frequencies from a point polarizable ion model. Their results are also listed in Table I. These are in amazingly good agreement with the experimentally obtained

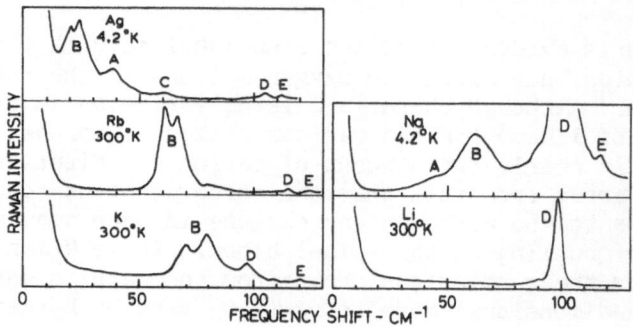

FIG. 4.   The low frequency E_{2g}, x(yx)z spectra of several β-Al$_2$O$_3$ isomorphs.

TABLE I: Raman and IR frequencies and calculated "attempt" frequencies. The ω_0 and ω_p listed under "theory" are from ref. 21.

Ion	Experiment			Theory	
	$\omega_r^{(18)}$ cm^{-1}	ω_i cm^{-1}	$\omega_0^{(18)}$ cm^{-1}	ω_0 cm^{-1}	ω_p cm^{-1}
Na$^+$	62	56$^{(1)}$	96	90	39
K$^+$	72, 82	80$^{(22)}$	91, 99	98	52
Rb$^+$	63, 69	86$^{(22)}$	73, 79	78	50
Ag$^+$	28	28$^{(1)}$	44	42	16

frequencies, considering that there are no adjustable parameters used in the calculations.

A second, and more crucial, factor is the question of the role of non-stoichiometry in the ion migration process. The corrected BR-site cation frequencies will represent attempt frequencies only if hopping proceeds by motion of isolated BR-site ions, through for instance, the sparsely occupied ABR sites. There is evidence from the temperature dependence of these cation vibrations that this is <u>not</u> the case. If that is the mechanism for conduction, there would be a potential barrier between the BR and ABR sites equal to the activation energy which is 0.17 eV for Na and 0.18 eV for Ag. Assuming, for simplicity, that the potential distributions are sinusoids of these amplitudes, we deduce from Slaters eigenvalues[2] that the vibrational frequencies should decrease by about 15% between 300 K and 850 K, where $kT \sim U/2$. The data shown in Figure (6) display just the opposite behaviour. Lorentzian fits to these data give increases in frequency of about 13% for Na and ten percent for Ag. Of further interest is the lack of harmonics of the cation frequency in the Raman data. This evidence suggests that the cation motions lead to fairly harmonic vibrations in a potential well much deeper than the measured U. Wang et al.[21] have expressed the same conclusion based on their model calculation, which predicts a potential barrier of ~ 2 eV for isolated BR-site cations. The observed increase in frequency can be accounted for on this basis, since the RMS vibrational amplitude of the cations increases much more rapidly than the thermal expansion of the hard β-alumina

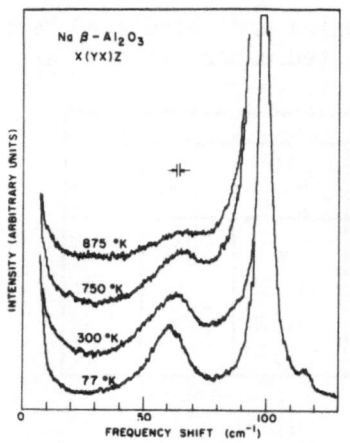 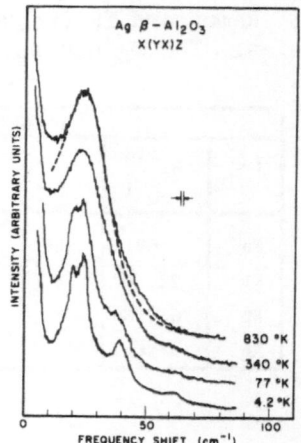

FIG. 5.   Temperature dependence of the cation vibrations of
Na and Agβ-Al$_2$O$_3$.

lattice.   Then the strong, short-range, repulsive potential
of the neighbouring O ions in the spinel block will increase
the force constants controlling the cation frequencies.
The Raman data are, therefore, in agreement with both the
cation frequencies and large energy barriers calculated
from the model of Wang, et al.[21].

Other conduction mechanisms can be devised which rely
on the occupation of the ABR and MO sites by the excess
cations.   Wang, et al.[21] have calculated attempt frequencies for the in-phase motion of pairs of cations
occupying neighbouring MO sites adjacent to an empty BR
site as shown in Figure (3), with local lattice relaxation
included.   Their results are listed as ω_p in Table I.   No
features with ~ 50% of the intensity of the B lines are
observed at these frequencies.   The only apparent effects
of the excess cation on the Raman spectra are the splittings
of the B lines, for which no satisfactory quantitative
explanation is available.   Crystal field calculations based
on the reported low-temperature super-lattices[22,23]
predict splittings considerably larger than those observed.

Some evidence of a distinct hopping mode is also
observed in Agβ-Al$_2$O$_3$.   The development of this feature at
high temperatures is shown in Figure 6.   The intensity of

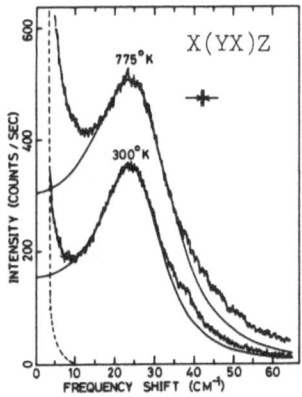
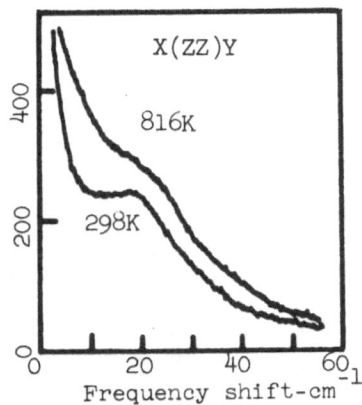

FIG. 6. Low frequency XY (left) and ZZ (right) spectra of
Agβ-alumina. The dashed line in the XY trace is the
response to unshifted laser light of the same intensity as
in the 775°K data. Smooth curves are lorentzian fits to
line B.

the additional low-frequency scattered light is comparable with that of the E_{2g} vibration. It is not clear at this
time if there are distinct $E_g(XY)$ and $A_{1g}(ZZ)$ modes or if the
mode responsible for the intense ZZ spectrum "leaks" into the
XY spectrum. Although a much more careful examination of
this spectrum is required before any more extensive information can be obtained, it is worth discussing as an example
the hopping behaviour to be expected in β-alumina. Referring
to Figure 3, the sites labelled ABR 2 and ABR 3 are equivalent to ABR 1 on the basis of the assumptions discussed
earlier, so there are five inequivalent sites in the unit
cell of one conducting plane. Intersite hopping leads to
three A_1 modes and one E_2 mode. (We use the C_{3v} symmetry of
a single plane since there is no hopping between planes.)
One of the A_1 modes is the "diffusion" mode discussed in
section IV. There remain two possible A_1 modes due to long
wavelength fluctuations in the relative occupations of the
BR, ABR and MO sites and a degenerate E_2 mode of the relative
occupations of the three MO sites. The A_1 and E_2 modes can
be observed separately by observations in ZZ and XY
geometries, respectively. Other hopping modes are possible
in doubly occupied cells, or if equation (1) is not applicable.

III. BRILLOUIN SCATTERING

The static and dynamic disorder in solid ionic conductors has important effects on the elastic constants,
particularly near order-disorder phase transitions. Recent
theoretical work in this area has indicated that some
rather useful information can be obtained by studying these
effects through the frequency and linewidth of the acoustic
phonons. Huberman and Martin[3] have recently examined the
dynamics of the mobile ions coupled to acoustic phonons and
have derived frequencies and line-widths for the coupled
modes. The phonons can interact with: (1) fluctuations in
ion density, the "diffusion mode"; (2) inter-site hopping
processes; (3) the ion current. The ion density fluctuations and hopping between inequivalent sites both affect
only the local bulk modulus and, therefore, longitudinal
phonons. Non-totally symmetric hopping modes and the ion
current couple to the shear elastic constants. These interactions are capable of altering the frequencies, widths and
intensities of the Brillouin scattering peaks in a manner
which is well known for the similar problem of acoustic and
molecular reorientation coupled modes in solids[24] and
liquids[25]. The observation of these coupled modes could
be used to obtain the dependence of the ionic chemical
potential of the various ion sites on the expansion and
deformation of the unit cell volume. An example of the
importance of this information is the study of Rice, et.
al.[26] who find that the dependence of Frenkel defect
energy on unit cell volume can lead to order-disorder transitions of first or second order, as well as more gradual
types such as those observed in the fluorite lattices.

In transparent materials, Brillouin scattering is an
ideal probe with which to study the coupled modes. The
only reported study of Brillouin scattering in an ionic
conductor is that of Harley et al.[5] on the fluorite
materials. The data they obtained for PbF_2 is shown in
Figure 7. From these measurements, it is apparent that the
bulk modulus, $(C_{11} + 2C_{12})/3$ decreases by about 30% over a
small temperature region which coincides with the order-
disorder transition near 800 K. Similar behaviour is
reported for $SrCl_2$ near 1000 K[5]. The intensities of the
longitudinal mode Brillouin peaks increase by about an
order of magnitude at this phase transition. The changes
in bulk modulus and intensity are both consistent with the

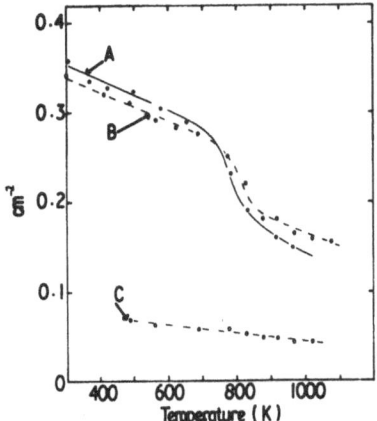

FIG. 7. Squared phonon frequencies for PbF_2: (A) $\bar{q}$ 11 $\langle 100 \rangle$, $\nu_{LA}^2 \propto C_{11}$; (B) $\bar{q}$ 11 $\langle 110 \rangle$, $\nu_{LA}^2 \propto C_{11} + C_{12} + 2C_{44}$; (C) $\bar{q}$ 11 $\langle 110 \rangle$, $\nu_{TA}^2 \propto C_{44}$.

type of coupled mode behaviour discussed by Huberman and Martin[3]. In this case the longitudinal phonons can couple to a mode involving fluctuations in the relative occupations of the normal and interstitial fluorine sites. It is the difference in polarizability of the unit cell when one or the other of these sites is occupied which can cause the increase in intensity of the Brillouin peaks. The large change in bulk modulus in this case could be due to the increasing density of interstitials, but in general, a fairly rapid change in the Brillouin frequency could also result from an increased hopping rate at higher T, which allows the ion density to relax within the period of the sound wave. In the latter case, the Brillouin lineshapes may become distorted by an overlap with the coupled hopping mode spectrum[24].

Some Brillouin scattering measurements have also been performed in $Na\beta-Al_2O_3$ and $Ag\beta-Al_2O_3$[27]. The elastic constants at room temperature are in agreement with those obtained by neutron scattering from higher frequency acoustic modes. No anomalies or significant changes are apparent between 77 K and 700 K. Although the room temperature shear elastic constant, C_{44}, is considerably affected by ion substitution[28], it is not influenced by the quali-

tative changes in cation distribution as a function of
temperature   observed by x-ray and neutron scattering[19].

IV. PHOTON CORRELATION SPECTROSCOPY

This technique is capable of measuring the spectral
distribution of quasielastically scattered light for frequency shifts between a few Hz and 1 MHz[27]. A prospective
application in ionic conductors is the direct measurement
of the diffusion coefficient using light scattered from ion
density fluctuations. For a wavevector transfer $\bar{q}$, the
scattered intensity due to a single diffusion mode has a
Lorentzian frequency distirbution of width $\gamma = Dq^2$, centered
at the incident light frequency. For laser light with
$\lambda \sim 5000$ Å and $D \sim 10^{-5}$ cm^2 sec^{-1}, γ is about 5×10^3 sec^{-1},
which is certainly accessible experimentally. However,
this type of spectroscopy is difficult if light scattered
from static imperfections is large compared with the quasielastic scattering, and it has so far only been applied
successfully to studies of diffusion of macro-molecules and
large particles for which the quasielastic scattering is
clearly dominant.

+   The author is grateful to Professor B. Bleaney for
fellowship support of the Science Research Council and the
use of the facilities of the Clarendon Laboratory and to
the Alfred P. Sloan Foundation for extended support.

*   Permanent address

(1)   S. J. Allen and J. P. Remeika, Phys. Rev. Lett. 33,
      1478 (1974).
(2)   J. C. Slater, Phys. Rev. 87, 807 (1952).
(3)   B. A. Huberman and R. M. Martin, Physical Review, to
      be published.
(4)   M. V. Klein, Proceedings of the Third International
      Conference on Light Scattering in Solids, p. 503,
      M. Balkanski, R. C. C. Leite, and S. P. J. Porto,
      editors, Flammarion Press (1976).
(5)   R. T. Harley, W. Hayes, A. J. Rushworth and J. F. Ryan,
      J. Phys. C. 8, L 530 (1975).
(6)   G. L. Bottger and C. V. Damsgaard, J. Chem. Phys. 57,
      1215 (1972).

(7) R. C. Hanson, T. A. Fjeldly and H. D. Hochheimer,
     Phys. Stat. Sol. (b) 70, 567 (1975).
(8) L. W. Strock, Z. Phys. Chem. 25, 441 (1934).
(9) W. Buhrer and P. Bruesch, Sol. St. Comm. 16, 155 (1975).
(10) M. J. Delaney and S. Ushioda, Bull. Am. Phys. Soc. 21,
     284 (1976).
(11) G. Burns, F. H. Dacol and M. W. Shafer, Bull. Am. Phys.
     Soc. 21, 284 (1976).
(12) B. A. Huberman and P. N. Sen, Phys. Rev. Lett. 33,
     1379 (1974).
(13) P. Fulde, L. Pietronero, W. R. Schneider and S.
     Strassler, Phys. Rev. Lett., 35, 1776 (1975).
(14) J. Hubbard and J. L. Beeby, J. Phys. C 2, 566 (1969).
(15) S. Hoshino, J. Phys. Soc. Japan 12, 315 (1957).
(16) W. J. Pardee and G. D. Mahan, J. Solid State Chem. 15,
     310 (1975).
(17) L. L. Chase, C. H. Hao and G. D. Mahan, Solid St.
     Comm. 18, 401 (1976).
(18) C. H. Hao, L. L. Chase and G. D. Mahan, Phys. Rev.
     B13, to be published, May (1976).
(19) W. L. Roth, General Electric Co. Technical Information
     Series Report No. 74CRD054, Schenectady, N.Y. (1974).
(20) A. S. Barker, Jr., J. A. Ditzenberger and J. D. Remeika,
     to be published.
(21) J. C. Wang, M. Gaffari and Sang-il Choi, J. Chem. Phys.
     63, 772 (1975).
(22) D. B. McWhan, S. J. Allen, Jr., J. P. Remeika and
     P. D. Dernier, Phys. Rev. Lett. 35, 953 (1975).
(23) Y. LeCars, R. Comes, L. Deschamps and J. Thery, Acta
     Cryst. A30, 305 (1974).
(24) An excellent example of recent results has been given
     by T. Bischoferger and E. Courtens, Phys. Rev. Lett.
     35, 1451 (1975).
(25) H. D. Dardy, V. Volterra and T. A. Litovitz, J. Chem.
     Phys. 59, 4491 (1973), and references cited.
(26) M. J. Rice, S. Strassler and G. A. Toombs, Phys. Rev.
     Lett. 32, 596 (1974).
(27) R. T. Harley and L. L. Chase, unpublished results.
(28) D. B. McWhan, S. Shapiro, J. P. Remeika and G. Shirane,
     J. Phys. C 8, L 487 (1975).
(29) H. Z. Cummins and H. L. Swinney in Progress in Optics,
     Volume VIII edited by E. Wolf, P. 133, North Holland,
     Amsterdam (1970).

NMR STUDIES OF SUPERIONIC CONDUCTORS, PRIMARILY BETA-ALUMINA

H.S. Story, W.C. Bailey, and I. Chung

Physics Dept., State Univ. of NY at Albany

1400 Washington Ave., Albany, NY   12222

and

W.L. Roth

General Electric Corporate Research & Development Center, Schenectady, NY   12301

INTRODUCTION

Nuclear magnetic resonance (NMR) is proving to be a valuable tool in the investigation of superionic conductors. This technique allows one to look into the material by using the conducting ion itself as a probe, and is most useful when used to complement primary methods like conductivity measurements, diffraction and structure studies. The nuclei of Na^+, Li^+ and F^- ions, and the hydrogen in NH_4^+ and OH_3^+ are particularly sensitive. It is fortunate that these ions are also among those of greatest interest in superionic conductors.

Although NMR studies of β''-alumina, CuI, $Li_2Ti_3O_7$ and other ionic conductors have been reported or are in progress, we shall discuss beta-alumina in detail and briefly refer to some of the other work.

NUCLEAR MAGNETIC RESONANCE

Some basic information about NMR relevant to superion-

317

ic conductors will be reviewed.[1] The nucleus of interest must have a spin I greater than zero. The spin system is conveniently described by a spin Hamiltonian:

$$\mathcal{H} = \mathcal{H}_Z + \mathcal{H}_D + \mathcal{H}_Q \tag{1}$$

$\mathcal{H}_Z$ is the interaction between the spins and external magnetic field. If a sample containing such nuclei is exposed to a magnetic field, the $2I+1$ magnetic energy levels of each nucleus will be split by the field. Transitions between these energy levels can be induced by radio frequency at a value (the Larmor frequency) equal to the level spacing divided by h. The RF may be continuous as in wide-line NMR or pulsed. The energy levels will be further perturbed by two important effects. $\mathcal{H}_D$ describes the magnetic dipole coupling between the spins, and with other spins in the lattice. It will broaden or further split these energy levels. The effect will dominate the spectra of spin $I=\frac{1}{2}$ nuclei like ^1H and ^{19}F in solids.

Nuclei of spin I greater than $\frac{1}{2}$ will generally have a nuclear quadrupole moment. $\mathcal{H}_Q$ describes the nuclear electric quadrupole interaction. The charge distribution in the crystal produces a crystal field at the nuclear site. The gradient in this crystal field will couple with the nuclear quadrupole moment to further modify the nuclear energy levels, Fig. 1. The effect of this coupling dominates the spectra for ^{23}Na and ^7Li. The quadrupole interaction can be characterized by five parameters; the coupling constant (C), the asymmetry parameter (η), and three angles which give the orientation of local principal axes relative to crystalline axes. C and η are determined by the electric charge distribution of ions in the crystal and by shielding due to the electrons in the host atom. A nucleus in a cubic environment will have C=0. η measures the departure of the local electric field from axial symmetry. If the environment of a nucleus has three-fold or higher symmetry, η will vanish. A study of the quadrupole parameters can help to elucidate the number of inequivalent sites, to determine the symmetries of these sites, to uncover phase changes, and to clarify the dynamics of ion motion. The angular parameters can in principle be determined only in a single crystal sample. The quadrupole coupling will give rise to line broadening and sometimes to a spectrum of resolved resonance lines.

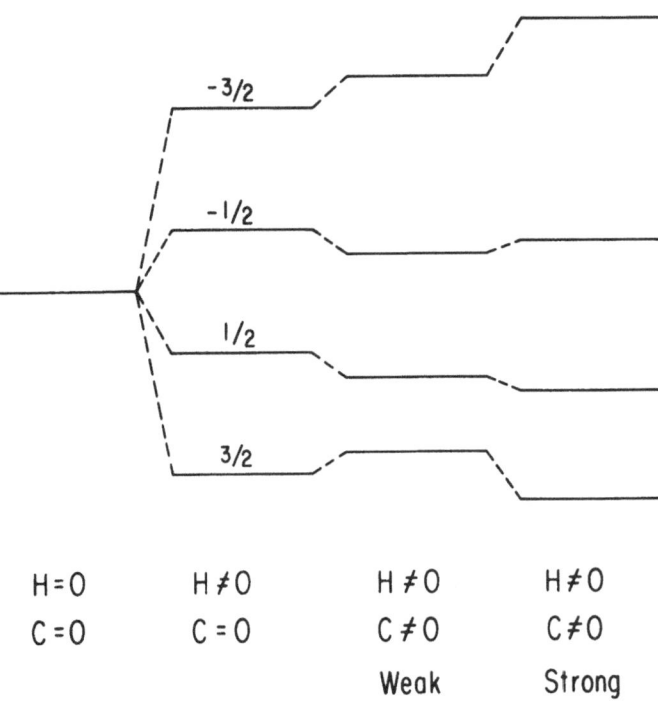

$$H=0 \qquad H\neq0 \qquad H\neq0 \qquad H\neq0$$
$$C=0 \qquad C=0 \qquad C\neq0 \qquad C\neq0$$
$$\qquad\qquad\qquad\qquad\qquad Weak \qquad Strong$$

Fig. 1. Schematic energy level diagram. H represents the external magnetic field. C is the quadrupole coupling strength. The energy level modification by C is small compared with the Zeeman splitting (C≈0) and depends on the direction of the magnetic field relative to the principal axis system.

Dynamic effects in NMR are crucial, especially so in superionic conductors, where diffusion carries the spins amoung different environments. Systems of spins are usually discussed in terms of a correlation time (or time of stay) τ for the spin in a particular environment. τ is

often related to temperature and an activation energy E
through:

$$\tau = \tau_o e^{E/kT} \tag{2}$$

τ affects NMR results in two important ways. Let us denote by $\Delta\omega$ a characteristic line width or line splitting
associated with the spectrum. If τ is sufficiently long
that its reciprocal, the correlation frequency, is less
than the static line broadening or splitting, $1/\tau \ll \Delta\omega$,
the static spectrum will be seen. Conversely, if $1/\tau \gg$
$\Delta\omega$, then the resonance line structure will collapse into
single line. As τ further decreases with increasing temperature, the width of the line W will decrease. This is
called motional narrowing. A plot of Log W (with corrections) versus reciprocal temperature yields an estimate of
an activation energy.

An important quantity which can be determined from NMR
experiments is T_1, the spin-lattice relaxation time. As
its name implies, it is a measure of the rate at which the
spin system, if polarized, decays to an unpolarized state
by transferring energy to the lattice. A shorter T_1 implies a faster decay rate.

The other important effect of τ will be on the spin-
lattice time. T_1 is related to the correlation time through terms like:

$$\frac{1}{T_1} = (\quad) \frac{\tau}{1 + \omega^2\tau^2} + (\quad) \frac{\tau}{1 + (2\omega)^2\tau^2} \tag{3}$$

where the factors in the parentheses would include the square of interaction strengths of the dipole or quadrupole
couplings, and angular dependences if they are present.
Fig. 2 shows a logarithmic plot of the first term of this
relation. Below the T_1 minimum (long τ) T_1 is proportional
to τ, above the minimum T_1 is proportional to $1/\tau$. Thus,
above or below the minimum, a plot of log T_1 versus $1/T$
will yield an activation energy. Both of these estimates
of activation energy may be in error if there is a wide
distribution in correlation time τ, instead of the assumed
single value at a given temperature.

Another relaxation time, the spin-spin relaxation time
T_2, is related to τ through terms like:

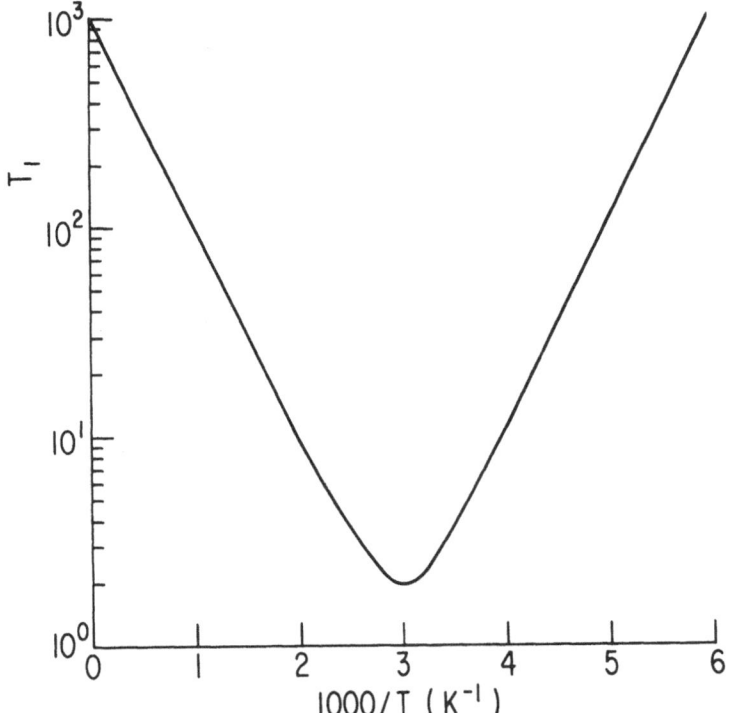

Fig. 2. Usual behavior of spin lattice relaxation T_1 time versus inverse temperature. From the slope of the straight portions, an activation energy can be found.

$$\frac{1}{T_2} = (\quad) \tau + (\quad) \frac{\tau}{1 + \omega^2\tau^2} + (\quad) \frac{\tau}{1 + (2\omega)^2\tau^2} \quad (4)$$

For a single homogenously broadened line, the linewidth W is proportional to $1/T_2$. The first term, proportional to τ, is responsible for the motional narrowing.

It is clear then, that there are thus two important crossover regions in NMR. One occurs when $1/\tau \approx \Delta\omega$, of the order of 10^5/s. The other corresponds to $1/\tau \approx \omega$ which is about 10^8/s.

In the next sections, we shall discuss some applications of NMR to β-alumina, first in single crystals, and then in polycrystalline material.

SINGLE CRYSTAL BETA ALUMINA

The most complete and clearest information comes from single crystals, and the place to start is at a temperature sufficiently low that ionic motion is slow. Bailey et al[2], studied the ^{23}Na resonance spectrum of single crystal β-alumina at 77 and 103K.

Fig. 3 shows the rotation pattern obtained as the magnetic field is rotated from the crystal c-axis direction toward an a-axis. The spectrum shows the presence of sodium at three-fold (η=0) sites – the curve F near θ=0. Since x-ray and neutron diffraction evidence places much of the sodium at BR sites, it is likely that this resonance may be assigned to such sites. This resonance line seems to broaden and disappear as θ increases, probably due to local variations in the crystal field. The resonance lines A, E are associated with well defined sites having a high asymmetry parameter; they are in sites which lack three-fold symmetry, for example sodium at a BR site with an aBR neighbor. Calculations of the field gradients from multipole expansions in somewhat simplified models are producing results which compare with experiment.[3] The remaining resonances come from other sites of low symmetry, perhaps other defects. Boilot et al.[4] found evidence of line structure at low temperature, but did not elaborate.

As the temperature is raised, τ decreases. At about 110K it reaches the point where the frequency 1/τ exceeds Δω in the static spectrum. The line structure then collapses to a single line. Boilot et al.[4] and Chung et al.[5] have studies the angular dependence of the splitting from H_O and line width of this line. Fig. 4 whows the line splitting as a function of θ for several temperatures. It is characterized by an almost constant quadrupole coupling over the temperature range from 110K to 500K. Over this range the average η uniformly diminishes, Fig. 5. Taken together, these facts are consistent with sodium ions that circulate in an environment with symmetry less than three-fold, and with very little change in the population densities of the various sites as the temperature changes. Only

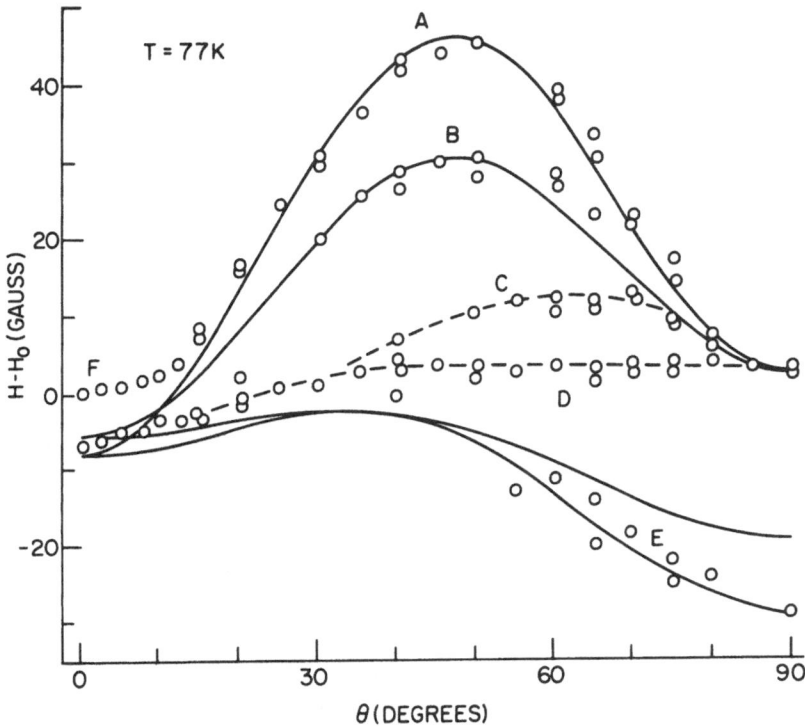

Fig. 3. Rotation pattern for a single crystal of sodium β-alumina at 77K. θ=0 when H is parallel to crystal c-axis; 90° when H is parallel to a.

the frequency of the motion increases with increasing temperature. Boilot's and Chung's results are in general agreement.

The presence of a single resonance line at all temperatures above about 110K indicates that all the sodium ions participate in the motion and are not "stuck" at any sites, at least on an NMR time scale, i.e. longer than 10^{-5} seconds. No additional sodium lines at room temperature which might come from immobile sodium have been found after careful search.

The line width of the single line decreases as the temperature and the correlation frequency increases, Fig. 6.

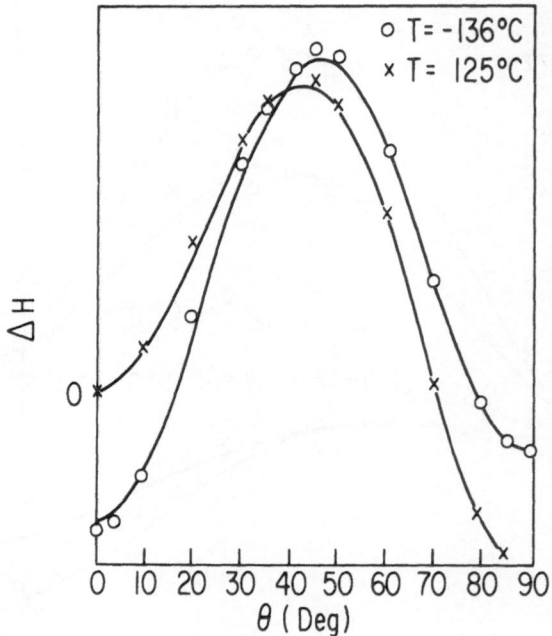

Fig. 4. Rotation pattern for the same crystal as in Fig.
3 at two higher temperatures. The ΔH scale is similar to
that in Fig. 3.

If the log of the line width (with certain corrections) is
plotted versus 1/T, an activation energy may be computed
from the slope. An estimate of about .1 eV is obtained.

Bailey found that at about 103K, the spectrum contained lines from the static spectrum together with the motion-
ally narrowed line.[2] There is therefore at this temperature
a wide distribution of correlation times probably extending

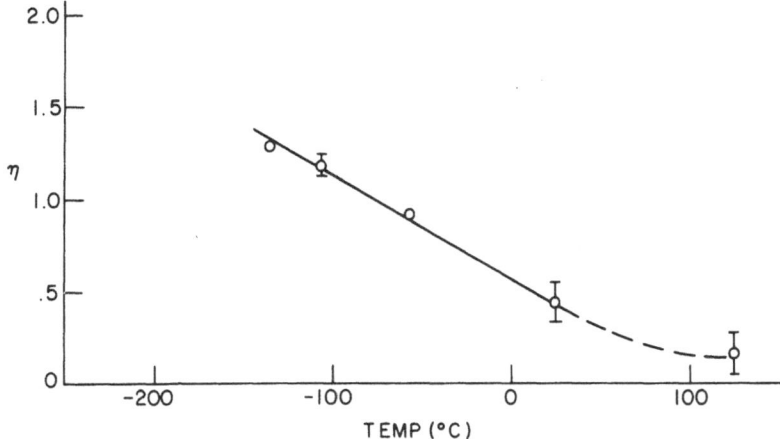

Fig. 5. Asymmetry parameter versus temperature. The asymmetry parameter decreases as the distinction between the inplane components of the field gradient tensor "blurs". In this respect, the planar motion becomes more "liquid-like".

over more than four orders of magnitude. This signifies a complicated hoping process amoung a number of sites with inequivalent environments. In a later section, we shall make further use of this wide distribution.

Walstedt et al.[5] have measured the spin-lattice relaxation time T_1 as a function of temperature in single crystals and found an activation energy which ranged from .04 to .1 eV in different crystals. They attribute the variation in T_1 behavior to variations in the defect structure in crystals of different origin.

POLYCRYSTALLINE BETA ALUMINA

Because single crystals of the desired materials are not always available, and since polycrystalline and ceramic materials are expected to be used in applications, they have been used in many studies. In such samples, there is assumed to be a random distribution of crystal orientations. The single resonance line in the single crystal gives rise

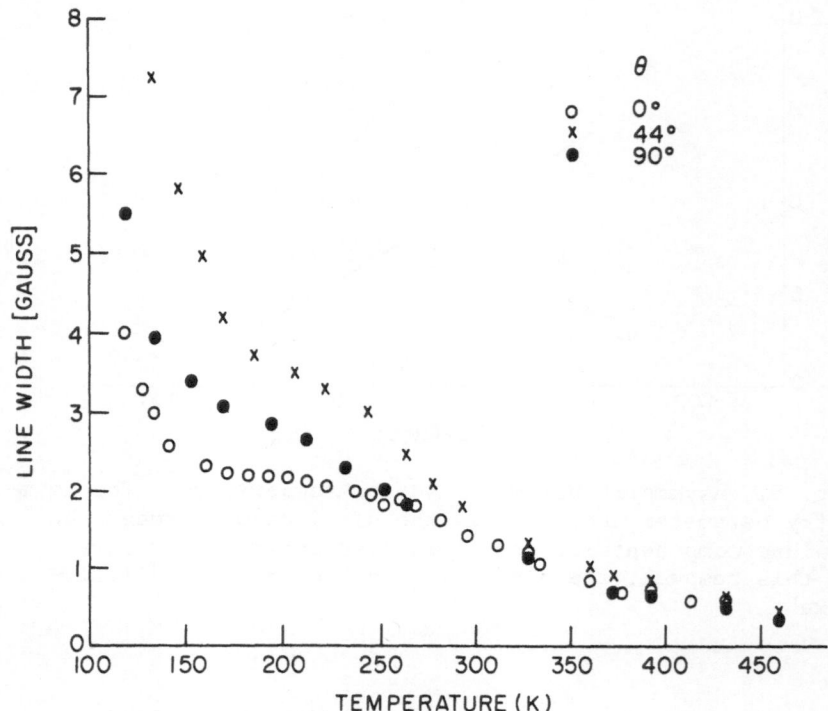

Fig. 6.   Linewidth at several angles as a function of T.
θ has the same meaning as in Fig. 3.   An activation energy
of about 0.1 eV is estimated from the line narrowing data.

to a powder pattern whose derivative leads to a two peak
spectrum, Fig. 7.   The effect of line narrowing with increasing temperature as the frequency of motion increases
is evident.

Kline et al[7] showed that the quadrupole coupling constant, as measured by the spacing between the peaks, is
reversibly reduced by exposure of the material to water
vapor.

Jermoe and Boilot[8] measured T_1 in polycrystalline β-
alumina as a function of temperature from about 100K to
700K, Fig. 8.   From the slope of their data on a logarithmic plot, they found two different activation energies of
about 0.1 and 0.2 eV and suggested the possibility of a

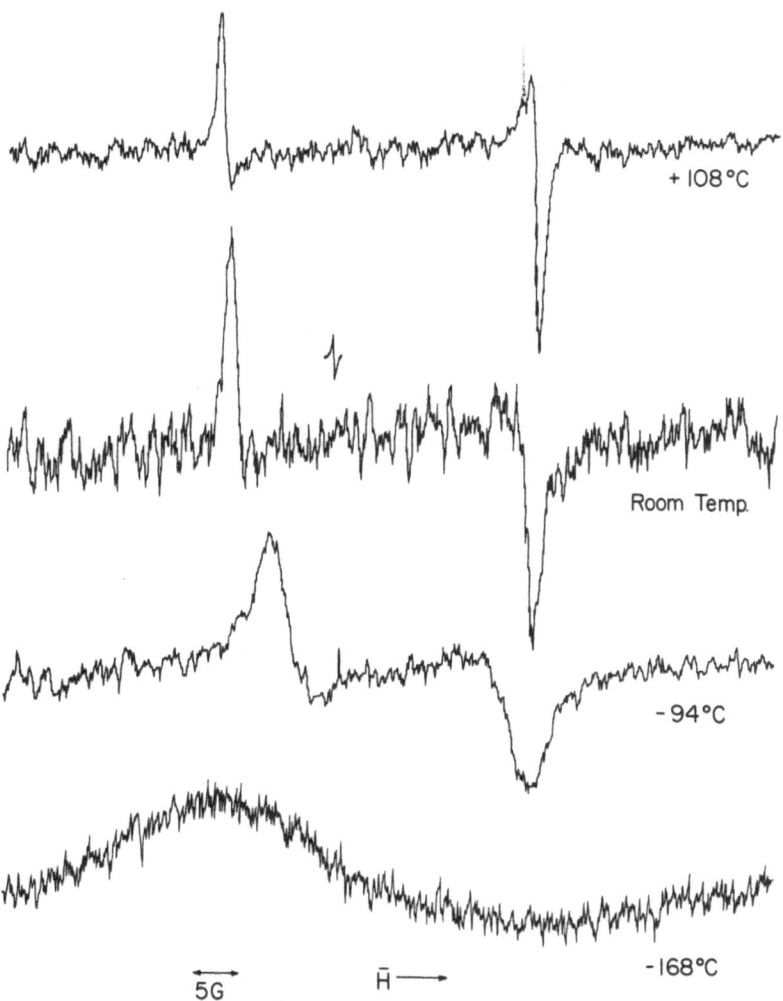

Fig. 7. Spectrum of a polycrystalline sample. C, η and W can be extracted but with less reliability than from a single crystal.

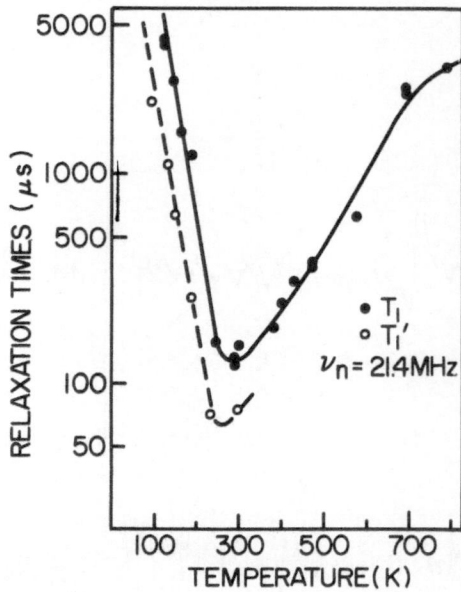

Fig. 8.   Spin-lattice relaxation time T_1 in polycrystalline
β-alumina from reference 8.   The authors extracted activation energies of .1 and .2 in the lower and higher temperature ranges, respectively.

multiplicity of processes in the conduction mechanism.

        A study of the effects on the NMR linewidth and coup-
ling parameters of sodium β-alumina containing certain
additives has been reported by Chung et al.[9] and by Roth et
al.[10]   Ceramic samples containing Li_2O, MgO, LiO, CaO and
Y_2O_3 in several different additive fractions were prepared.
The addition of Li, Mg and Ni up to 0.3 atom ratio (additive cation to sodium ratio) produced a reduction in NMR
linewidth that is correlated with increased conductivity,
and a large effect on the coupling constant Fig. 9.   The addition of yttrium produced no effect on the NMR sodium spectrum presumably because it goes into the grain boundaries.
It is believed that calcium forms a compound in the conduction plane and inhibits the conduction process.   Walstedt
and Remeika[11] have found an increase in activation energy
accompanying increasing potassium content in the mixed

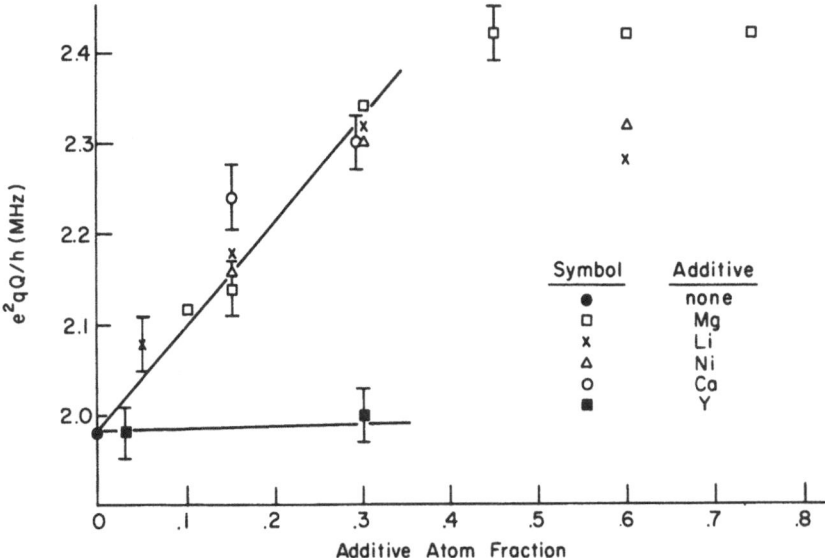

Fig. 9. Effect on the coupling constant of several different additives in polycrystalline β-alumina. Mg, Ni and Li
probably go into the spinal block of the structure but they
do directly modify the conduction plane. In addition to
the effect shown in the figure, they also produce additional
line narrowing. Y produces no effect and goes into the grain
boundaries. Ca broadens the linewidth.

sodium-potassium β-alumina.

We have made some NMR measurements of [7]Li and ^{1}H spectra in lithium and ammonium β-alumina respectively. In each
of these cases as well as in Na, there exists a moderately
narrow temperature range where both the broad resonance
characteristic of slow motion and the narrow resonance characteristic of fast motion can be simultaneously seen. As
the temperature is raised through this region, the intensity
of the broad line diminishes as the narrow one grows. Resing[12] has shown that this "apparent phase transition" behavior is a consequence of a broad distribution in the
correlation time. By using the temperature at the "center"
of this distribution, T_C and quantities drawn from NMR with
a reasonable pre-exponential factor, an estimate of average
activation energy can be obtained. This result for the

three β-aluminas is shown in Table 1, and indicates good
agreement with the activation energy from dielectric loss
data[13]. It is likely that in the presence of a wide distribution in correlation time, the various logarithmic estimates will underestimate the activation energy. The use
of the "apparent phase transition" data may be more reliable, at least for the intercomparison of different samples.

SUMMARY OF BETA ALUMINA

Briefly, we may summarize the principal results of NMR
studies of Na β-alumina as follows: 1. At low temperatures
(77 Kelvin) sodium occupies several different kinds of lattice sites or configurations; 2. At higher temperatures,
sodium ions move among these sites with all of the sodium
ions participating in the motion; 3. The activation energy
can be estimated in several ways from the NMR data - various estimates range from .04 to .20 eV and it may depend on
the defect concentration; 4. There is a wide range in the
frequencies of ionic motion - at 100K it probably extends
over at least four orders of magnitude, giving rise to so-
called "apparent phase transition" behavior; 5. The moving
ions see an average environment which has a symmetry lower
than three-fold; 6. Water reversibly penetrates the conduction planes inhibiting the sodium motion; 7. The inclusion of particular additives during the synthesis of ceramic β-alumina alters the NMR spectrum; the intragranular
and intergranular effects of certain additives have been
identified.

OTHER MATERIALS

NMR studies of other materials have been reported or
are in progress. Boyce and Huberman[14] have measured T_1 in
CuI as a function of temperature. The expected increase in
T_1 above the temperature of the T_1 minimum did not occur.
They attributed this to an increasing density of defects as
the temperature approached that of the superionic conductor
transition. Follstaedt and Biefield[15] have measured T_1 in
for ^7Li in $LiAlSiO_4$ glass ceramic samples and suggested an
order-disorder transition at about 460°C. Boyce and Mikkelsen[16] have investigated $Li_2Ti_3O_7$ and again found anomolous
behavior of T_1. A series of compounds of the composition

A_xTiS_2 where A = Li or Na, and x ranges from zero to one, were studied by Silbernagel and Whittingham[17] using NMR, electrochemical and x-ray techniques. In this case, NMR Knight shifts as well as quadrupole interactions and spin-lattice relaxation times were measured. The presence of a Knight shift indicates an electronic contribution to the conductivity.

Bailey[18] is also studying the ^{23}Na spectrum in sodium β"-alumina. He finds results which parallel those found in β-alumina.

Because of the limited space, we have not attempted to include the hydrides, fluorine systems and materials showing large Knight shifts.

Table I

Mobile Ion	$\Delta\omega$ (s^{-1})	ω_0* (s^{-1})	T_C (K)	E (eV)	E* (eV)
Na^+	2×10^5	1.9×10^{12}	100	.14	.16
Li^+	4×10^4	3.8×10^{14}	170	.34	.38
H in NH_4^+	6×10^4	1.7×10^{14}	270	.51	.48

ω_0 is taken from Ref. 13, the activation energy E is computed from the first three columns by:

$$E = KT_c \ln \left(\frac{\omega_0}{\Delta\omega} \right)$$

E* is from Ref. 13, and may be compared with the NMR estimate.

1. For general NMR background, see for example: A. Abra-
    gram, The Principles of Nuclear Magnetism (Oxford
    University, London, 1961).

2. W. Bailey, S. Glowinkowski, H.S. Story, and W.L. Roth
    (to be published J. Chem. Phys. May 1976).

3. W. Bailey and H.S. Story, in preparation.

4. J.P. Boilot, L. Zuppipoli, G. Delplanque, and D. Jer-
    ome, Phil. Mag. 32, 343 (1975).

5. I. Chung, H.S. Story, and W.L. Roth, J. Chem. Phys.
    63, 4903 (1975). Also Bull. Am. Phys. Soc. Series II,
    19, 202 (1974).

6. R.E. Walstedt, R. Dupree and J.R. Remeika, Bull. Am.
    Phys. Soc. Series II, 21, 285 (1976).

7. D. Kline, H.S. Story and W.L. Roth, J. Chem. Phys. 57,
    5180 (1972).

8. D. Jerome and J.P. Boilot, J. Phys. (Paris) Lett. 35,
    L-129 (1974).

9. I. Chung, this conference.

10. W.L. Roth, I. Chung and H.S. Story, in preparation.

11. R.E. Walstedt and J.P. Remeika, Bull. Am. Phys. Soc.
    Series II, 20 (1975).

12. H.A. Resing, J. Chem. Phys. 43, 669 (1965).

13. R.H. Radzilowski, Y.F. Yao and J.T. Kummer, J. Appl.
    Phys. 40, (1969).

14. J.B. Boyce and B.A. Huberman, Bull. Am. Phys. Soc.
    Series II, 20 (1975).

15. D.M. Follstaedt and R.M. Biefeld, Bull. Am. Phys. Soc.,
    Series II, 21, 285 (1976).

16. J.B. Boyce and J.C. Mikkelsen, Bull. Am. Phys. Soc.,
    Series II, 21, 285 (1976).

17. B.G. Silbernagel and M.S. Whittingham, Bull. Am. Phys.
     Soc. Series II, <u>21</u>, 285 (1976).

18. W. Bailey, this conference.

ADDITIVES AND RESISTIVITY IN BETA-ALUMINA

John H. Kennedy

Department of Chemistry, University of
California
Santa Barbara, California, 93106

It is well known that the resistivity of β-alumina is
sensitive to composition. However, the situation is complicated by the fact that the constituents of the material
being studied may play various roles in affecting the
measured resistivity. First, β-alumina conducts via
mobile sodium ions, and replacement of these sodium ions
by other metal cations will, naturally, change the resistivity. Second, the resistivity of the β-alumina will
depend on the concentration of the mobile sodium ions, and
additives which will increase that concentration will decrease the resistivity. This assumes that these additives
play no other role in the conduction process. Third,
most studies of β-alumina have used polycrystalline samples,
and sintering to near theoretical density is required.
Additives or impurities which aid or inhibit the sintering
process will affect the measured resistivity of the sintered
polycrystalline material. Fourth, β-alumina may also adopt
somewhat modified structures, e.g., β″-alumina, which exhibit different resistivities compared to the parent
material. Additives which produce the new structures
will, of course, affect the resistivity. Fifth, the
aluminum ions may be completely replaced by other +3 ions
leading to analogs of β-alumina. These materials may
show resistivities which are considerably different from
the parent material. This review will examine briefly
each of these five areas. Some investigators have reported their results in resistivity terms while others
have used conductivity. In order to compare more easily,
all results will be presented in conductivity units.

REPLACEMENT OF SODIUM

One of the properties of β-alumina recognized early on by investigators is the ability to replace sodium ions by other metal cations using ion exchange techniques. A classic study was carried out by Yao and Kummer (1) in which sodium ions were exchanged in molten salts by Ag^+, K^+, Rb^+, Li^+, and Cs^+. Some experiments were also carried out with divalent ions, and some exchange was noted with Sr^{2+}, Pb^{2+}, Fe^{2+}, Ba^{2+}, Sn^{2+}, Mn^{2+}, and Ca^{2+}. However, in a number of these cases, the single crystals fractured which was attributed to a shortening of the a-axis with M^{2+} exchange.

TABLE I

Ions Substituted for Na^+ in β-Alumina
(Single Crystal Samples)

Ion	E_a, kcal/mole		$\sigma(ohm\text{-}cm)^{-1}, 25^\circ C$
	D	σ	
Na	3.81	3.78	140×10^{-4}
Ag	4.05	3.94	67×10^{-4}
K	5.36	6.78	0.65×10^{-4}
Rb	7.18	-	-
Tl	8.22	8.19	0.02×10^{-4}
Li	8.71	8.54 high temp 4.30 low temp	1.3×10^{-4}

Other ions reported: NH_4^+, In^+, Ga^+, Cu^+, NO^+, H_3O^+

Cs^+ (partial)

Ca^{2+}

Eu^{2+} (conduction plane and spinel block)

Conductivity depends on mobility which in turn depends on diffusion. Tracer diffusion coefficients were determined for a number of the +1 substituted β-aluminas as a function of temperature (1), and the activation energies are given in Table I. Few studies have been carried out to determine the actual conductivity of these materials. Whittingham and Huggins determined the conductivities of β-alumina single crystals substituted with various +1 ions as a function of temperature (2). Conductivity at 25°C and activation energy for conductivity are given in Table I. It can be seen that the transport process for diffusion and conductivity must be similar because of the close agreement between diffusion and conductivity activation energies. Only in the case of K^+ is there any appreciable difference. It can also be seen from Table I why most research has been carried out using sodium β-alumina. Aside from Ag^+ all other M^+ β-aluminas exhibit conductivities more than two orders of magnitude smaller. A detailed study of silver β-alumina was carried out by Whittingham and Huggins (3), and the results are shown in Fig.1.

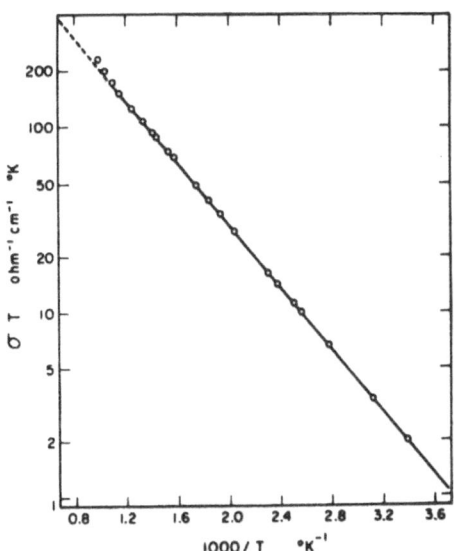

FIGURE 1: The conductivity of Ag^+ β-alumina from ref.3.

Other metal ions which have been substituted for Na^+ are mentioned in a review by Kummer (4) and are given in Table I. It is interesting to note that exchange by Eu^{2+} has been reported (5,6) although Mossbauer results (5) indicated that Eu^{2+} may be in both the conduction plane (Na^+ site) and in the spinel block (Al^{3+} site). This points out that it is not always obvious where an additive may reside in β-alumina confusing the issue of its role in the conductivity process. Roth has shown the location of several additives by observing ^{23}Na NMR line width (7), and the results are given in Table II. Note that Li^+ may also reside in the conduction plane and the spinel block.

CONCENTRATION OF CONDUCTIVE SODIUM IONS

Beta-alumina has the nominal formula $NaAl_{11}O_{17}$ but always contains excess sodium ions. The conductivity is sensitive to the concentration of excess sodium ions, and therefore, attempts have been made to incorporate additives which will increase the sodium ion concentration and thereby the conductivity. Charge balance in undoped β-alumina has been attributed to Al^{3+} vacancies and to O^{2-} interstitials in the conduction plane (7). If a metal

TABLE II

Effect of Additives in Beta-Alumina on Sodium Motion

Motion is Shown by Narrowing the Width of
^{23}Na NMR Spectrum

Additive	Line Width (gauss)	Location of Additive
None	2.0	
Mg	1.4	Replaces Al(2) in spinel block[a]
Ni	1.4	In spinel block
Li	1.5	In spinel block and conduction plane
Ca	4.0	In conduction plane
Y	2.0	Forms a second phase

[a] Al(2) is tetrahedral site.

From: W. L. Roth
       Trans. Amer. Cryst. Assoc., 11, 51 (1975).

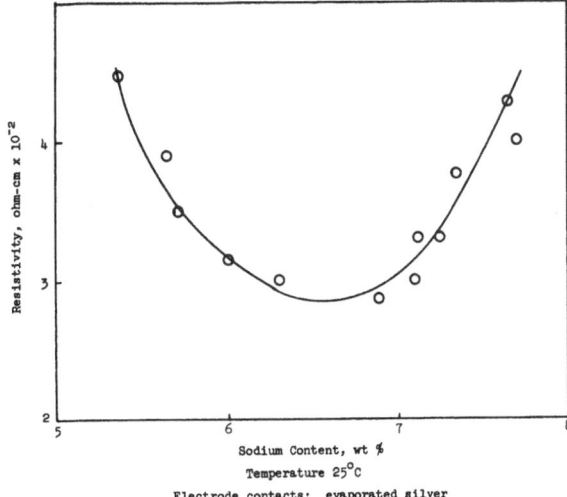

FIGURE 2: Resistivity as a function of sodium content for β-alumina containing 1% MgO from ref.8.

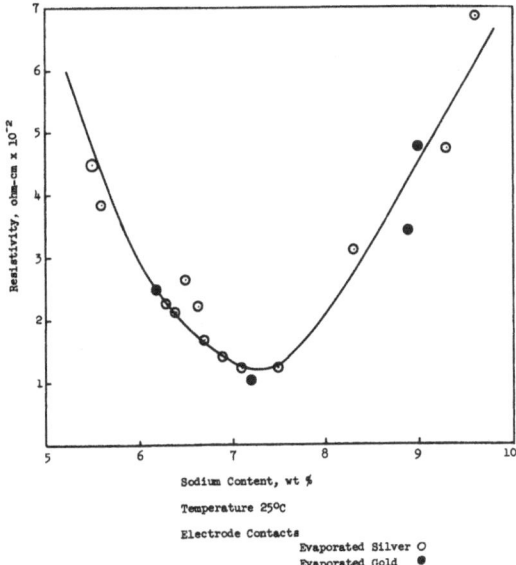

FIGURE 3: Resistivity as a function of sodium content for β-alumina containing 2% MgO from ref.8.

TABLE III

Comparison of Calculated Sodium Content with Sodium
Content at Minimum Resistivity

MgO, wt %	Calculated Sodium content, wt % Na$_2$O	Observed Sodium Content, wt % Na$_2$O (minimum resistivity)
1	6.1 + 0.75 = 6.85	6.8
2	6.1 + 1.5 = 7.6	7.4
4	6.1 + 3.0 = 9.1	9.5

ion having a charge less than +3 replaced an aluminum ion
then additional sodium ions could be incorporated for
charge balance.   Sodium content and conductivity as a
function of MgO doping level were studied by Kennedy and
Sammells (8).   It was found that the optimum concentration
of sodium (minimum resistivity) depended on MgO content.
Fig.2 and Fig.3 show that as the MgO content increased the
optimum sodium content also increased.   The optimum sodium
content could be predicted by assuming that each Mg^{2+}
added incorporated one extra Na^+ compared to the undoped
material (Table III).

TABLE IV

Conductivity of 2 w/o Transition Metal-Doped β-Alumina
∞ Frequency vs. Electrolyte-Thickness

Dopant	Oxidation State	Conductivity (ohm-cm)$^{-1}$, 25°C
None	-	5.5 x 10^{-4}
Cr	+3	4.7 x 10^{-4}
Fe	+3	6.7 x 10^{-4}
Co	+2	8.0 x 10^{-4}
Mn	+2	10.4 x 10^{-4}
Ni	+2	11.5 x 10^{-4}
(Mg)	+2	80 x 10^{-4}

Doping with other +2 metal ions also increased conductivity compared to similar doping with +3 metal ions
(9) as shown in Table IV.   However the effects were smaller
than with Mg^{2+}.   The situation became more complicated
when higher doping levels were used.   Table V shows the
conductivity of iron-doped β-alumina which remained close
to undoped material until > 4 w/o iron was present.   A
rapid increase in conductivity then was observed which may
include an electronic conductivity component in addition to
the ionic conductivity (9).   Many other metal ions have
been added to β-alumina which appear to have increased the
conductivity and are summarized in Table VI (10-20).   Many
of these reports are from patents and do not give sufficient
detail to prove the effectiveness of the additive or its
role in increasing the conductivity.

SINTERING TO HIGH DENSITY

Many of the additives proposed in Table VI may not be
increasing the concentration of mobile sodium ions or their
mobility in the bulk crystals but aid sintering so that
grain boundary resistance was decreased.   This appears to
be the major effect with Y_2O_3 (19,20) and TiO_2 (17,18).

TABLE V

Conductivity of Iron-Doped β-Alumina
∞ Frequency vs. Electrolyte Thickness

w/o	Conductivity (ohm-cm)$^{-1}$, 25°C	Δσ
0	5.5×10^{-4}	-
1	7.4×10^{-4}	1.9×10^{-4}
2	6.7×10^{-4}	1.2×10^{-4}
4	8.0×10^{-4}	2.5×10^{-4}
6	16.4×10^{-4}	10.9×10^{-4}
8	28.2×10^{-4}	22.7×10^{-4}
15	43.5×10^{-4}	38.0×10^{-4}

TABLE VI

Reports of Metal Oxide Additions Increasing
β-Alumina Conductivity

Metal Oxide Additives	w/o	$\sigma(\text{ohm-cm})^{-1}, 300^\circ C$	Investigators
CuO, NiO ZnO	7.2	17×10^{-2}	Harada, Ohta, and Imai
CuO	0.9-20 m/o		Imai, Harada, Ogawa, Hasegawa
Co_2O_3 $Cr_2O_3, Fe_2O_3, Nb_2O_5,$ TiO_2	0.1-6	18×10^{-2}	Inoue
MgO, MnO_2, ZnO mixtures	2-25		Ogino, Kodama, Miyake
MnO_2/ZnO	6/5	25×10^{-2}	
CaO, Co_2O_3, CuO, Fe oxide, $Li_2O, MgO,$ Mn oxide, NiO TiO_2, ZrO_2			Tretzakov
B_2O_3, BaO, CoO, Li_2O			Terazaki
TiO_2	4		Kovachev
Y_2O_3	0.1 0.5	4×10^{-2} 6×10^{-2} (?)	Imai, Watanabe, Ogawa
Y_2O_3	1	$10-20 \times 10^{-2}$	Charles, Mitoff, Morris

The effects at grain boundaries with additives may be so spectacular that changes in the bulk may be masked. Calcium, like magnesium, might be expected to substitute for aluminum and increase the concentration of mobile sodium ions, but NMR results indicated it resided in the conductive plane (7) and therefore should decrease the conductivity in proportion to its concentration. However, De Jonghe showed (21) that even 1% calcium addition increased the resistance at $300^\circ C$ by 50-100 times (Fig.4). The effect was attributed to calcium accumulating in the grain boundary region.

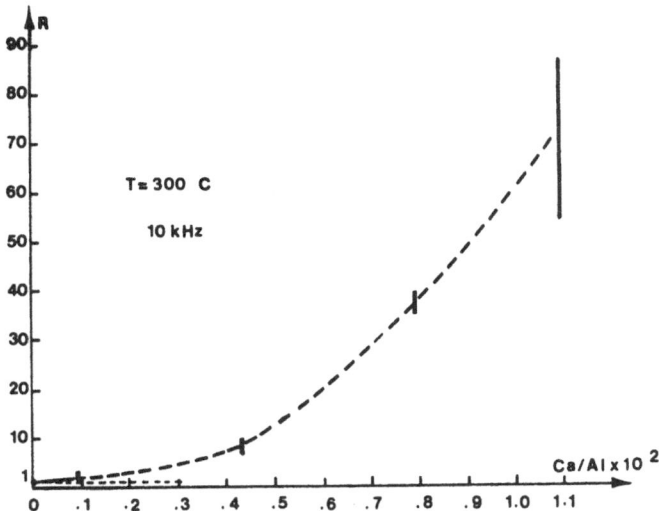

FIGURE 4:   Specific impedance ratio R $(\Omega_{doped}/\Omega_{pure})$ as a
            function of calcium content at $300°C$ and 10 kHz
            from ref.21.

        Wynn Jones and Miles studied additives with special
attention given to sintering characteristics and final
density (22,23).   They found that 1% additions of Ca, Sr,
Ti, and Zr enhanced sintering.   High conductivity was
found with β-alumina containing Ti, Zr, Li, and Mg.   They
noted that Li and Mg stabilized compositions containing up
to 9% Na_2O.   However, Ca and Sr hindered conductivity
even though highly dense material could be obtained.   Some
of the results are summarized in Table VII.

        Addition of SiO_2 and B_2O_3 to aid sintering and thereby
achieve high conductivity in polycrystalline material was
noted in the early patents on β-alumina (24).   More
recently, De Jonghe (21) showed that larger concentrations
(0.5 w/o) of SiO_2 decreased conductivity appreciably, and
SiO_2 content should be kept to about 500 ppm for best
results.   Conductivity as a function of SiO_2 was consistent with Kummer and Weber's results given in Table
VIII.   De Jonghe (21) also showed that 5% sodium-
aluminum oxide sintering additive lowered sintering

TABLE VII

Additives to Aid Sintering of β-Alumina

Additive to Al_2O_3 w/o	Sintering Temp °C	Density g/cm^3	Best Conductivity (ohm-cm)$^{-1}$ (20°C)
5.8 Na_2O	1750-1870	2.93-3.19	6.7 x 10^{-4}
8.0 Na_2O, 2.0 MgO	1670-1830	2.96-3.19	27.0 x 10^{-4}
7.0 Na_2O, 1.1 CaO	1690-1790	3.05-3.07	0.1 x 10^{-4}
9.0 Na_2O, 0.7 Li_2O	1620-1720	3.03-3.14	5.0 x 10^{-4}
6.7 Na_2O, 2.4 ZrO_2	1690-1790	3.19-3.22	12.5 x 10^{-4}
6.8 Na_2O, 1.0 SrO	1690-1790	3.00-3.07	0.6 x 10^{-4}

From: I. Wynn Jones and L. J. Miles
Proc. Brit. Ceram. Soc., 19, 161 (1971).

temperature to 1600-1650°C, and conductivity increased from 1.0×10^{-2} (ohm-cm)$^{-1}$ at 300°C for undoped β-alumina to 2.4×10^{-2} (ohm-cm)$^{-1}$. An added heat treatment at 1375°C increased the conductivity to 3.8×10^{-2} (ohm-cm)$^{-1}$ as shown in Fig.5.

Hot-pressing allows high density β-alumina to be achieved at lower temperatures and without special additives. Duncan, for example, showed that $ZrSiO_4$ used to aid sintering of β-alumina was not needed when hot-pressing was employed (25). Clendenen and Olson (26) used hot-pressing to obtain β-alumina with a conductivity of $7-9 \times 10^{-2}$ at 290°C. Hot-pressed material is significantly oriented and presents an additional variable to be considered since β-alumina is so anisotropic.

The ideal approach to study the effect of additives on bulk conductivity of β-alumina would be to use single crystals. Other than the ion exchange studies this approach has not been taken. A recent paper on edge-defined film-fed growth of β-alumina and Mg-doped β-alumina, including β″-alumina single crystals by Morrison et al. (27) may open the door to these studies.

TABLE VIII

Effect of SiO_2 and B_2O_3 on β-Alumina Conductivity

w/o	SiO_2		Investigators
	σ (25°C)	σ (300°C)	
0.05	13.3 x 10^{-4}	3.6 x 10^{-2}	Kummer and Weber
		5.5 x 10^{-2}	De Jonghe
0.3	1.7 x 10^{-4}	1.7 x 10^{-2}	Kummer and Weber
0.5		0.2 x 10^{-2}	De Jonghe

w/o	B_2O_3		
	Other Additives, w/o	σ (25°C)	σ (300°C)
0.5-1	6.37 Na_2O, 0.69 SiO_2	0.45 x 10^{-4}	0.5 x 10^{-2}
0.16	5.75 Na_2O, 0.04 SiO_2	1.3 x 10^{-4}	1.4 x 10^{-2}
0.5	β-Alumina		3.1 x 10^{-2} (312°C)

From: J. T. Kummer and N. Weber
        U.S. Patent No. 3,404,036, October 1, 1968.

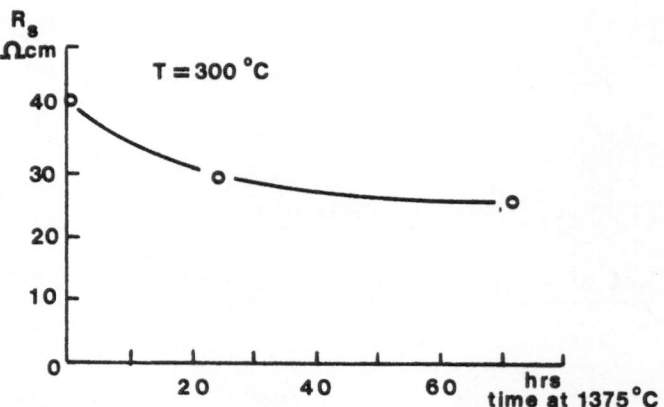

FIGURE 5: Decrease in specific resistance, R_s determined
at $300^\circ C$ and 10 kHz, of electrolyte sintered
with eutectic additive, upon subsequent
annealing at $1375^\circ C$ in air from ref.21.

β''-ALUMINA

The existence of other structures similar to β-alumina
in the $Na_2O-Al_2O_3$ system has been known for many years
(4) but only in the last few years have the structure and
conductivity properties been understood in detail.   The
primary difference between β-alumina and β''-alumina is
that the conduction plane is not a mirror plane in β''-
alumina leading to two identical Na^+ sites.   Also, β'' is
stabilized by Mg^{2+} or Li^+ and, in the case of Mg^{2+}, has
the ideal formula $Na_2MgAl_{10}O_{17}$, and both sites should be
filled.   The material is Na^+ deficient, but because it
contains a higher concentration of mobile Na^+ the conductivity of β'' is 3-5 times greater than β-alumina (4,
28-30).   One of the problems in obtaining good conductivity values for β''-alumina is that it often contains
some β-alumina.   Virkar, et al. (29) showed that after
hot-pressing, annealing at $\overline{1450}^\circ C$ eliminated any β-alumina
present and increased the conductivity.   Similarly, Whalen,
et al. (28) showed that the conductivity increased from
$\overline{10} \overline{x} 10^{-2}$ to 20×10^{-2} (ohm-cm)$^{-1}$ at $300^\circ C$ with longer
singering time at $1585^\circ C$.   Weiner (30) hot-pressed β''-

alumina at 1440°C (4260 psi), and the conductivity was
13 x 10^{-2} (ohm-cm)$^{-1}$ at 300°C initially. However, after
annealing for 22 hr at 1300°C the conductivity was 19 x
10^{-2} (ohm-cm)$^{-1}$ at 300°C and reached 40 x 10^{-2} (ohm-cm)$^{-1}$
at 300°C when annealed at 1450°C for 20 hr.

REPLACEMENT OF ALUMINUM

The gallium analogs of β-alumina, $NaGa_{11}O_{17}$ and
$KGa_{11}O_{17}$ were prepared by Foster and Stumf in 1951 (31).
A more recent study was carried out by Foster and
Scardefield (32). The only conductivity information
reported for these compounds has been the work of
Brinkhoff (33) who studied gallium substitution in β-
alumina from $MAl_{11}O_{17}$ to $MGa_{11}O_{17}$ where M was Na^+ or K^+.
The results are shown in Fig.6 and show that conductivity
for the sodium compounds was not affected appreciably by
gallium substitution while a marked decrease was observed
for the potassium compounds. Gallium is an interesting
element since Ga^{3+} may substitute for Al^{3+} while Ga^+ can
replace Na^+ (4).

The iron analogs of β-alumina are known for potassium,
$KFe_{11}O_{17}$ (34,35). The stoichiometry for $K_2O \cdot \underline{n}Fe_2O_3$ ranges

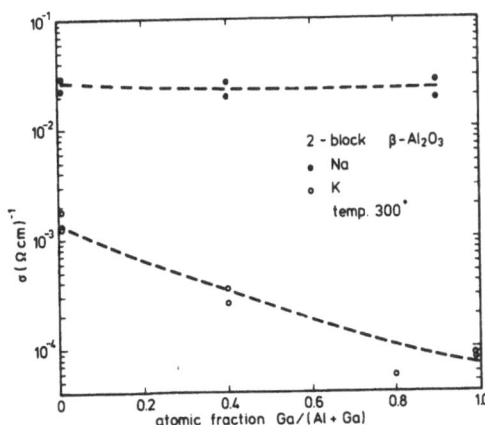

FIGURE 6:   Conductivity as a function of Ga substitution
            in 2-block β-alumina from ref.33.

from n = 5-11.   Roth and Romanczuk (34) showed that
potassium ferrite was a mixed conductor with an activation
energy of 6.1 kcal/mole for ionic conductivity and 2.2
kcal/mole for electronic conductivity.   The ionic
activation energy agreed reasonably well with the potassium
diffusion value of 7.2 kcal/mole and also agreed with the
activation energy for conduction in potassium β-alumina
(6.8 kcal/mole).   Takahashi, et al. also reported a mixed
conductivity of 0.5 (ohm-cm)$^{-1}$ at 100°C for $K_2O \cdot 6$-7 Fe_2O_3.
The possibility that an additive to β-alumina may lead to
electronic conductivity would prevent its use in any
battery application using β-alumina as the electrolyte (as
part of the anode or cathode structure it could, however,
be an advantage).   Little data exist except for the low
levels of electronic conductivity reported by Whittingham
and Huggins for silver β-alumina (3) and a study by Galli,
et al. which showed the electronic transport number to be
essentially zero for undoped, 0.5 w/o magnesium-doped,
3.8 w/o Fe_2O_3-doped, and 3.57 w/o CoO - 1.28 w/o TiO_2-
doped β-alumina (36).   More work needs to be done in this
area.

Another area in which additives may be deleterious is
strength and life.   Whalen, et al. (28) showed that as
the conductivity increased with sintering time, the fracture
strength decreased from 28,000 psi to 12,000 psi.   Fally,
et al. (37) reported that cell life was shorter for magnesium-doped β-alumina compared to undoped β-alumina.
Weiner (30) also found that degradation was greater for
β-alumina containing 0.7% Li_2O than for β-alumina containing 0.25% Li_2O.   Thus, we reach the conclusion that
additives do play several important roles in the properties
of β-alumina and can be employed to increase the conductivity several-fold from that of undoped β-alumina.
However, before a doped β-alumina is recommended for a
practical application the effect of the additive on several
other properties such as sinterability, electronic conductivity, strength, and cell life must be investigated.

ACKNOWLEDGMENT

The author thanks the National Science Foundation for
financial support of his research in this area (Grant No.
DMR 73-07507 A02) and to James Akridge for his contribution to this review.

REFERENCES

1. Y. F. Y. Yao and J. T. Kummer, J. Inorg. Nucl. Chem.,
   29, 2453 (1967).
2. M. S. Whittingham and R. A. Huggins, (R. S. Roth and
   S. J. Schneider, Ed.), Nat. Bur. Stand. (U.S.), Spec.
   Publ., No.364, 139 (1972).
3. M. S. Whittingham and R. A. Huggins, J. Electrochem.
   Soc., 118, 1 (1971).
4. J. T. Kummer, (H. Reiss and J. O. McCaldin, Ed.),
   Prog. Solid State Chem., 7, 141 (1972).
5. R. L. Cohen, J. P. Remeika, and K. W. West, J. Phys.
   (Paris), Colloq., 6, 513 (1974).
6. D. B. McWhan, S. J. Allen, Jr., J. P. Remeika, and
   P. D. Dernier, Phys. Rev. Lett., 35, 953 (1975).
7. W. L. Roth, Trans. Amer. Cryst. Assoc., 11, 51 (1975).
8. J. H. Kennedy and A. F. Sammells, J. Electrochem. Soc.,
   119, 1609 (1972).
9. J. H. Kennedy and J. R. Akridge, Electrochem. Soc.
   Meeting, Paper 84, Washington, D.C., May 2-7, 1976.
10. M. Harada, T. Ohta, and A. Imai, Japan. Patent No.75-
    05, 384, March 3, 1975.
11. A. Imai, M. Harata, Y. Ogawa, and S. Hasegawa, U.S.
    Patent No.3,671,324, June 20, 1972.
12. K. Inoue, Japan. Patent No.75-22,720, August 1, 1975.
13. I. Ogino, T. Kodama, and Y. Miyake, Japan. Patent No.
    73-93,609, December 4, 1973.
14. Yu. D. Tret'yakov, Vestn. Mosk. Univ., Khim., 15, 643
    (1974).
15. Yu. D. Tret'yakov, Tezisy Dokl.-Vses. Soveshch.
    Elektrokhim., 5th 1974, 1, 98 (1975).
16. M. Terazaki, Japan. Patent No.75-00,436, January 9,
    1975.
17. I. Kovachev, Teor. Tekhnol. Spekaniya, Dokl.,
    Mezhdunar. Kollok., 2nd, 249 (1974).
18. K. Nishimura, M. Hasegawa, and M. Takagi, Japan.
    Patent No.73-43,646, December 20, 1973.
19. S. Imai, R. Watanabe, and Y. Ogawa, Japan. Patent
    No.73-31,731, October 1, 1973.
20. R. J. Charles, S. P. Mitoff, and W. G. Morris, U.S.
    Patent No.3,607,435, September 21, 1971.
21. L. C. De Jonghe, EPRI Rept.252, July 1975.
22. I. Wynn Jones and L. J. Miles, Proc. Brit. Ceram.
    Soc., 19, 161 (1971).

23. I. Wynn Jones and L. J. Miles, Brit. Patent No.
     1,375,167, November 27, 1974.
24. J. T. Kummer and N. Weber, U.S. Patent No.3,404,036,
     October 1, 1968.
25. J. H. Duncan, Brit. Patent No.1,287,571, August 31,
     1972.
26. R. L. Clendenen and E. E. Olson, U.S. Patent No.
     3,795,723, March 5, 1974.
27. A. D. Morrison, R. W. Stormont, and F. H. Cocks,
     J. Am. Ceram. Soc., $\underline{58}$, 41 (1975).
28. T. J. Whalen, G. J. Tennenhouse, and C. Meyer, J. Am.
     Ceram. Soc., $\underline{57}$, 497 (1974).
29. A. V. Vrikar, G. R. Miller, and R. S. Gordon, Amer.
     Ceram. Soc. 28th Pacific Coast Regional and Nuclear
     Div. Meeting, 2-FC-75P, October 29-31, 1975.
30. S. A. Weiner, RANN Rept., Contract NSF-C805(AER-73-
     07199), July 1975.
31. L. M. Foster and H. C. Stumf, J. Am. Chem. Soc., $\underline{73}$,
     1591 (1951).
32. L. M. Foster and J. E. Scardefield, J. Electrochem.
     Soc., $\underline{123}$, 141 (1976).
33. H. C. Brinkhoff, J. Phys. Chem. Solids, $\underline{35}$, 1225
     (1974).
34. W. L. Roth and R. J. Romanczuk, J. Electrochem. Soc.,
     $\underline{116}$, 975 (1969).
35. T. Takahashi, K. Kuwabara, and Y. Kase, Denki
     Kagaku, $\underline{43}$, 273 (1975).
36. R. Galli, P. Longhi, T. Massini, F. A. Tropeano,
     Electrochim. Acta, $\underline{18}$, 1013 (1973).
37. J. Fally, C. Lasne, Y. Lazennec, and P. Margotin,
     J. Electrochem. Soc., $\underline{120}$, 1292 (1973).

THE INTER- AND INTRA-GRANULAR RESISTIVITY OF BETA-ALUMINA

R. W. Powers

General Electric Research and Development Center

P. O. Box 8, Schenectady, New York 12301

The electrical properties of polycrystalline beta-alumina involve two features not encountered with single crystals--tortuosity and grain boundaries. The former, the elongation of the conduction path arising from the random orientation of the individual grains, will only be touched upon in this paper. Rather, the main concerns are with a model for separating measured impedances into intergranular (grain boundary) and intragranular (crystal) components, with practical methods for determining the parameters characterizing these components, and with variations in parameter values with composition.

THE MODEL

The equivalent circuit shown in Figure 1 has been discovered and rediscovered over the years by a number of different workers in the course of studies on a wide variety of polycrystalline conductors such as composition carbon resistors, nickel zinc ferrites, calcium-stabilized zirconia, and beta-alumina. (1-6) Their approaches to this model have differed widely and the results of their studies are scattered throughout the scientific literature. In Fig. 1, r_c represents the resistivity of the grain interiors while r_b represents the extra resistivity in ceramic specimens associated with intergranular material. The capacitor C_b models the dielectric properties of the intergranular component.

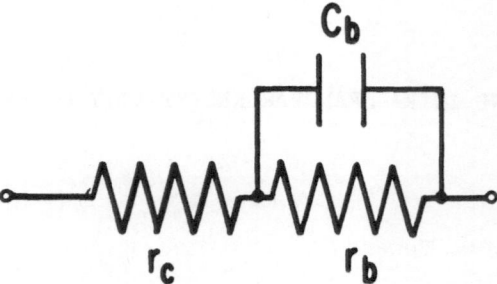

Fig. 1   Simplified model for the electrical properties of polycrystalline beta-alumina

It is now clear that the utility of the model has nothing to do with the nature of the conductor. The model works in that limiting case where the intergranular resistance shunting a grain greatly exceeds that of the grain interiors. Figure 2 may be helpful in understanding the point under discussion. In a polycrystalline conductor, each grain is, of course, completed surrounded by grain boundaries except for those grains at external surfaces or those bounding pores. Actual grain boundaries might be resolved into those aligned perpendicular to the field (hatched on the figure) and others parallel to the field (cross-hatched on the figure). In general, a suitable equivalent circuit would need take account of both grain boundary resistance in series and in parallel with that of intragranular material.   However, in the limiting case where the grain boundary shunting resistance is much larger than the crystal resistance, the former can be neglected, and we obtain the essential features of the equivalent circuit shown in Figure 1. The limiting case under discussion in this paper is the extreme opposite of the more familiar one encountered in some of the alkali metal and silver halides where shorting of grain interiors by grain boundaries is observed at lower temperatures. (7)

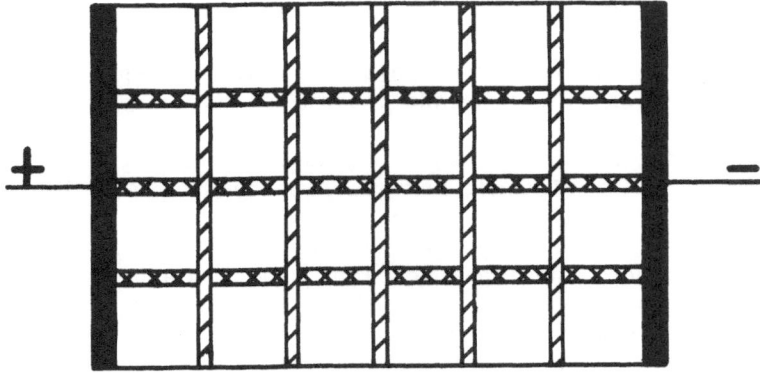

Fig. 2   Schematic showing basis of simplified model

In their pioneering study on beta alumina, N. Weber
and J. T. Kummer found that the conductivity of single
crystals was very much greater than that of polycrystalline
ceramic fabricated from single crystals. (8)   They reported
single crystal resistivity values of about 30 and 3.5 ohm
cm at room temperature and 300° respectively, while comparable values for a specimen of ceramic were 250 and 18 ohm
cm.   They attributed the difference to interfacial resistance between crystals in the ceramic.   Other findings such
as the higher activation energy observed for the conductivity
of polycrystalline specimens and the curvature noted on
Arrhenius plots supported this interpretation. (9, 10)

PROCEDURES FOR OBTAINING PARAMETERS CHARACTERIZING THE
SIMPLIFIED MODEL

Since the procedures used for analyzing measurements
of the electrical properties of beta-alumina depend somewhat
on the form of these measurements, the experimental

arrangement will be described first. A simple four-point AC
arrangement as shown in Figure 3 has been used to eliminate
complications from electrode polarization. The specimen is
placed in series with a decade resistance box and the two
are energized by a sine wave generator. The voltage wave-
form appearing across the inner contacts to the specimen and
that across the resistance box are displayed on a dual beam
oscilloscope. The decade resistance box is adjusted so that
the two voltage amplitudes are the same. In this way, the
impedance between the inner contacts can be measured to an
accuracy of about one percent with a little care. The phase
relationship between the current and potential can be
obtained from the relative displacement of the two waveforms.
Specimens were usually in the form of cylindrical tubes, 1
cm in diameter, about 8 cm in length, and of 1 mm wall
thickness. Suitability of contacts to specimens was determined by the phase relationship between current and potential
at low frequencies, e.g., 10 hz. They seldom differed in
phase by more than a couple degrees. Furthermore, the
impedance data were strictly ohmic. A more detailed
description of experimental techniques is given elsewhere (6).

Impedance measurements have been used in two very
different ways in this Laboratory to separate the contribution of grain boundaries from that of intragranular

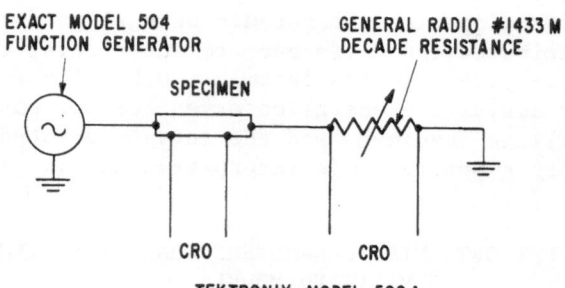

Fig. 3   Four point arrangement for measuring components of
         electrolyte impedance

material. The first and perhaps more obvious technique is based on the frequency dependence of the impedance. The second method is related to the fact that the parameter r_b depends on the grain size. Measurements are made on a series of specimens of widely different r_b values and an extrapolation is made to zero contribution from r_b to obtain the crystal component of resistivity, r_c.

For a specimen whose electrical properties can be described by the equivalent circuit shown in Figure 1, the frequency dependence of the impedance, z, is given by the expression

$$z = (r_c + r_b) \left[\frac{1 + (\frac{r_c}{r_c + r_b})^2 (r_b C_b \, 2\pi f)^2}{1 + (r_b C_b \, 2\pi f)^2} \right]^{1/2} \tag{1}$$

There are two limiting plateau values of impedance. The low frequency value amounts to the sum of r_c and r_b. The high frequency limit is r_c. We can define a cross-over frequency, f_c, as that at which the impedance equals the geometric mean of the low and the high frequency limiting impedances. The parameter C_b can be obtained using the relationship

$$C_b = \frac{1}{2\pi f_c r_b} \left[\frac{r_c + r_b}{r_c} \right]^{1/2} \tag{2}$$

Impedance and phase measurements on a beta-alumina ceramic containing 7.2 percent soda and 1.0 percent magnesia, 1.0 percent zirconia, and 0.5 percent yttria have been published by S. P. Mitoff and this writer. (6) Data on one specimen at 52°C are presented on Figure 4 to show the adequacy of the simplified model in describing the electrical properties of beta-alumina ceramics. To be noted is the fact that the measured frequency interval between the low and the high frequency impedance plateaus is considerably wider than that calculated using a single set of (r_b, C_b) values. Similarly, the dispersion of the measured phase angles occurs over a wider frequency range than expected for a single (r_b, C_b) set. These discrepancies are those expected if the grain boundaries are not homogeneous, i.e., if they all do not have the same resistive and dielectric properties. Of course, a better fit to experimental data could be made by

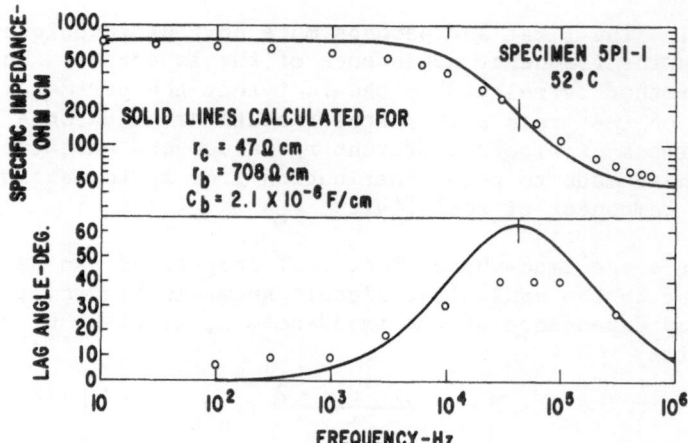

Fig. 4   Impedance and phase angle vs. frequency curves
         showing deviations from behavior of simplified
         model

dividing the total grain boundary resistivity, r_b, into
components, r_{b1}, r_{b2}, r_{b3}, etc. each shunted with its
respective capacitor, C_{b1}, C_{b2}, C_{b3}, etc.

Impedance measurements at several temperatures on this
same specimen, No. 5P1-1, are presented on Figure 5. The
low frequency plateau is better defined at higher temperatures whereas the high frequency plateau is more closely
approached at lower temperatures. Nonetheless, $(r_c + r_b)$
values could still be obtained in many cases by extrapolation of low frequency z vs. f plots to f → 0 and r_c values
from extrapolation of high frequency z vs. 1/f plots to
1/f → 0. The variation of both r_c and r_b with temperature
is presented on Figure 6 in the form of log T/r vs. 1/T
plots.

The values of r_c from two different specimens are seen
on Figure 6 to lie along the same Arrhenius line to a very
good approximation. The temperature variation of r_c
corresponds to an activation energy of 4.4 kcal/mole, close
to but nonetheless significantly higher than the value of
3.8 kcal/mole found by both Yao and Kummer and by
Whittingham and Huggins for single crystal beta alumina.
(11, 12) The causes of this discrepancy will be discussed

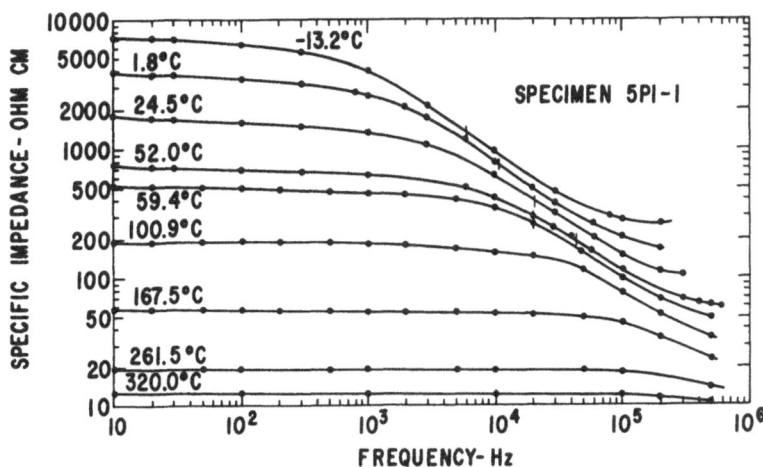

Fig. 5   Specific impedance vs. frequency curves for
          specimen 5P1-1

below.   On the other hand, r_b values on the two specimens
differ significantly as the grain sizes were not the same.
Nonetheless, the temperature dependence of the separate
sets of r_b values is identical and corresponds to an
activation energy of 6.6 kcal/mole, a value much higher
than that for r_c.   In contrast to r_b and r_c, C_b values for
both specimens were found to be independent of temperature.

The variation of r_b at $25^{\circ}C$ with the reciprocal of the
grain size is brought out in Table 1.   While the author
feels that the correlation is sufficiently good to indicate
an association between the electrical parameter r_b and
grain boundaries, the relationship deviates significantly
from the ideal simple proportionality.   A number of causes
for deviations can be cited.   Quantitative ceramography
with beta-alumina is difficult mainly because all boundaries
do not etch at the same rate.   Uncertainties in the quantity,
number of grain boundaries intersections per unit length,
are estimated to be about ± 25 percent.   Furthermore, it is
very difficult to change the grain size without changing
the porosity in addition.   In a recent study on zirconia,

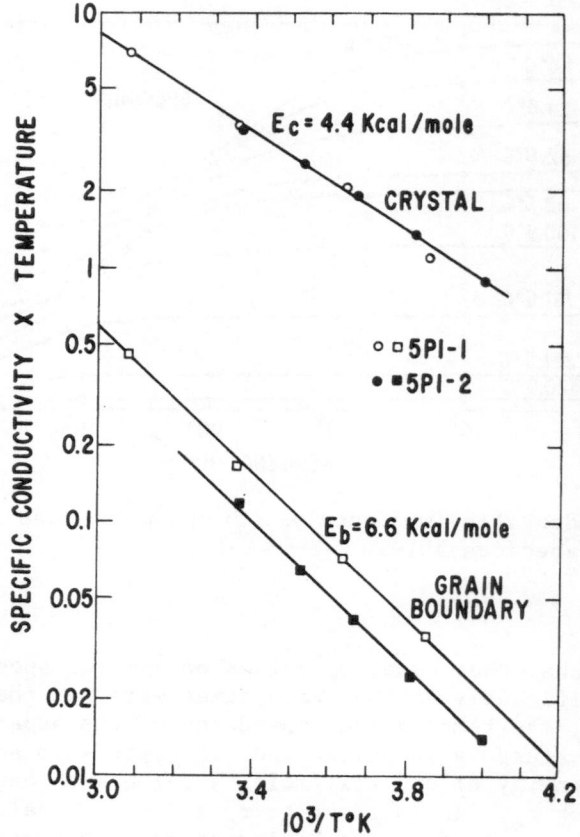

Fig. 6   Arrhenius plots for components of resistivity

the parameter r_b was found to correlate better with
porosity than with grain size in poorly sintered specimens
containing 5-25 percent porosity. (13)   This author's casual
observations that r_b can vary at room temperature by tens of
Kohm cm from specimen to specimen in poorly sintered beta-
alumina ceramics would tend to confirm this claim.   While
porosity was less than five percent in nearly all specimens
on which data are given in this paper, still variations in

Table 1

Correlation of Parameter r_b with Number of Grain Boundary
Intersections per Unit Length

Specimen	r_b (kohm cm)	Intersections per cm $\cdot$ 10^{-3}	r_b/Intersections per cm
5P1-1	1.33	0.81	1.64
5P1-2	1.88	2.25	0.84
5P1-3	2.2	2.75	0.80
5P1-4	6.9	3.5	1.97

porosity are likely to affect r_b values to some extent in
specimens of small grain size. On the other hand, when
specimens of beta-alumina are sintered close to theoretical
density, exaggerated grain growth often occurs. A few
grains may grow to dimensions 100 times the average grain
size. Such grains can seriously perturb current flow in
their vicinity and present a problem as to proper averaging
of the reciprocal grain size.

In ceramics without the additive yttria, values for the
parameter C_b are much smaller with the result that the
cross-over frequency is moved toward higher frequencies.
r_c values for such ceramics are not attainable at any temperature. It might seem obvious that use of instrumentation
with a higher frequency capability would solve these problems.
Dr. G. C. Farrington in this Laboratory developed a pulse
technique with a rise time near 10^{-9} sec. (14) However,
effects which appear as extra capacitances arise in the
elapsed time region around 10^{-8} sec. Farrington has interpreted these as evidence for localized motion of sodium ions
between non-equivalent sites in the conduction plane. However
interesting this finding, it practically precludes use of
higher frequencies to determine r_c values. (15) Another
procedure had to be developed.

As indicated above, the temperature comparison method for obtaining the parameters r_b and r_c involves measurements on a set of specimens with widely varying r_b values. Impedance measurements must be made over a range of temperatures usually from room temperature to above $400^{\circ}C$. However, they need be carried out only at low frequencies, e.g., 10 hz. Consequently, no information about C_b is obtained. A major advantage of this technique is that knowledge of the relationship between r_b and the grain size is not required.

All the major assumptions on which the technique is based are supported by experimental findings presented above. These assumptions are:

1. that $r = r_c + r_b$.

2. that all specimen in the set are assumed to have the same r_c values.

3. that the temperature variation of r_c is assumed to follow the relationship $T/r_c = \sigma_o \exp(-E_c/RT)$. Here, σ_o is the pre-exponential constant; E_c, the activation energy; R, the gas constant; and T is the absolute temperature.

4. that, whereas r_b values will vary from specimen to specimen within the set, they are assumed to have the same temperature dependence. E_b, the activation energy associated with r_b, can vary with temperature, however.

These assumptions lead to the equation

$$r_T = r_{CT} + (T/L) \exp\left[\frac{E_b}{R}\left(\frac{1}{T} - \frac{1}{L}\right)\right] r_{bL} \qquad (3)$$

Here T refers to a temperature, e.g., $300^{\circ}C = 573.2^{\circ}K$, at which measured values of the resistivity, r_T, are available. The crystal component of the resistivity at T is indicated as r_{CT}. L is a lower or reference temperature and throughout this paper is always taken to be $26.8^{\circ}C$ or $300^{\circ}K$. The grain boundary component of resistivity at L is r_{bL}.

Consider now a plot in which resistivity measurements

at T for a set of specimens are compared against measurements at the temperature L. Points should lie along a
straight line if the assumptions stated above are valid.
A schematic plot is shown on Figure 7 to illustrate the
geometric significance of equation (3). Assume for the
moment that r_c is known at the temperature L. A vertical
line, $r_L = r_{cL}$, is erected as shown by the dashed line on
Figure 7. The intersection of this line with that through
the points establishes the value of r_c at the temperature
T, common to all the specimens. For example, for the
hypothetical specimen whose (r_T, r_L) values are indicated
by the symbol ▲ on the figure, r_T is the line segment ac,
of which the r_{cT} component is bc and the r_{bT} component is
ab. Equation (3) really implies that, if a value of r_c is
known at one temperatures, values at other temperatures can
be obtained. Plots such as Figure 7 have been called resistivity distribution plots because they indicate at a glance
the distribution of the measured resistivity r between its
r_c and r_b components.

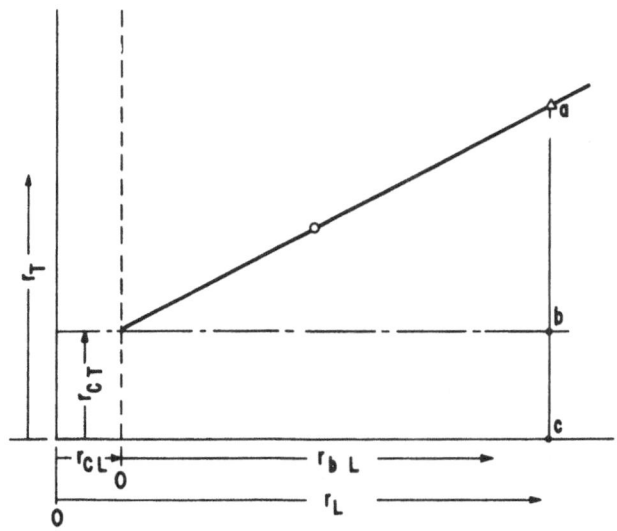

Fig. 7 Schematic resistivity distribution plot

The grain boundary activation energy, E_b, averaged
between the temperatures L and T, can be obtained from the
slope, m, of the line through the data points. E_b is found
from the expression

$$E_b = \frac{R \ln (mL/T)}{(1/T - 1/L)} \tag{4}$$

It is more useful to display resistivity distribution
plots for several temperatures on the same figure. Such plots
are shown in Figure 8 for ceramic containing 7.2% soda.
Actual resistivity measurements were made at room temperature
and at about 75°, 125°, 175°, 225°, 275°, 325°, 375°, and 415°C. The
values at the even temperatures shown on plots were obtained
by interpolation. Data points were fitted to the straight
lines shown using the least squares method. Most data points
for 150°C lie off the graph and consequently are not shown on
Figure 8. The activation energy for the grain boundary
resistivity averaged between 26.8°C and the temperatures in
question were calculated and are shown on Figure 8. For this
ceramic, they only range from 7.23 to 7.48 kcal/mole.

Actually, r_c values at 26.8°C are not known but can be
determined using a minimum variance principle. A trial value
of $r_{c26.8°C}$ is assumed and hence r_c values at other temperatures are derived as explained above. Such temperatures are
usually 150°, 200°, 250°, 300°, 350°, and 400°C. The set of r_c
values at these temperatures and at 26.8°C are fitted to the
straight line

$$\ln T/r_{cT} = I + M (1/T)$$

using a least squares procedure. Here the ordinate intercept
I equals $\ln \sigma_0$ and the slope M equals $(-E_c/R)$. The goodness
of fit of the data points to the best straight line is
measured by the variance. This process is repeated for other
trial values of $r_{c26.8°C}$ until a minimum variance is obtained. The trial value of $r_{c26.8°C}$, together with the r_c values
derived at other temperatures corresponding to this minimum
variance is assumed to be the best set of crystal resistivity
values. A more detailed description of the temperature
comparison method is given elsewhere. (16)

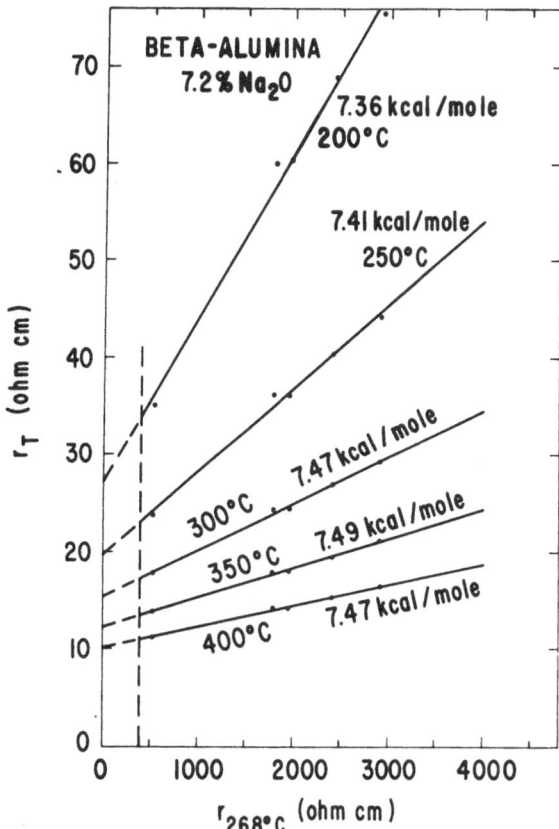

Fig. 8 Resistivity distribution plots for beta-alumina ceramic containing 7.2 percent Na_2O

THE VARIATION IN PARAMETER VALUES WITH CERAMIC COMPOSITION

The temperature comparison method has been used to obtain resistivity parameter values for a number of different ceramic compositions. One composition studied was that one on which parameter values were also available using the first method described in this paper. The activation energy for r_c was found to be 4.4 kcal/mole by both methods. Those for r_b differed only slightly: 6.6 kcal/mole by the first method and 6.4 by the temperature comparison method. Since activation energies for r_b by this second method are averaged values, values at temperatures between 150° and $400^\circ C$ were calculated and an extrapolation was made to $10^\circ C$, the midpoint of the temperature interval over which measurements were carried out by the first technique. The r_c values at $26.8^\circ C$ were found to be 77 and 103 ohm cm by the first and second methods respectively. This agreement in parameter values is considered very satisfactory.

Resistivity parameter data for several beta-alumina ceramic compositions are presented in Table 2. The variation in the crystal resistivity with temperature for some of these compositions is shown on Figure 9 in the form of log T/r_c vs. $1/T$ plots. The soda content of the ceramic is seen to have very marked effects on the resistivity of the crystal component. An increase from 6.2 to 7.2 percent Na_2O concentration leads to an increase in both the pre-exponential constant and the activation energy with the result that the resistivity at 7.2 percent Na_2O is less than that at 6.2 percent above $150^\circ C$ and greater below this temperature. A further increase in soda concentration from 7.2 to 8.4 percent has an almost negligible effect. The presence of magnesia at the 1.0 percent level also gives rise to an increase in both the pre-exponential constant and the activation energy.

The activation energy found for r_c in ceramic with 6.2 percent Na_2O, 3.88 kcal/mole, does not differ greatly from the single crystal values of 3.77 kcal/mole found by Yao and Kummer and the value 3.79 reported by Whittingham and Huggins. If the r_c value at $25^\circ C$ for ceramic with 6.2 percent Na_2O is divided by 1.34, the tortuosity factor as calculated recently by S. P. Mitoff in this Laboratory, an equivalent single crystal resistivity of 205 ohm cm is obtained. (17) This can be compared with values of 75 ohm cm reported by Whittingham and Huggins and 33 ohm cm, by Yao and Kummer. The result of the present study exceeds that of Whittingham and Huggins by roughly the same factor as that of Yao and Kummer is lower. Admittedly, data on resistivity distribution plots are much more scattered for the ceramic with 6.2 percent Na_2O than those from any other ceramic composition--a fact consistent, however, with the rapid change in resistivity parameters between 6.2 and 7.2 percent soda. Still the differences are much greater than any likely measurement error. In this author's opinion, these result from unknown subtle differences that arise in

Table 2

Resistivity Parameters for Various Beta-Alumina Ceramics

Parameter	Low soda Ceramic	Standard Ceramic	Soda-enriched Ceramic	1% MgO Ceramic
Composition % Na_2O % other additives	6.2	7.2	8.4	7.2 1.0%MgO
r_c, crystal (intragrain) resistivity at 300°C-ohm cm	22.8	17.0	17.0	11.0
r_c at 26.8°C	266	391	418	327
σ_o, pre-exponential factor	761	2151	2294	4288
E_c, activation energy for r_c, kcal/mole	3.88	4.73	4.81	5.04
E_b, grain boundary activation energy averaged between 26.8°C and				
150°C - kcal/mole	7.19	7.23	7.48	7.60
200	7.49	7.36	7.67	7.78
250	7.87	7.41	7.89	7.92
300	8.5	7.47	8.11	7.99
350		7.49	8.36	7.99
400		7.47	8.70	7.99

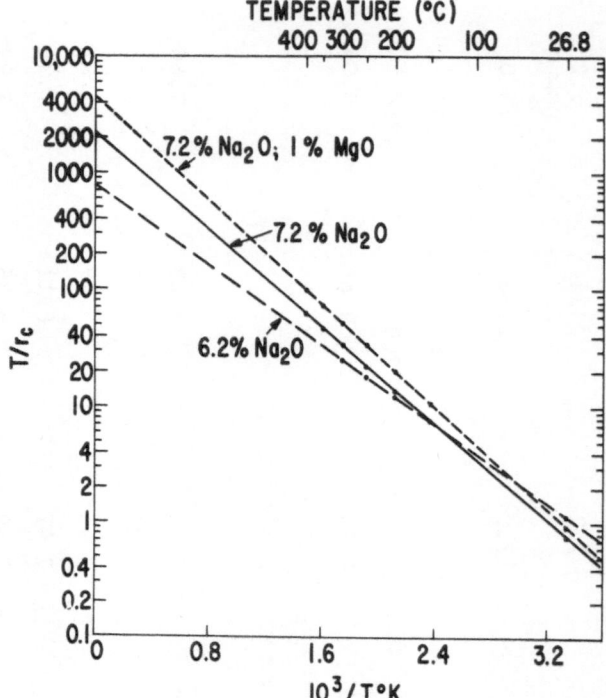

Fig. 9   The variation of r_c with temperature for three beta-
alumina ceramic compositions

specimen preparation. The single crystals, on which results
were cited above, were large grains removed from cast blocks
of a commercial beta-alumina designated as Monofrax H by
the manufacturer, the Carborundum Corporation. They contain
about 6.0 percent Na_2O. On the other hand, specimens used
in the present study were made from a semi-commercial beta-
alumina powder designated as XB-2 by Alcoa, the manufacturer.
The soda content of this powder is 7.2 percent. The powder

was formed into tubes by electrophoretic deposition and sintered at temperatures between 1700 and 1775°C. After sintering, specimens were heated in air at 1575°C for about one day to reduce the soda content to approximately 6.2 percent.

In Table 2, the averaged activation energies for the grain boundary resistivity are shown to increase with temperature. While the increase may be of marginal significance in the standard ceramic and that with 1% MgO, it is regarded as real for the other two ceramics. A change in an apparent activation energy with temperature is often attributed to heterogeneities in the rate controlling process. In the present case, these might be compositional variations within the grain boundaries. It is interesting that the increase of activation energy with temperature is least in the standard ceramic with 7.2 percent Na_2O, one in which no additives were incorporated and in which the soda content was not altered from that of the as-received powders. The incorporation of additives gives rise possibly to a small increase in activation energies with temperature. However, the largest effects appear associated with alterations in the soda content from that of the as-received powders. Additional work will be needed to determine how sensitive are values for the activation energies for grain boundary resistivities to details of specimen preparation.

ACKNOWLEDGEMENTS

The author is grateful to R. N. King and E. Szymalak for experimental assistance with the preparation of specimens. Discussions with S. P. Mitoff were very helpful.

REFERENCES

1. O. S. Puckle, The Wireless Engineer, 12, 303 (1935)

2. F. E. Terman, Radio Engineers Handbook, p. 42, McGraw-Hill, New York, 1943

3. C. G. Koops, Phys. Rev., 83, 121 (1951)

R.W. POWERS

4. J. E. Bauerle, Phys. Chem. Solids, 30, 2657 (1969)

5. S. P. Mitoff, chapter in Fast Ion Transport in Solids,
    W. vanGool, Editor, North Holland American Elsevier
    Publishing Co., New York (1973)

6. R. W. Powers and S. P. Mitoff, J. Electrochem. Soc., 122,
    226 (1975)

7. W. G. Johnston, Phys. Rev., 98, 1777 (1955)

8. N. Weber and J. T. Kummer, Power Source Conference
    Proc., Red Bank, N.J., 1967, p. 37

9. A. Imai and M. Harata, Abstract No. 277, p. 673,
    Electrochemical Society Extended Abstracts, Spring
    meeting, Los Angeles, Calif., May 10-15, 1970

10. I. Wynn Jones and L. J. Miles, Proc. Brit. Ceram.
    Soc., No. 19, p. 161 (1970)

11. Y. F. Yao and J. T. Kummer, J. Inorg. Nucl. Chem., 29,
    2453 (1967)

12. M. S. Whittingham and R. A. Huggins, J. Chem. Phys.,
    54, 414 (1971)

13. M. V. Inozemtsev, M. V. Perfilev, and A. S. Lipilin,
    Elektrokhimiya 10, 1471 (1974)

14. G. C. Farrington, J. Electrochem. Soc., 121, 1314 (1974)

15. G. C. Farrington, J. Electrochem. Soc., in press

16. R. W. Powers, manuscript to be submitted to J. Am.
    Ceramic Soc.

17. S. P. Mitoff, manuscript to be submitted to J. Am.
    Ceramic Soc.

Nuclear Relaxation and Barrier Height Distribution in Na β-Alumina.

R. E. WALSTEDT, R. DUPREE*, AND J. P. REMEIKA, Bell Laboratories.--Studies of nuclear relaxation times T_1 of ^{23}Na in several specimens of Na β-alumina give results which vary markedly among specimens and are apparently correlated with the method of crystal preparation. In particular, melt-grown crystals (Union Carbide Co.) exhibit a distribution of barrier heights $g(U)$ for hopping motion centered below the measured bulk value, U_{bulk}, whereas flux-grown material gives a $g(U)$ centered as much as 40% higher and well above U_{bulk}. Gaussian models for $g(U)$ ($\propto \exp[-U-U_0)^2/\Delta U^2]$) yield widths $\Delta U \sim 0.5 U_0$, where U_0 is the peak value. Attempt frequencies are consistently an order of magnitude lower than the infrared and Raman scattering peak values for vibrational motion. This analysis and the implication of our results will be discussed.
*On leave from University of Warwick, Warwick, England.

NMR Study of Sodium Ion Motion in β"-Alumina.

W. C. BAILEY AND H. S. STORY, SUNY at Albany AND W. L. ROTH, General Electric Corporate Research and Development.--A study has been made of the temperature dependence of the nuclear magnetic resonance (NMR) of ^{23}Na(I=3/2) in polycrystalline Li_2O stabilized β"-alumina. The spectra correspond to quadrupolar perturbation of the central magnetic transition. The features of the spectra are consistent with a model which assumes planar motion of sodium ions among lattice sites related by threefold symmetry.[1,2] Within the framework of this model, measurements were made of the quadrupole interaction parameters eQV_{zz}/h and η. At room temperature, eQV_{zz}/h is 2.98 MHz with variations of no more than 3% over the temperature range studied. η on the other hand, decreases sharply with increasing temperature from 0.98 at -94C, to 0.37 at room temperature, to 0.21 at 115C.

At temperatures below -94C, η appears to continue increasing, but the spectral features have become very broad and weak, making it difficult to extract values. It may be that the frequency of the motion is becoming comparable to the quadrupole splitting frequency, yielding broad resonance lines and signaling the onset of a "static" spectrum.

Although difficult to measure quantitatively, line narrowing characteristics of ionic motion is clearly evident with increasing temperature.

The NMR results are consistent with a model in which the motion is among the vertices (sodium sites) of a planar honeycomb network, and in which diffusion occurs via a vacancy mechanism. The presence of vacancies destroys the local threefold symmetry of the occupied sodium sites, yielding a non-zero average $\eta \cdot \eta$ increases with decreasing temperature as the frequency of the motion decreases because the averaging becomes less complete.

[1]I. Chung, H. S. Story, and W. L. Roth, J. Chem. Phys. 63, 4903 (1975).

[2]J. P. Boilot, L. Zuppipoli, G. Delplanque, and D. Jerome, Phil. Mag., 32, 343 (1975).

Sodium NMR Study of the Effects of Additives in Sodium Beta-Alumina Ceramic Samples. I. CHUNG AND H. S. STORY, State University of New York at Albany, AND W. L. ROTH, General Electric Corporate Research and Development.--Some microscopic effects of various additives in sodium beta-alumina have been observed in the ^{23}Na NMR spectrum. Polycrystalline ceramic samples with varying fractions of the additive such as magnesium, nickel, calcium, lithium, and yttrium were studied.

Two effects on the NMR spectrum were noted. Generally, increasing the additive fraction increases the nuclear quadrupole coupling constant, indicating that the conduction plane is directly affected. The resonance linewidth is also affected; the addition of magnesium, nickel, and lithium enhances the motional narrowing of the sodium resonance spectrum, while calcium broadens the line.

We interpret these results in terms of the removal of defects in the conduction plane. It appears that some additives may remove defects which are partially blocking the conduction pathways in pure beta-alumina.

Calculation of the Phonon Spectra of β-Alumina.* WILLIAM Y. HSU, University of Illinois at Urbana-Champaign. --The phonon spectra and density of states of β-alumina ($1.3\ M_2O \cdot 11\ Al_2O_3$) have been computed assuming that: (1) only the dynamics of the monovalent metal ions M^+ need be treated, the other ions are approximated by an elastic potential derived from the calculation of Wang, Gaffari and

Choi[1]; (2) the screened Coulomb interaction among M^+ ions
in a single layer is taken into full account, but different
layers are assumed to be non-interacting; and (3) overall
charge neutrality is maintained by a uniform, negatively-
charged background. The equations of motion, within the
harmonic approximation, are obtained by standard methods[2]
and the dynamical matrix corresponding to the Coulomb interactions is treated exactly by a generalization[3] of the
Ewald method.[2] Two ideal lattice structures are considered:
(1) the stoichiometric structure; and (2) an extended superlattice structure. Allowance for non-ideal structures is
made by using a self-consistent-perturbation method; disorder
is treated as a set of randomly distributed static defects,
the defects-induced phonon self-energy is assumed to be
independent of momentum and effects of localized phonon
modes are ignored. The observed infrared reflection spectra[4]
and Raman spectra[5] are in close correspondence with the
computed vibrational density of states and features in these
optical spectra are thus identified and ascribed to specific
ionic normal modes.

*Research supported by the National Science Foundation under
 grant NSF-DMR-72-03026.
1J. C. Wang, M. Gaffari and S. Choi, J. Chem. Phys. 63, 772
 (1975).
2See, for example, W. Cochran in Phonons in Perfect Lattices
 and in Lattices with Point Imperfections, ed. by R.W.H.
 Stevenson (Plenum, New York, 1966), p.61.
3C. Clark, Phys. Rev. 109, 1133 (1958) and references therein.
4S. J. Allen, Jr., F. DeRosa and J. P. Remeika, Bull. Am. Phys.
 Soc. 21, 284 (1976) and private communication.
5C. H. Hao, L. L. Chase and G. D. Mahan, to be published and
 private communication.

The Thermal Conductivity of β-Alumina.* P. J.
ANTHONY AND A. C. ANDERSON, University of Illinois.--The
thermal conductivity of Na, Ag, Li, and K β-alumina single
crystals have been measured from 0.1 to 100 K in an attempt
to obtain information about the low frequency excitation
spectra of the ions. All but one sample was measured with
the heat flow perpendicular to the c-axis. The Na data,
taken both parallel and perpendicular to the c-axis, indicate
that the conductivity is isotropic to within experimental
scatter. Near 10 K, the thermal conductivity of K β-alumina

is a factor of 10^2 larger than for Li. The results can be fit assuming boundary scattering of the thermal phonons and a temperature dependent mean free path for the phonons very similar to that used to explain the thermal conductivity of glass.

*Supported by NSF Grant DMR 72-03026.

A Collective Ionic Model for Na β-Alumina--Evidence From Conductivity Measurements Between 10^7-10^{13} Hz.

U. STROM, P. C. TAYLOR, S. G. BISHOP, T. L. REINECKE AND K. L. NGAI, Naval Research Laboratory.--It has recently been proposed[1] that the ionic conductivity in β-aluminas involves a broad distribution of low energy modes and associated relaxation times. It is suggested that this distribution of modes arises from the strong ion-ion interactions known to exist in the β-aluminas. The model predicts features which are similar to those seen in amorphous materials: in particular, (i) a low temperature specific heat enhanced above that expected from T^3 Debye behavior; (ii) an a-c conductivity which, at temperatures sufficiently low to inhibit diffusion, rises as $\sigma \sim \omega^\alpha$ (where $\alpha \sim 1$), which has a power law temperature dependence, and which saturates at high frequencies; (iii) a predicted coupling to the phonon modes associated with the mobile cations. The model is shown to compare well with recent experimental results of the ionic conductivity in Na β-alumina from 10^7 to 10^{13} Hz at 4.2 to 700K. There is evidence for all features noted in (ii) above including saturation at $\sim 10^9$ Hz and a temperature dependence ($\sigma \sim T^2$) between 30 and 650K (for $10^9 \lesssim \omega \lesssim 10^{11}$ Hz). The large 23 cm^{-1} width of the Na ion phonon mode observed at 65 cm^{-1} is also consistent with the model.

[1]U. Strom, P. C. Taylor, S. G. Bishop, T. L. Reinecke and K. L. Ngai, Phys. Rev. B 13 (1976) in press.

Far Infrared Spectra of Sodium-β-Alumina and Its K^+, Rb^+, Cs^+, Ag^+ and Tl^+ Analogs.*

WILLIAM M. RISEN, JR. AND WAYNE M. BUTLER, Brown University.--The far infrared spectra (20-450 cm^{-1}) of powdered forms of $xNa_2O \cdot 11\ Al_2O_3$ ($x \approx 1.2$-1.8) and of K^+, Cs^+, Ag^+, and Tl^+-exchanged analogs have been determined. The reflectance spectra of the Na, K, Rb, and Cs forms have been obtained from the (001) face of large crystals, and the low frequency polarized Raman spectra of crystals of the Na, K, Ag, and Tl forms have been obtained. The analog forms were prepared by the exchange methods of Yao and Kummer, using Na-β-alumina experimental samples obtained from Alcoa Corporation and Harbison Carborundum Company. Analyzed experimental Na and Rb forms, prepared by O. Muller of General Electric, were studied, also, and samples of intermediate Na content were prepared by the method of Harata.

Employing a site symmetry interpretation, the cation-β-structure bands corresponding to vibrations at the BR, aBR, and mO (mid oxygen) sites have been identified and assigned. For Na-β-alumina and its K analog, BR and mO vibrations are found, while for the Ag and Tl analogs, BR, aBR, and mO bands are observed. The spectra will be discussed.

*This work was supported in part by the Office of Naval Research.

The Lattice Location of Conducting Atoms in Ag, K β-Alumina by Particle Channeling. L. C. FELDMAN, J. P. REMEIKA, AND P. J. SILVERMAN, Bell Laboratories, AND K. KOMAKI, Rutgers University and Bell Laboratories.--Particle channeling is a direct method of determining the lattice positions of atoms in a solid. In this technique we measure the yield of MeV He$^+$ ions Rutherford backscattered from the different atoms of the crystal as a function of the angle between the incident beam and major symmetry directions. This technique has been applied to the determination of the sites of Ag and K in Ag, K-β-alumina. The site is determined by (1) observing the change in the measured distributions as a function of temperature and (2) comparing the measurements to detailed channeling calculations. The lattice sites are generally consistent with the most recent results of X-ray refinements. For the case of Ag β-alumina, the conducting ions are observed to migrate under the influence of the probing beam. The systematics of these phenomena and a suggested model for the motion will be discussed.

Structural Characteristics and Non-Stoichiometry of
β-Alumina Type Compounds.* H. SATO, Y. HIROTSU AND J. K.
McCOY, Purdue University.--Structural characteristics of
β-alumina type compounds are investigated by comparing them
with those of isomorphous compounds in the $BaO-Fe_2O_3-MeO$
system (where Me represents a spinel-forming divalent ion)
and by observing the lattice images of a series of β-alumina
type compounds by transmission electron microscopy. The
results suggest that the stoichiometry of β-alumina is fixed
but is far from the idealized formula $Na_2O \cdot 11Al_2O_3$ and that
the existence of the solubility range is due to its
structural characteristics to accommodate mixed periods in
the structure rather than to a change in the Na composition
in the "defect" layer. The results also suggest that the
charge distribution in the crystal should be balanced locally. The Beevers-Ross structure for β-alumina has been
reexamined from this point of view to understand the charge
compensation involved in the excess Na ions.
*This work was supported by the National Science Foundation
 Grant DMR 7502959 and MRL Program DMR 7203018A4.

Planar [00.1] Disorder in Sodium Beta Alumina.
L. C. DeJONGHE, Cornell University.--$00.\ell$ lattice images are
used to show the presence of [00.1] disorder in β alumina.
It is shown that both stacking sequence faults and β" intergrowth are present. The faults are believed to be introduced in the early stages of sintering when neighboring
particles combine to form a larger crystallite, or when
spinel block ledges are added out of sequence. The observation shows that the planar disorder observed by Bevan,
et al. cannot be attributed solely to β" or β"' intergrowth.

Latest Results of "Laboratoire de Chimie Appliquée
de l'état solide de l'ENSCP" on Compounds of β Alumina Type.
J. ANTOINE, J. P. BOILOT, G. COLLIN, R. COLLONGUES,
D. GRATIAS, A. KAHN, J. LIVAGE, J. THERY, AND D. VIVIEN.--
The results mainly deal with (1) influence of the type of
A^{3+} cation on the existence, stability and ionic conductivity
of β and β" aluminas in $Na_2O-A_2O_3$ systems. (2) Syntaxy
between β and β" types of structures in $Na_2O-Ga_2O_3$ and
$Na_2O-Al_2O_3$ systems (electron microscope investigation).

(3) Localization of impurities in β alumina (E.P.R. study).
(4) Influence of doping ions contained in the starting
material on the orientation of the preparative reaction for
β and β" phases and on their electric conductivity.
(5) Structure determination of β alumina type compounds:
sodium gallate, potassium and thallium aluminates.
     More results were obtained together with the
"Laboratoire de Physique des Solides" (Paris-Sud University):
(1) NMR approach (Dr. Jerome and M. Zuppiroli). (2) Study
by X-Ray Diffusion (see Dr. Comes Communication).
     These results regard the local arrangement of the
conducting ions.

Powder Neutron Diffraction Analysis of Sodium Silver
and Deuterium Beta Aluminas. B. C. TOFIELD* A. J. JACOBSON
AND W. A. ENGLAND, Harwell*, and Inorganic Chemistry, Oxford.
--The main features of the beta-alumina structure are known
from X-ray diffraction studies on the Na, K, Ag and Tl compounds[1,2] but some anomalies remain. In particular, three
separate mechanisms for the compensation of the excess soda
have been presented recently involving vacancies on aluminum
sites or extra oxygen near the mid-oxygen position.[1,3]
Also, the structural basis for the anomalous lattice parameters[4] of Li and H beta-aluminas is not known. We have
prepared a pure beta-phase ceramic containing a significantly
higher excess of soda ($1.47Na_2O:11Al_2O_3$ by atomic absorption
analysis for Na and Al) than found in the single crystals
studied hitherto and have so far studied the crystal structures of this and the ion-exchanged Ag and D compounds by
powder neutron diffraction using profile analysis. Data
are presented for these compounds at low temperature and
for the Ag compound at room temperature. Refinements nearing
completion show that i) atoms in the mirror plane have significant shifts in position for anisotropic relative to
isotropic temperature factor refinements, ii) quite good
agreement is found for the Ag refinement at room temperature
compared to that done with X-rays, iii) the nonstoichiometry
is almost certainly compensated by extra oxygen in the mirror
plane, and iv) the location of D in the mirror plane is
quite different to that of the cations in the other compounds,
being found between pairs of $O(2)$ and $O(4)$ oxygens.
[1]P. D. Dernier and J. P. Remeika, J. Solid State Chem. 17,
245 (1976), and references therein.

[2]T. Kodama and G. Muto, J. Solid State Chem. 17, 61 (1976).
[3]J. Antoine, D. Vivien, J. Livage, J. Thery and R. Collongues,
 Mat. Res. Bull. 10, 865 (1975).
[4]J. T. Kummer, Prog. Solid State Chem. 7, 141 (1972).

Ammonium Ion Reorientation in NH_4^+ β-Alumina.*
J. D. AXE, L. M. CORLISS, J. M. HASTINGS, Brookhaven National Laboratory, O. MULLER** AND W. L. ROTH, General
Electric Corporate Research and Development.--Neutron inelastic scattering techniques have been used to study the
motion of NH_4^+ ions in NH_4^+ β-Alumina. The results strongly
suggest jump rotational reorientation of the NH_4^+ ion
between equivalent configurations. The translational diffusion coefficient of the NH_4^+ ion appears to be less than
2.5×10^{-6} cm^2/sec at 473 K.
*Research at Brookhaven National Laboratory performed under
 the auspices of the U.S. Energy Research and Dev. Admin.
**Present address, Xerox Corporation, Webster, N. Y.

X-Ray Diffuse Scattering From Alkali and Europium
β-Aluminas. D. B. McWHAN, P. D. DERNIER, C. VETTIER* AND
J. P. REMEIKA, Bell Laboratories.--The diffuse scattering
of $MoK\alpha$ X-rays from $(1+x)M_2O \cdot 11Al_2O_3$ with M = Li^+, Na^+,
K^+, Rb^+, Ag^+, or $1/2$ Eu^{++} and x ≈ 0.26 has been measured
on crystals grown from the melt and from a flux. The
modulation of the diffuse scattering in the [001] direction
is related to the nature of the compensating defect and the
correlations between defects and diffusing ions. The data
are consistent with a defect model of extra bridges made up
of excess oxygen ions and aluminum ions which move from an
octahedral position in the spinel block into the extra
tetrahedral bridge positions.[1] The correlation length
measurements[2] in the diffusing plane as a function of ion
and temperature have been extended to the melt grown crystals. There is a strong correlation between the ionic
conductivity and the degree of order between the compensating
defects. The melt grown crystals have a conductivity a
factor of four higher than the flux grown material[3] and
considerably less correlation among the compensating defects.
As a result of this the overall in plane correlation lengths
do not increase as dramatically with the size of the diffusing

ion as was observed in the flux grown material.
[1]W. L. Roth, Trans. Am. Cryst. Assoc. 11, 51 (1975) and
W. L. Roth, F. Reidinger, and S. LaPlaca, this conference.
[2]D. B. McWhan, S. J. Allen, Jr., J. P. Remeika and P. D.
Dernier, Phys. Rev. Letters 35, 953 (1975) and 36, 344
(E) (1976).
[3]A. S. Barker, Jr., J. A. Ditzenberger, and J. P. Remeika,
to be published.
*Also at Laboratoire de Magnetisme, B. P. 166,
38042 Grenoble, CEDEX, France.

Surface Studies of Na-β-Alumina. M. L. KNOTEK,
Sandia Labs.--Studies of the thermal desorption of ions
from β-alumina (a-c plane) have been extended to the case
of an ion-etched surface and a Cu-contaminated surface.
The desorption spectra from the ion-etched surface have
both time and temperature dependence, both of which suggest
either ion interactions with adsorbates on the surface or
that the ions are mobile on such a surface, which differs
from an "untreated" surface. A series of K desorption
spectra are presented to illustrate the effects of doping
the surface. Doping the surface with Cu results in marked
changes in the K spectra. Certain K sites appear to be
totally undisturbed while others are either displaced by
Cu or transformer by the presence of a Cu atom into a more
weakly binding site. Implications of such results on the
study of surface contamination of solid electrolytes in
technological are that the details of a specific contaminant
effect on a surface are readily measurable in detail. Due
to the complexity of the data, considerably better surface
definition is needed before a detailed model can be derived.
Auger electron spectroscopy (AES) and low energy electron
diffraction (LEED) are now being used to assure a controlled
and well defined surface at all points in the experiment.
LEED results on the cleaved a-b plane of β-alumina have
shown that this surface is well ordered and much more stable
than the corresponding α-alumina surface. Changes in LEED
patterns due to deposited ions are also readily apparent
and suggest long range ordering of the ions.

Transport Properties of Single Crystal Beta Alumina*.
N. CHOUDHURY[1], J. N. MUNDY, AND S. PURI[2], Argonne National
Laboratory.--A comparison of ionic conductivity measurements
with sodium tracer self diffusion measurements allows the
calculation of the "Haven ratio." Both measurements are
being made on single crystals of beta alumina so that an
accurate comparison can be made. The ionic conductivity
measurements have been made with an a.c. method using
sodium electrodes over a temperature range 25-400°C. The
measurement of conductivity shows a small frequency
dependence that appears to result from the sodium electrode/
beta alumina interface. The activation energy is in good
agreement with previously measured values. The self-
diffusion measurements will be made with a non-destructive
method using positron annihilation techniques.

*Work supported by the U.S. Energy Research and Development
 Administration
[1]Present Address: Systems Research Laboratories, Inc.
[2]Also Northeastern Illinois University

Other Superionic Conductors

SOME SUPER IONIC CONDUCTORS AND THEIR APPLICATIONS

Takehiko Takahashi
Department of Applied Chemistry
Faculty of Engineering
Nagoya University
Nagoya 464, Japan

This paper presents the general survey of some special types of solid electrolytes and their applications.
1. Anion Conducting Solid Electrolytes
1.1 Fluorite- and tysonite-type fluorides have been known to exhibit the high fluoride ion conductivity.

As fluorite-type fluoride ion conductors, we have alkali metal fluorides and lead fluoride doped with mono- or tri-valent metal fluorides. Recently, beta-lead fluoride doped with 2 mole% potassium fluoride has been found to have the fluoride ion conductivity of $10^{-3}\Omega^{-1}cm^{-1}$ at room temperature (1).

Lanthanum or cerium(III) fluoride doped with di-valent metal fluoride has been known as a tysonite-type fluoride ion conductor. Here, some characteristics of tysonite-type fluoride ion conductor based on cerium fluoride are described. As the dopant, calcium fluoride has been recognized to be most effective to give the high fluoride ion conductivity to cerium fluoride. The charge carrier in this solid solution has been found to be the fluoride ion which migrates through fluoride ion vacancies presented in the crystal. In Fig. 1, the conductivity of cerium fluoride containing 5 mole % calcium fluoride is shown together with those of the other fluoride ion conductors. The fluoride ion conducting

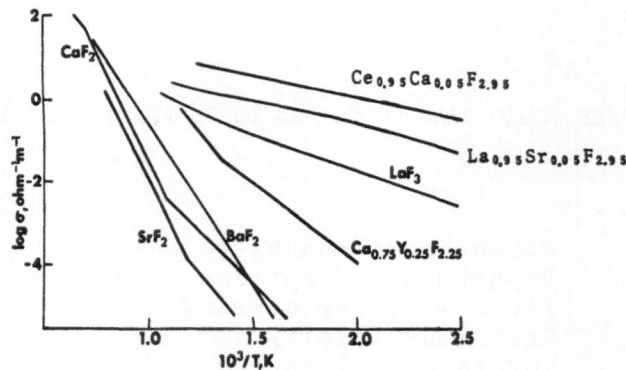

Fig. 1.   Conductivity as a function of temperature for some
fluoride ion conductors.

solid electrolytes have the possibilities of
being applied to the electrolyte membrane for
fluoride ion selective electrodes or to the
electrolyte for high power density solid batteries.

1.2.   Oxide Ion Conductors

As oxide ion conductors, stabilized zirco-
nias, thorias and cerias which have the fluorite-
 type lattice structure have vastly been developed and applied practically to various uses,
for example, an oxygen analyser or sensor and
the concentration determination of oxygen in
metals.

These oxide ion conductors have also been
tested to be applied to the electrolyte for
electrolysis of water, fuel cells and lambda
sensor for automobilies.

In this paper, some properties of perovskite-
type oxides and solid solutions based on bismuth
oxide which have the high oxide ion conductivities at high temperature are mentioned.

1.2.1.   Perovskite-Type Oxides (2)

Perovskite-type oxides ABO_3, where A and B
are di- or tri-valent and tetra- or tri-valent

cations, respectively, have been found to exhibit the high oxide ion conductivity at high
temperature when A or B is substituted partially
with the cation of lower valency giving rise to
oxide ion vacancies. In Fig. 2, the electrical
conductivities in air related to perovskite-type
oxides are shown which have been sintered at
1400~1500°C. The ion transference numbers
measured by the EMF of oxygen gas concentration
cell have indicated that these solid solutions
are mixed ionic and electronic conductors under
ordinary atmosphere of oxygen and under the
sufficiently low partial pressure of oxygen,
they are regarded as highly oxide ion conductors.
The latter situation would be realized in solid
electrolyte fuel cells at anode side and the

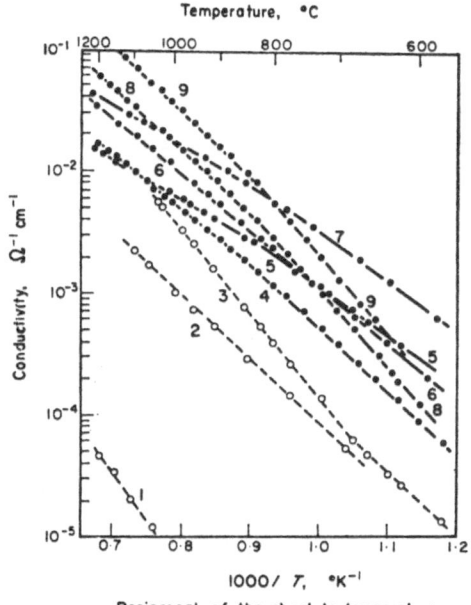

Fig. 2. The conductivity of various perovskite-type oxides
(in air): 1, $LaAlO_3$; 2, $CaTiO_3$; 3, $SrTiO_3$; 4, $La_{0.7}Ca_{0.3}$-
$AlO_{3-\alpha}$; 5, $La_{0.9}Ba_{0.1}AlO_{3-\alpha}$; 6, $SrTi_{0.9}Al_{0.1}O_{3-\alpha}$;
7, $CaTi_{0.95}Mg_{0.05}O_{3-\alpha}$; 8, $CaTi_{0.5}Al_{0.5}O_{3-\alpha}$; 9, $CaTi_{0.7}Al_{0.3}$
$O_{3-\alpha}$.

perovskite-type oxides have successfully been tested as the electrolyte for fuel cells at high temperature.

1.2.2. Oxide Solid Solutions Based on Bismuth Oxide

Bismuth sesqui-oxide transforms from α-phase to δ-phase above 730°C. The δ-phase has a defect fluorite-type structure and exhibits the high oxide ion conductivity which is attributed to the migration of a number of oxide ion vacancies. This oxide, however, melts at 820°C and in order to obtain an oxide conductor which is stable over a wide temperature range, especially below 730°C, di- , tri-, penta- and hexa-valent metal oxides have been added to bismuth oxide to form solid solutions. The stabilizer oxides which have been successfully introduced in Bi_2O_3 are SrO, BaO , La_2O_3 , Y_2O_3, Gd_2O_3 , V_2O_5, MoO_3 and WO_3 . For example, the Bi_2O_3-Y_2O_3 system has been sintered at 800 ~ 1000°C to obtain a solid solution, the conductivity of which in air is shown in Fig. 3. The face centered cubic

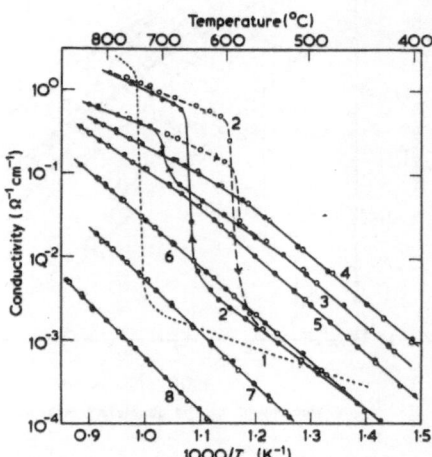

Fig. 3. Conductivity of $(Bi_2O_3)_{1-x}(Y_2O_3)_x$ in air. The values for $x(x \times 100 = mol\%)$ are: 1 -0; 2 - 0·05; 3 - 0·20; 4 - 0·25; 5 - 0·33; 6 - 0·425; 7 - 0·50; 8 - 0·60.

phase has been found in a composition range of
25~40 mole% Y_2O_3 over a wide range of temperature which showed high oxide ion conductivities
many times higher than those of stabilized zir-
conias at corresponding temperatures. High ion
conduction in this phase has been attributed to
the migration of oxide ion vacancies which have
been confirmed by density measurements. The
electronic conduction in this phase is negligibly small under ambient atmosphere (3).

These oxide ion conductors based on Bi_2O_3
have a deficiency of exhibiting mixed ionic
and electronic conduction in reducing atmosphere
which may be due to the reduction of Bi_2O_3.

2. Cation Conducting Solid Electrolytes
2.1 Silver Ion Conductors (4)
In order to stabilize α-AgI at room temperature, the substitution of iodide or silver ion
in AgI by other anion, cation or both ions has
been examined.
2.1.1. Substitution of Iodide Ions by Anions
By this time, S^{2-}, CrO_4^{2-}, MoO_4^{2-}, PO_4^{3-},
AsO_4^{3-}, VO_4^{3-} and $P_2O_7^{4-}$ have been successfully
introduced into AgI to obtain high silver ion
conductivity compounds at room temperature. Of
these silver ion conductors, the compound which
gives the highest silver ion conductivity at
room temperature is $Ag_{19}I_{15}P_2O_7$, the conductivity
of which is $9.0·10^{-2}\Omega^{-1}cm^{-1}$ at 25°C.

Most of the silver ion conductors of this
group have a simple cubic structure and though
the precise structural analyses have not yet
been carried out, the small activation energies
for silver ion conduction of 14~16 KJ/mole will
suggest that these silver ion conductors will
have a so-called average structure.

2.1.2 Substitution of Silver Ions by Cations

Alkali metal ions such as Rb^+, K^+, NH_4^+ and
substituted ammonium ions, sulfonium ions and
carbonium ions have effectively been introduced
into the lattice of silver iodide to obtain
solid silver ion conductors stable at room temperature. Of these solid electrolytes, $RbAg_4I_5$

Table 1.   Onium iodide-silver iodide solid electrolytes
           (QAg_nI_{n+m})

Q	n	m	Ionic conductivity at 25°C ($\Omega^{-1}cm^{-1}$)
Tetramethylammonium (Me$_4$N)	6.5	1	0.04
Pyridinium (Py)	5	1	0.08
Dimethylpyrrolidinium	7	1	0.07
Benzyltrimethylammonium	8	1	0.01
Methane-1,1-bis-methyldiethylammonium	12	2	0.05
Octamethyldiethylenetriammonium	22	3	0.06
1-Methyl-1-thioniacyclohexane	7	1	0.06
Isopropyldimethylsulfonium	7	1	0.05
Tropyllium	4	1	0.006

has been well known to show the silver ion con-
ductivity of $2.7 \cdot 10^{-1} \Omega^{-1} cm^{-1}$ at 24°C which is
the highest silver ion conductivity ever found
in the solid state.

In Table 1, the conductivities of typical
examples of QAg_nI_{n+m} are shown where Q is the
onium cation and n the mole ratio of AgI to QI_m.

2.1.3 Substitution of Silver and Iodide Ions
       by both Cations and Anions

Mercury(II) and chalcogen ions and cyanide
and alkali metal ions have been proposed as the
effective substituents. For example, $Ag_8HgS_2I_6$
has the silver ion conductivity of $1.47 \cdot 10^{-1} \Omega^{-1}$
cm^{-1} at 25°C and in a unit cell, 2.25 cations
are statistically distributed over 42 sites,
thus taking an average structure. The conduction
by mercury(II) ion or electron has not been
found.

In the inorganic solid silver ion conductor
having an average structure a linear relationship has been found between the thermoelectric
power and the reciprocal of the absolute
temperature and the heat of transport

of mobile silver ions calculated from this relation has been found to be almost the same value
as the activation energy for electrical conduction which has been explained by the free ion
model of solid electrolytes (5). But, the heat
of transport in alkyl-ammonium ion substituted
silver iodide has been found to be somewhat
smaller than the activation energy for electrical
conduction which will indicate that the number
of mobile silver ions in these solid electrolytes
is not so large as in the inorganic solid silver
ion conductors.

2.2   Applications of Silver Ion Conductors
      These silver ion conductors have been tried
to use as the electrolyte for batteries, energy
storage devices or electrochemical functional
elements. For example, a proportional timer
with electrical read-out or an electrical inte-
grator with long term characteristics has been
developed and commercially available. One of
the practical configurations is shown in Fig. 4.

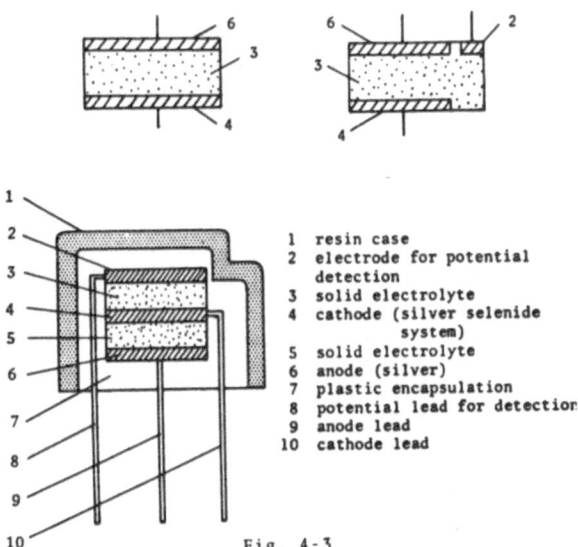

1 resin case
2 electrode for potential
   detection
3 solid electrolyte
4 cathode (silver selenide
               system)
5 solid electrolyte
6 anode (silver)
7 plastic encapsulation
8 potential lead for detection
9 anode lead
10 cathode lead

Fig. 4-3

Fig. 4.   Cross section of electrochemical functional elements.

This is actually a small capacity secondary
battery. One electrode is silver and the counter electrode and the reference or voltage detection electrode are the silver ion-electron
mixed conductor. As the mixed conductor,
$(Ag_2Se)_{0.925}(Ag_3PO_4)_{0.075}$ is used, which has the
α-silver selenide-like structure. Upon passing
a direct current from the mixed conductor to
the silver electrode a definit amount of silver
is transferred from the mixed conductor phase
to the silver electrode, decreasing the cation-
to-anion ratio in the mixed conductor. At this
time, the output voltage appears between the
mixed conductor electrode and the voltage detection electrode which rises almost linearly
with the number of coulombs passed through the
cell. This charging process and the reverse
discharging process show highly reproducible
voltage change between 0mV and 120 mV, thus the
quantity of electricity can be memorized in
the form of voltage. For example, the rated
capacity of 50μAh element can be charged with
40 mA in a few seconds which can act as a timer
for many hours. This element can be operated
over a wide temperature range from -20°to +60°C
and for up to 10^6 cycles of charge and discharge.
For most practical applications, the input
quantity of electricity is read by the output
voltage using a proportional linear-relationship
between them which deviates from the correct
input number of coulombs by 5.5% at the largest.
This element is recommended also to use as an
integrating   or memory device which can detect
temperature, noise, light and other physical
quantities.

     Another application of silver ion conductors
is to use them as the electrolyte for variable
resistor with memory so-called memistor. This
device is composed of
     Ag|Silver Ion Conductor|Silver Thin Film
or
     Ag|Silver Ion Conductor|Mixed Conductor.

As the silver ion conductor and the mixed conductor, $Ag_6I_4WO_4$ and $(Ag_2S)_{0.69}(Ag_{1.7}Te)_{0.285}$ ⎯

$(Ag_4P_2O_7)_{0.025}$ have been tested, respectively.
On passing direct current across the cell from
right to left, the resistivity of the silver
thin film or mixed conductor increases and when
the direction of the current is reversed, the
resistivity decreases. A linear relationship
has been found between the reciprocal of resistivity and the time of turning on a current when
the current density is held constant, and the
electricity passed across the cell is evaluated
by measuring the resistivity of the silver thin
film or mixed conductor. The element using a
mixed conductor as one electrode functions also
as batteries and is expected to be applied to
an element in electronic circuits. Recently,
a solid state display system has been developed
using solid silver ion conductors. In this
system, one electrode is a transparent electrode
and by applying a pulse current to electrolyze
the ionic conductor so as to produce silver on
the transparent electrode, the display can be
read which can be erased by passing a current
in the reverse direction.

2.3 Copper(I) Ion Conductors

In order to stabilize at room temperature an
average structure of copper(I) halide which
appears at high temperature, the substitution
of the constituent ions by various ions has been
attempted. Up to the present, the most effective
substituent ions have been the organic substituted ammonium ions. N-hydro or N-alkyl hexa-
methylenetetramine (6), N,N'-hydro or N,N'-alkyl
triethylenediamine (7) (8), N-hydro or N-methyl
quinuclidinium,N-methyl pyridinium (9), N-methyl
piperidinium and N-methyl morpholinium (10)
halides have been introduced into copper(I)
halides to obtain high copper(I) ion conductivity solids. For example, the temperature
dependence of the electrical conductivity of
the N,N'-hydro or N,N'-alkyl triethylenediamine
halide-copper(I) halide system is shown in Fig.5.
The charge carriers in these solids have been
proved to be the copper(I) ion with negligibly
low electronic conductivity.

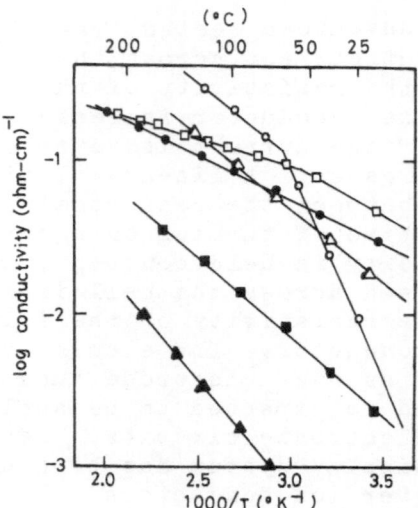

Fig. 5. Temperature dependence of the electrical conduc-
tivity of $17CuCl \cdot 3C_6H_{12}N_2 2HCl$ (o), $4CuCl \cdot C_6H_{12}N_2 2CH_3Cl$ (Δ),
$7CuBr \cdot C_6H_{12}N_2 2HBr$ ($\square$), $47CuBr \cdot 3C_6H_{12}N_2 2CH_3Br$ ($\bullet$), $17CuBr \cdot$
$3C_6H_{12}N_2 2C_2H_5Br$ ($\blacktriangle$), and $17CuI \cdot 3C_6H_{12}N_2 2CH_3I$ ($\blacksquare$).

These copper(I) ion conductors will be
applied to the same objects as silver ion conductors. For example, the galvanic cell of the
following type has been studied

$$Cu | 7CuBr \cdot C_6H_{12}N_4 CH_3Br | Chalcogen.$$

When the cathode is selenium, the open-circuit
voltage is 0.373 V at 25°C and this cell can
be operated up to 150°C above which the electrolyte decomposes. The energy density of 4.5 Wh/
kg is obtained at room temperature.

2.4 Alkali Metal Ion Conductors

2.4.1 Lithium Ion Conductors

Lithium ion conductors which exhibit sufficiently high conductivity at room temperature
have not yet been found. $LiNb_6O_{15}F^-$,
$Li_2Ge_7O_{15}^-$ (11) and $Li_2CuFe(CN)_6^-$ (12) type
compounds have been found to have the lithium
ion conductivities of $10^{-7} \sim 10^{-4} \Omega^{-1} cm^{-1}$ at room

temperature and the reaction product between 60
mole% lithium iodide and 40 mole% alumina has
been recognized to exhibit the lithium ion conductivity of $10^{-5}\Omega^{-1}cm^{-1}$ at room temperature (13).
    The high lithium ion conductivity has been
found at relatively high temperature in $Li_2O\cdot$
$3TiO_2$. When this sintered oxides is cooled
suddenly to freeze its high temperature phase,
the lithium ion conductivity of $10^{-2}\Omega^{-1}cm^{-1}$ is
obtained at 400°C. The conduction in this compound has been recognized to be due to the
migration of lithium ion through a tunnel in
the crystal. Further, the partial substitution
of lithium in lithium titanate with titanium,
aluminum or magnesium has been found to give a
stable lithium ion conductor at relatively high
temperature. For example, $Li_{1.1}Al_{0.9}TiO_3$ shows
the lithium ion conductivity of about $10^{-3}\Omega^{-1}cm^{-1}$
at 500°C, and the lithium ion in these conduc-
tors has been estimated to move through the
lithium ion vacancies which would be produced in
lithium ion layer in the crystal. Similar
results have been obtained when the lithium ion
in lithium stannate or the yttrium ion in
$LiYSiO_4$ is substituted partially by magnesium
or calcium ion.

2.4.2 Sodium Ion Conductors

    The most popular sodium ion conductors are
β- and β''-alumina. Besides these sodium ion
conductors, a solid solution of sodium sulfate
with 10 mole% calcium sulfate has been found to
show the sodium ion conductivity of $9.4\cdot10^{-3}\Omega^{-1}$
cm^{-1} at 300°C and in the systems sodium sulfate
and cadmium sulfate or yttrium sulfate the
sodium ion conductors have also been found.

2.4.3 Potassium Ion Conductors

    In the system $K_2O-Fe_2O_3$, potassium β- ferrite
which is mainly an electronic conductor has been
found, and when the di-valent metal oxide such
as zinc oxide was added to potassium β-ferrite,
for example, to form $K_2O\cdot5.2Fe_2O_3\cdot0.8ZnO$, the
potassium ion conductivity of $1.8\cdot10^{-2}\Omega^{-1}cm^{-1}$

has been found at 300°C with negligibly low
electronic conductivity.

As the potassium ion conductor, the hollandi-
te-type oxide such as $K_x Mg_{x/2} Ti_{(8-x/2)} O_{16}$ $(1.6 \lesssim x \lesssim 2.0)$ has been reported to have the potassium
ion conductivity of $1.7 \cdot 10^{-2} \Omega^{-1} cm^{-1}$ at 25°C (14).
This value was evaluated by dielectric and
capacitance measurements. The direct conductivity measurements, however, have indicated the
lower values of $10^{-5} \sim 10^{-4} \Omega^{-1} cm^{-1}$ at $200 \sim 400°C$.
The charge carrier has been determined to be the
potassium ion which migrates through the potassium ion vacancies in the tunnel which runs along
the direction of C axis in the hollandite crystal. Similar conduction has been obtained in
$K_x Al_x Ti_{(8-x)} O_{16} (1.6 \lesssim x \lesssim 2.0)$, the potassium ion
conductivities of which are $10^{-5} \sim 10^{-6} \Omega^{-1} cm^{-1}$ at
400°C. Furthermore, β-alumina type potassium
gallate has been found to be a potassium ion
conductor.

2.5   Proton Conductors

As proton has small size and an electronic
charge cloud is lack, it will not be subjected
to the usual ion repulsion effects which make
other ions separate individuals. Consequently,
it seems likely that protons in solids would be
transported with an associated ion.

For example, $H_3 OClO_4$ has been found to have
the proton conductivity of $3 \cdot 10^{-4} \Omega^{-1} cm^{-1}$ at 25°C.
But, as it decomposes to $H_2 O$ and $HClO_4$ at about
50°C and is hygroscopic, its practical application in the solid state must be limited

In the systems triethylenediamine- and
hexamethylenetetramine-sulfuric acid, the high
conductivity proton conductors have been found
(15). In these systems $C_6 H_{12} N_2 \cdot 3/2 H_2 SO_4$ and
$C_6 H_{12} N_4 \cdot 2 H_2 SO_4$ exhibit relatively high proton
conductivities. For example, $C_6 H_{12} N_2 \cdot 3/2 H_2 SO_4$
has a proton conductivity of $2.5 \cdot 10^{-5} \Omega^{-1} cm^{-1}$ at
100°C and is stable up to 200°C. The trans-
ference number of proton in this compound has
been found to be nearly unity by the electrolysis
method and the proton conduction mechanism has
been explained on the basis of hydrogen bonds

in the crystal.

Recently, heteropolyacids represented by the chemical formula $H_3(AB_{12}O_{40})nH_2O$ have been found to have high proton conductivities. For example, the hydrate of phosphotungstic acid (PWA) $H_3(PW_{12}O_{40})29H_2O$ has been proved to have the proton conductivity of $4 \cdot 10^{-2}\Omega^{-1}cm^{-1}$ at room temperature. The transference number of proton has been determined to be unity by the electrolysis method. This compound is composed of the anion $(PW_{12}O_{40})^{3-}$ and the cation $(H_3 \cdot 29H_2O)^{3+}$ forming two interpenetrating diamond lattice complexes, the crystal structure of which is shown in Fig. 6 (16). The proton in this crystal has been assumed to migrate along the hydrogen bonds between water molecules. The higher hydrate of phosphomolybdic acid (PMA) $H_3(PMo_{12}O_{40})30H_2O$ exhibits also the high proton conductivity. In Fig. 7, the comparision of the proton conductivities of various proton conductors with those of the oxide ion conductor is shown. $H_3(PMo_{12}O_{40})30H_2O$ has been tested as the electrolyte for hydrogen-oxygen

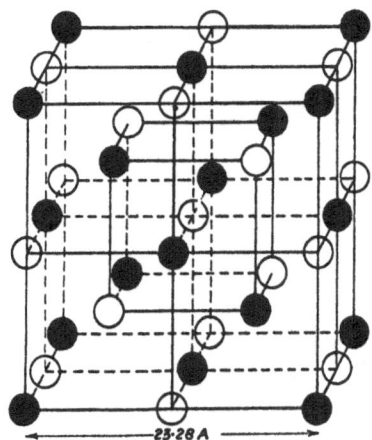

Fig. 6. Anions $(PW_{12}O_{40})^{-3}$ and cations $(H_3.29H_2O)^{+3}$ shown as black and white circles.

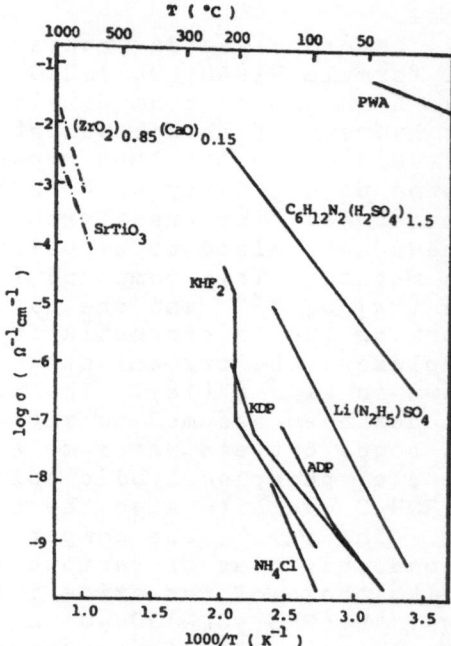

Fig. 7.   Temperature dependence of specific conductivities
of some proton and oxide ion conductors.

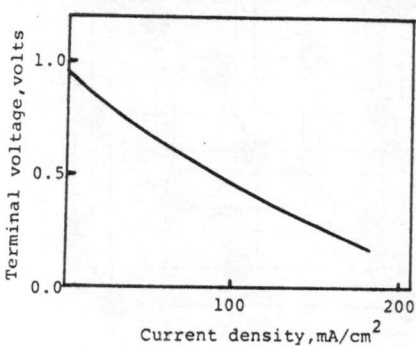

Fig. 8.   Performance of fuel cell using PMA as solid
electrolyte.

fuel cells, the discharge curve of which is
illustrated in Fig. 8 when the active area and
the thickness of the electrolyte are 0.38 cm^2
and 0.1 cm, respectively, and platinum black is
used as the electrode material (17).

3. Summary

In order to explain the conduction mechanism
in superionic conductors, free ion model,
domain model and others have been proposed.

However, to develop the field of superionic
conductors or solid-state ionics still more,
further accumulation of a great deal of data in
this field must be necessary.

In a general way, when mono-valent ions are
charge carriers in solid electrolytes, higher
conductivity would be obtained than in the case
of the charge carriers being poly-valent ions.
And in beta-alumina type superionic conductors,
for example, the activation energies for ionic
diffusion coefficients have been known to be
a function of the ionic radius of the charge
carrier.

These results suggest that in order to develop further the studies on solid superionic
conductors, not only the effects of their crystal
structures on the ionic conductivities, but
also the influences of the ionic value, ionic
radius and polarizability of mobile ions and
the interaction between the mobile ions and the
counter ions on their mobilities must be clari-
fied. The limiting ionic conductivity of solid
electrolytes has been estimated to be $4 \sim 10 \Omega^{-1}$
cm^{-1} and the superionic conductors the conductivities of which are higher than those of the
solid electrolytes ever found may be expected
to discover in the future.

References

(1) J.H. Kennedy, R.C. Miles, J. Electrochem.
     Soc. 123, 47 (1976).
(2) T. Takahashi, H. Iwahara, Energy Conv. 11,
     105 (1971).
(3) T. Takahashi, H. Iwahara, T. Arao,

J. Appl. Electrochem. $\underline{5}$, 187 (1975).

(4)   T. Takahashi, ibid. $\underline{3}$, 79 (1973).

(5)   M.J.Rice, W.L. Roth. J. Solid State Chem.
      $\underline{4}$, 294 (1972).

(6)   T. Takahashi, O. Yamamoto, S. Ikeda,
      J. Electrochem. Soc. $\underline{120}$, 1431 (1973).

(7)   T. Takahashi, O. Yamamoto, ibid. $\underline{122}$,
      83 (1975).

(8)   M. Lazzari, R.C. Pace, B. Scrosati,
      Electrochim. Acta $\underline{20}$, 331 (1975).

(9)   A.F. Sammells, J.Z. Gougoutas, B.B. Owens,
      J. Electrochem. Soc. $\underline{122}$, 1291 (1975).

(10) T. Takahashi, N. Wakabayashi, O. Yamamoto,
      ibid. $\underline{123}$, 129 (1976).

(11) W.L. Roth, O. Muller, NASA CR-134610 SRD-
      74-034 (April 1974).

(12) R.A. Huggins, NTIS AD-773 972 (Nov. 1973).
      ONR Techn. Rept. No. NR 056-555 (March
      1974).

(13) C.C. Liang, J. Electrochem. Soc. $\underline{120}$, 1289
      (1973).

(14) J. Singer et al, NASA Techn. Note D-7157
      (1973).

(15) T. Takahashi, S. Tanase, O. Yamamoto,
      S. Yamauchi, J. Solid State Chem. in press.

(16) A.J. Bradley, J.W. Illingworth, Proc. Roy.
      Soc. A, $\underline{157}$, 113 (1936).

(17) O. Nakamura, H. Kodama, M. Uemura, Y. Miyake,
      Ext. Abst. 16th Symp. Batteries (Japan)
      p. 59 (1975) Tokyo.

FLUORITE-TYPE OXYGEN CONDUCTORS

A.S. Nowick and D.S. Park

Henry Krumb School of Mines
Columbia University
New York, New York 10027

INTRODUCTION

Solid electrolytes which conduct by oxygen-ion transport have been of great interest for use in high-temperature
fuel cells, as well as for the measurement of thermodynamic
and kinetic properties of systems containing oxygen. By
far, the greatest activity in the study of oxygen-ion
conductors, both fundamental and applied, has centered about
oxides which possess the fluorite structure, notably solid
solutions of ZrO_2, HfO_2, ThO_2 and CeO_2. In all cases, the
host oxide is doped with divalent or trivalent cations
(e.g. alkaline earth ions, rare-earth ions, or Y^{3+}) thereby
introducing oxygen vacancies as compensating defects. Fig.
1 shows part of the unit cell of the fluorite structure
containing both an aliovalent cation and an oxygen vacancy.
If these two defects are nearest neighbors to each other,
as shown in the figure, they are said to be associated; if
farther apart, the defects are generally regarded as dissociated. The anion vacancy model has been well established
by two methods: a) simultaneous density and lattice parameter measurements, and b) x-ray intensity measurements
[1-4]. In the case of ZrO_2 and HfO_2, the pure oxides do
not possess the fluorite structure, which is only stabilized
when $\sim$ 10% di- or trivalent cation is present. Accordingly,
these two oxides cannot be studied in the fluorite structure
in the dilute range of doping.

The fluorite oxides doped with aliovalent cations are
very good ionic conductors at temperatures above $\sim$ 700°C.
This fact is quite remarkable when the high melting point

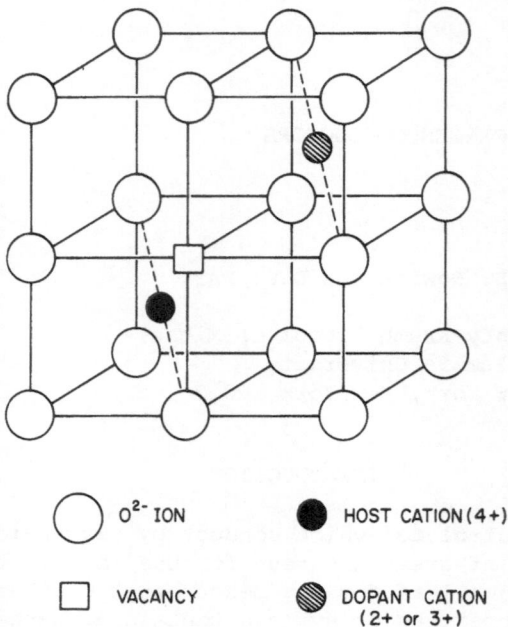

○ O^{2-} ION   ● HOST CATION (4+)

□ VACANCY   ◓ DOPANT CATION
                                      (2+ or 3+)

Figure 1.   Schematic diagram showing half a unit cell
of a fluorite-type oxide which contains an oxygen vacancy
and an aliovalent impurity.

($\sim$ 2700°C) of these materials is considered. The reasons
for this relatively high conductivity can be related to two
important properties of the fluorite structure: (1) the
ability of this structure to accomodate a high degree of
disorder and (2) the high anion-vacancy mobility in the
fluorite structure, probably due tc the openness of this
structure.   To illustrate item (1), we note that the
solubility of aliovalent ions in the fluorite structure is
very high compared to the alkali halides.   To illustrate
item (2), we note that in CaF_2 (and related fluorides) it
has been inferred, both experimentally and theoretically,
that fluorine-ion vacancies move even faster than fluorine-
ion interstitials [5].   This is contrary to the usual
situation.

For a material to function as a good solid electrolyte,
not only must it have a high defect mobility, but also

electronic conduction must be minimal. A convenient
measure of the relative importance of the two processes is
the ionic transference number $t_i \equiv \sigma_i/(\sigma_i + \sigma_e)$ where σ_i is
the ionic conductivity and σ_e the electronic conductivity.
Since electrons or holes generally have much higher
mobility than atomic defects, electronic conduction can
dominate over ionic even when a relatively low electron
concentration is present. Electronic defects are generally
introduced when a compound departs from stiochiometry.
Thus, for example, PrO_2 and TbO_2, which also have the
fluorite structure, are very easily reduced (even for p_{O_2}
~ 1 atm) with the introduction of oxygen vacancies and
electrons [6], while UO_2 readily oxidizes to produce oxygen
intersititials plus electron holes [7]. Such nonstoichio-
metric materials behave as semiconductors and are therefore
not suitable as solid electrolytes.

The oxides which we consider as solid electrolytes
also show some degree of nonstoichiometry, depending on the
oxygen partial pressure, p_{O_2}. According to the application
of interest, one can tolerate some departure from a t_i value
of unity. It has become conventional, however, to define
the electrolytic domain as the region in a p_{O_2} - T diagram
in which $t_i \gtrsim 0.99$. This domain can be determined by
various methods [8]; perhaps the simplest is to utilize the
fact that σ_i is independent of p_{O_2} while the electronic
conductivity, σ_e, varies as a power of p_{O_2}. Figure 2 shows
the electrolytic domain as determined in this way for
CeO_2:5% Y_2O_3. (The figure also shows the boundaries along
which $t_i = 0.5$ and 0.9.) Table 1 shows a comparison of data
for three important oxide conductors. It is clear that
with respect to the p_{O_2} range of the electrolytic domain,
CeO_2:CaO is the poorest. Nevertheless, this may not too
severely limit the usefulness of doped CeO_2 for fuel cell
applications where one side of the electrolyte is subjected
to a highly reducing atmosphere, but the other is in an
oxidizing (high p_{O_2}) environment. Also, the high conductivity of a CeO_2-based electrolyte is a factor in its
favor [11].

The amount of previous work on the oxides of the
fluorite structure and their electrical properties is considerable. Fortunately, there exist several major reviews
[11-13] which survey and compile much of the existing data.
Accordingly, in the present paper, we do not feel compelled

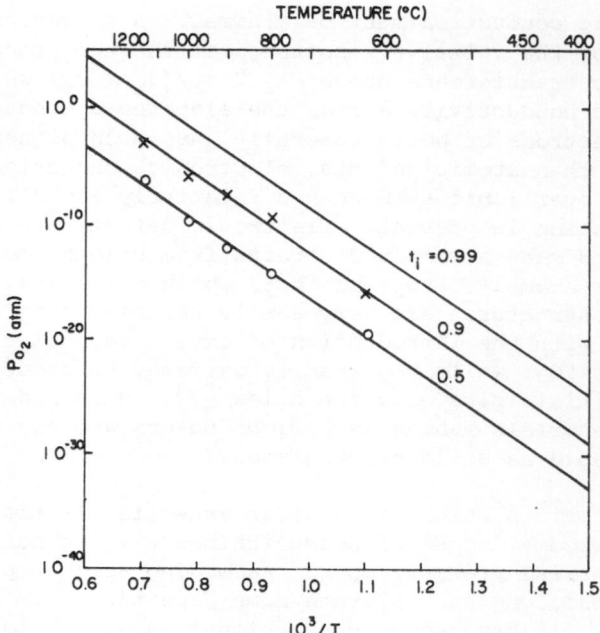

Figure 2. Plots of p_{O_2} vs. T^{-1} at fixed values of t_i for $CeO_2:5\%$ Y_2O_3. The upper curve marks the boundary of the electrolytic domain ($t_i \gtrsim 0.99$). After Tuller and Nowick [9].

Table 1

Limits of electrolytic domain at 800°C for three oxygen conductors.

Electrolyte	Upper p_{O_2} (atm)	Lower p_{O_2} (atm)	Ref.
$CeO_2:5\%$ Y_2O_3	> 1	10^{-6}	9
$ZrO_2:15\%$ CaO	> 1	10^{-24}	8
$ThO_2:8\%$ Y_2O_3	10^{-7}	10^{-40}	10

to summarize the data on these systems; rather it is our
intention to review critically the existing information in
order to determine what is and is not understood about the
defect structure and defect interactions in these oxides.

The importance of defect interactions is best indicated
by the existence of a maximum in ionic conductivity as a
function of the concentration of aliovalent cations. From
a simple vacancy model, one would expect the conductivity
to increase monotonically with doping level, since the concentration of vacancies continuously increases. Instead,
numerous investigators have shown that σ goes through a
maximum in the range between 3 and 8% oxygen vacancies in
virtually all of these systems. Clearly, then, defect
interactions must be playing a major role at the higher
concentrations.

In approaching the problem of attempting to understand
the electrical behavior of these oxides, it is convenient
to divide the study up into two ranges. In the "dilute"
range ($\lesssim$ 3% vacancies) a theory involving relatively simple
defect interactions, or perhaps none at all, should apply.
On the other hand, in the "concentrated" range, more complex
interaction effects surely become important. The next two
sections deal, in turn, with these two ranges.

THE "DILUTE" RANGE

Ionic conductivity has been studied most extensively
in materials such as the alkali halides, silver halides and
alkaline earth fluorides [14,15,5]. Analysis of such systems generally begins by considering the three stages shown
schematically in Fig. 3. In all three stages the conductivity obeys the semi-empirical formula

$$\sigma = (A/T) \exp(-H/kT) \tag{1}$$

where H is the effective activation enthalpy, A is a pre-
exponential constant, and k and T have their usual meanings.
To derive the empirical constants A and H, we begin from
the basic relation

$$\sigma = n_V q \mu_V \tag{2}$$

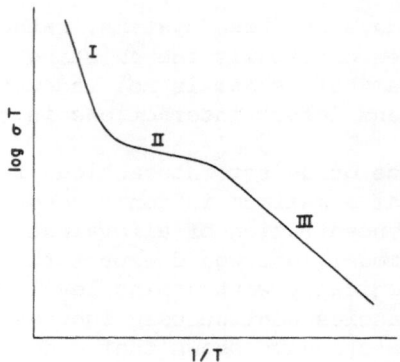

Figure 3. Schematic diagram showing the three stages
in the conductivity plot, log σT vs. 1/T.

where n_v is the concentration (no./vol.) of the dominant
charge-carrying defect, q its effective charge (relative to
the crystal) and μ_v its mobility. (For present purposes,
we consider the defect to be a vacancy.) The mobility μ_v
is given by

$$\mu_v = (B/T) \exp(-H_v^m/kT) \tag{3}$$

where H_v^m is the activation enthalpy for motion and the constant B is given by

$$B = \alpha v' a^2 q/k \tag{4}$$

Here α is a geometrical factor, a is the lattice parameter
and v' a frequency factor which includes the attempt frequency and entropy of activation. In the highest temperature region, stage I of Fig. 3, we obtain intrinsic
conductivity, where the enthalpy H_I involves an enthalpy of
formation as well as that for motion. At intermediate
temperatures, particularly when the material is deliberately
doped with aliovalent impurities, we obtain stage II,
called the extrinsic region. In this stage n_v is a constant, determined by the doping level, and thus

$$H_{II} = H_v^m \tag{5}$$

Finally, at still lower temperatures association between

the aliovalent impurity and the compensating defect (in the manner shown, for example, in Fig. 1) becomes significant. Then the concentration of dissociated defects n_V and of pairs, n_p, are related through the mass action equation

$$\frac{n_p}{n_V^2} = \frac{n_o - n_V}{n_V^2} = K_A = \frac{z}{N_O} \exp\ (E_A/kT) \tag{6}$$

where N_O is the concentration of sites on which the defects can reside, n_o is the total defect concentration introduced by the dopant, while z is the number of orientations and E_A the association energy (really free energy) of the pair. In stage III association is almost complete; thus $n_V \ll n_p \simeq n_o$. By combining Eqs. (2) - (6) we then obtain

$$H_{III} = H_V^m + \frac{1}{2} E_A \tag{7}$$

and

$$A = \alpha v' a^2 q^2 (n_o N_O/z)^{\frac{1}{2}} k^{-1} \tag{8}$$

There are other refinements of this theory that can be considered, notably the application of Debye-Hückel type corrections [14]. For our purposes these considerations can be neglected.

In examining the fluorite-type oxides from the viewpoint of the above equations, it first becomes apparent that since the doping level is much higher than that used in the various halides, and the measuring temperatures are far below the melting temperatures, we are not likely to encounter stage I. Experimentally, a substantial linear region of the conductivity plot (log σT vs. $1/T$) is consistently reported for these oxides. A particularly striking example is shown in Fig. 4, where a straight line is obtained for CeO_2:10% CaO over the temperature range from 100° to 900°C and over 8 decades in σ. The question then arises as to whether it is stage II or stage III that is observed. Strangely enough this question has not been dealt with clearly in the literature. While some authors have mentioned the possibility that the measured H may involve an association energy, the vast majority of workers assume that H represents H_V^m and that there must be a linear dependence of σ on dopant concentration in the dilute range. This dependence is characteristic of stage II; for stage III,

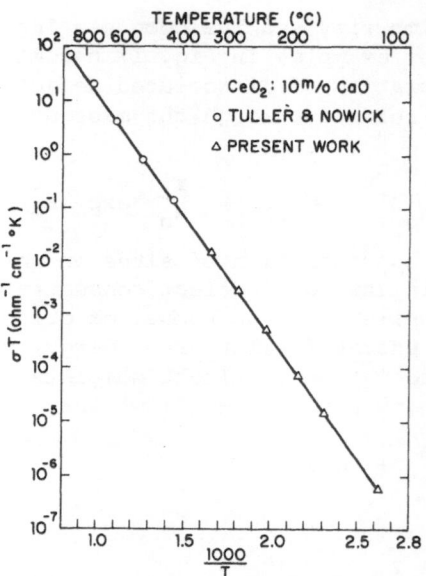

Figure 4.   The plot of log σT vs. T^{-1} for CeO_2:10% CaO illustrating a linear relation over a wide range of temperature.   (The high temperature points are from ref. [9] .)

on the other hand, $\sigma \propto n_o^{\frac{1}{2}}$ (see Eq. (8)).

We shall give reasons for believing that data such as that shown in Fig. 4 must represent stage III:

a) In alkali halides and alkaline earth fluorides, association energies E_A are typically $\sim$ 0.3 - 0.5 eV [15]. Recalling that E_A originates in the Coulomb interaction between the excess (or deficient) charge of the aliovalent ion and that of the compensating defect, and considering that in a system such as CeO_2:CaO the effective charges are 2e (where e is the electronic charge), it is anticipated that values of E_A should certainly be no lower than $\sim$ 0.3 eV in these oxides.   In Fig. 5 we have plotted the temperature of the break between stages II and III as a function of mole % CaO for a fluorite-type oxide, with E_A as a parameter.   (The break point is defined arbitrarily as that at which $n_v = n_o/2$ in Eq. (6).)   It is clear that for reasonable E_A values, the break for oxides with several % CaO

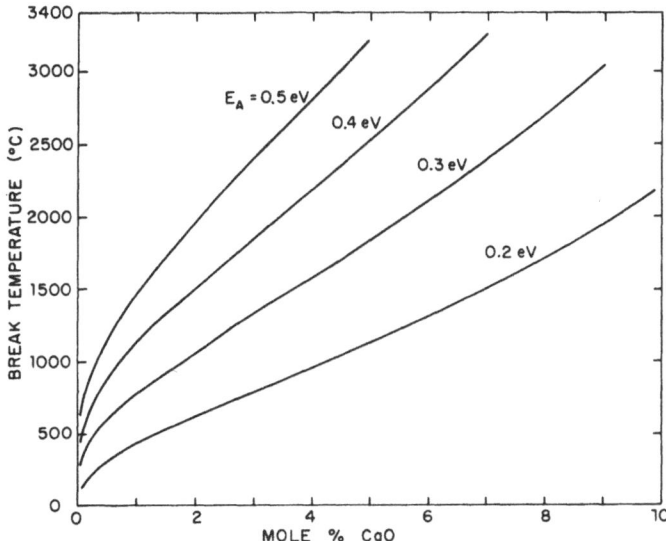

Figure 5. Calculated curves of the temperature of the break between stages II and III as a function of CaO concentration in a fluorite-type oxide, for various E_A values.

must occur well above 1000°C. There are numerous examples in the literature where the conductivity plot shows a deviation from linearity going to a lower slope at high temperatures [16,17]. This might indicate the onset of the break which we are discussing. However, in view of disagreements over the occurrence of such breaks in a given material, more work seems to be necessary before any definitive conclusion can be reached. (Some factors that can give false breaks include electrode effects and other types of polarization phenomena, which serve to make σ frequency dependent.)

    b) If the usual observations fall into stage II, it is clear from Eq. (5) that H_{II} should be independent of the dopant used. Yet, it has been consistently reported that H is lower for trivalent dopants such as Y^{3+} than for divalent Ca^{2+}. This point can be reasonably well explained if we interpret H as H_{III}, which involves E_A as expressed in Eq. (7). This is true because we can expect a lower E_A in

Table 2

Summary of results from conductivity measurements
on five well studied materials, including activation
enthalpy H, preexponential factor A, and
calculated frequency factor ν'.

Material	% $V_O^{..}$	σ $(\Omega\text{-cm})^{-1}$	T $(^\circ K)$	H (eV)	A $(10^5 \Omega^{-1} cm^{-1} {}^\circ K)$	ν' $(10^{13} sec^{-1})$
$ThO_2 : 2.6\% \ Y_2O_3$	1.3	3.2×10^{-3}	1273	1.2	2.1	2.0
$CeO_2 : 2\% \ CaO$	1.0	6.7×10^{-7}	500	0.92	6	6.2
$CeO_2 : 5\% \ Y_2O_3$	2.4	1.1×10^{-5}	500	0.83	11	7.3
$ZrO_2 : 12\% \ CaO$	6.0	5×10^{-2}	1273	1.1	14	5.6
$ZrO_2 : 10\% \ Y_2O_3$	4.5	10×10^{-2}	1273	0.85	2.9	1.3

the case of, for example, Y^{3+}-doped samples than in the
case of Ca^{2+}-doped samples because of the lower effective
charge of Y^{3+}. Some values of H are listed in Table 2,
illustrating the difference for di- and trivalent dopants.
Included are only materials which have been well investigated and for which there is general agreement concerning
the value of H [12]. (The results for ZrO_2 unfortunately
do not fall in the truly "dilute" range, because the
fluorite structure is not stable until sufficient dopant is
added, but are included because this material has received
so much attention.)

c) There is no substantial doping range over which σ
vs. n_O is linear. Figure 6 shows, however, that for the
$CeO_2 : CaO$ system an $n_O^{\frac{1}{2}}$ relation is well obeyed.

d) Dielectric relaxation, which will be described
below, shows the presence of electric dipoles of about the
right number to be consistent with the occurrence of nearly
complete association in the vicinity of room temperature.

In view of these arguments, there seems to be little
doubt that the conductivity results fall into stage III and
therefore that, in the dilute range, H should be set equal
to H_{III} and interpreted according to Eq. (7). This conclusion also suggests that we examine the preexponential

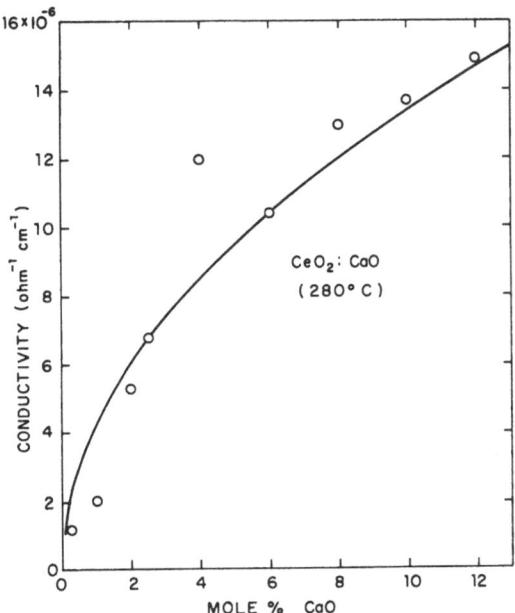

Figure 6. The plot of conductivity of CeO_2:CaO vs.
mole % CaO at 280°C. Circles are experimental values,
while the solid curve represents a square root dependence.

constant A, in terms of Eq. (8). For the fluorite structure,
the constant α is 4, for nearest neighbor pairs z = 8, and
finally q = 2e for oxygen vacancies. Table 2 lists the
empirical values of A for each of the materials considered,
and in the final column, gives the value of ν' determined
from Eq. (8). The results obtained are quite reasonable
since it is to be expected that $\nu' \sim 10^{13} - 10^{14} sec^{-1}$ when
both the attempt frequency and the entropy of activation
factor are taken into account. Even for the ZrO_2:CaO case
there is no indication that A is anomalous, as had been
suggested earlier [18]. This is of interest because a conduction mechanism involving cooperative transport [19] has
been invoked based in part on the belief that A is anomalously large.

    It has been long known that dielectric and anelastic
relaxation studies provide strong supplements to conductivity

measurements in ionic systems; while σ measures the dissociated defects, relaxation studies measure the behavior of associated pairs. Detailed reviews of these techniques and the information derivable from them are available [20, 21]. Suffice it to say that dielectric relaxation measures the additional time-dependent polarization due to preferential reorientation of the defects in the presence of an electric field. Correspondingly anelastic relaxation measures a time dependent strain in response to a stress field. The activation enthalpy for relaxation, H_r, is that for the jump of a defect between two bound states (e.g., between adjacent bound states of the vacancy in Fig. 1). It is therefore expected that H_r will not be very different from the activation enthalpy H_V^m for the motion of a free vacancy. Dielectric and anelastic relaxation measurements have been made on dilute ThO_2:CaO by Wachtman [22] who obtained $H_r = 0.95$ eV, and in CeO_2:CaO by Lay and Whitmore [23] who obtained $H_r = 0.86$ eV. The fact that both of these values are lower than the corresponding activation enthalpies in Table 2 is consistent with the earlier conclusion that the enthalpy obtained from the conductivity is H_{III}. The present authors have extended the work of Lay and Whitmore on the CeO_2:CaO system to higher concentrations in order to study the onset of more complex interaction effects that mark the "high" concentration range. In this work, dielectric relaxation has been measured by the ionic thermocurrent (ITC) method [24]. In this measurement the material is first polarized at a selected temperature under an electric field and then cooled to low temperatures to freeze in the state of polarization. Upon heating slowly, after removing the field, the polarization is released in the form of an "ionic thermocurrent" which shows a peak in a current-versus-T plot. Figure 7 shows such curves for CeO_2:CaO with varying amounts of CaO. While the 1 and 2% samples show a well defined peak which is reasonably ascribed to Ca^{2+}-V_O pairs, the 8% CaO sample (4% vacancies) shows considerable broadening and the development of additional relaxation peaks. Actually the broadening sets in at about 4% CaO (2% vacancies), which must then be regarded as roughly the end of the "dilute" range.

From the area under the curves in Fig. 7 the total number of dipoles can be calculated [24,25]. As already mentioned above, these results for CeO_2 containing 1 and 2% CaO are consistent with the idea that virtually all of the

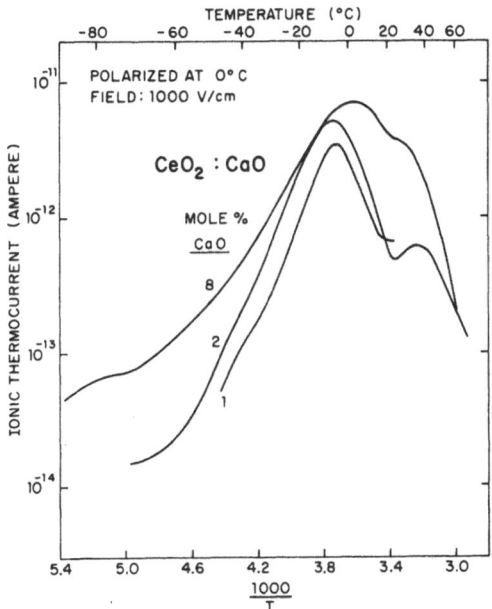

Figure 7. Curves of the ionic thermocurrent vs. T^{-1} for CeO_2:CaO with three different CaO concentrations. In all cases, the electrode area was 1 cm^2, and the sample was polarized at 0°C under a field of 1000 V/cm. (The apparent peak at 40°C is not a true relaxation peak.)

defects are in the form of bound pairs at these temperatures.

THE "CONCENTRATED" RANGE

For dopant concentrations well outside of the "dilute" range, there is considerable evidence that much more complex interactions than pair formation take place. The treatment of this range must be more qualitative until the nature of these defect interactions is better known. Nevertheless, this range is a very interesting one, as well as important for practical applications.

The most striking demonstrations of interaction effects

comes from the occurrence of a maximum in σ vs. dopant concentration and from the broadening of the dielectric relaxation peak, as mentioned earlier. In some cases (e.g. for $ZrO_2:CaO$ [3,18] and $ZrO_2:Y_2O_3$ [26]) it has been established that the apparent activation enthalpy H, obtained from conductivity plots, increases substantially with increasing concentration. In other cases, however, an increase in H is not apparent over a moderate range of concentrations [27,28]. Finally diffraction techniques, including diffuse scattering, give evidence for complex interactions, as will be discussed below.

A key factor in analyzing the behavior of the concentrated oxides is the low rate of cation diffusion in these oxides. It has been well established that both cation self diffusion and dopant diffusion take place orders of magnitude more slowly than oxygen diffusion in these oxides [12]. One might then expect that a high concentration solid solution cooled moderately rapidly from high temperatures can be regarded as containing essentially a random distribution of the aliovalent cations. Since each oxygen site has 4 nearest-neighbor cation sites, we might then think of vacancies as distributed in sites with 0, 1, 2,...nearest dopant cations, in accordance with the appropriate association energies. This would constitute a trivial extension of the model for the dilute range, but it could not adequately explain the various experimental observations. Let us then see what other modifications need to be introduced.

First, as the concentration of vacancies increases to the point where vacancy-vacancy distances become small, we must consider that vacancies will tend to stay apart from each other. Barker [29] and O'Keeffe [30] have recognized this point and showed that, based entirely on geometrical considerations, a maximum in the conductivity as a function of vacancy concentration is predicted. However, this theory does not take into account the possibility that the activation enthalpy for motion of a vacancy may increase. Such an increase comes about because at high concentrations the jumps of a single vacancy inevitably take it from an energetically favorable to a less favorable position. Alternatively, at sufficiently high concentrations, coordinated motion of more than one vacancy (cooperative diffusion) may be required to produce jumps.

The second modification to a simple geometrical theory
comes from the onset of ordering phenomena. Such processes
are limited by cation migration and, therefore, are sluggish
below $\sim$ 1500°C, but there is good evidence to show that
they occur. In fact the occurrence of a slow long-range
ordering process in ZrO_2:CaO (especially for concentration
$\sim$ 18 - 20% CaO) was reported by Tien and Subbarao [3] and
later studied in greater detail by Carter and Roth [18].
The onset of order is accompanied by as much as a 50%
decrease in conductivity (an aging effect). The sluggish-
ness of the reaction is indicated by the fact that ordering
requires several days at 1000°C. On the other hand, at
1400°C and above, the ordered structure is unstable and
goes back to the disordered state. It has recently been
shown [31] that the ordered structure has the composition
$CaZr_4O_9$ and is similar to that proposed for $CaHf_4O_9$ [32].

Very recently, Ray and Stubican [31] have demonstrated
that a sluggish long-range ordering also takes place in
ZrO_2:Y_2O_3 to form a rhombohedral structure of composition
$Zr_3Y_4O_{12}$. Such a 7:12 structure is very commonly observed
in compounds containing both tri- and quadrivalent cations
[33] and is characterized by chains of oxygen vacancies
running along a <111> direction of the original fluorite
structure. This ordering reaction also occurs sluggishly
near 1000°C, and the ordered phase is unstable above 1300°C.

In addition to these observations of long-range order,
evidence for short-range order in these oxides has been
obtained from both Bragg and diffuse scattering of neutrons,
as well as from electron diffraction. Although different
investigators disagree on details, there seems to be general agreement that short-range order takes the form of
local regions in which the atomic arrangement resembles
that of a long-range ordered structure which is nearby in
composition. Most often this means that the ordered
regions have a higher-than-average concentration of aliovalent dopant. Along different lines, recent work by
Steele and Fender [34] on neutron diffuse scattering in
CeO_2:Y_2O_3 shows evidence for striking departures from a
random vacancy distribution, in relatively dilute solutions
($\sim$ 2.5% vacancies). In fact, these authors interpret their
results as due to a cluster containing a pair of vacancies
along a <111> direction together with either three or four
Y^{3+} ions. While there is a need for more work along these

lines (e.g. a dilute oxide doped with divalent Ca^{2+} has not yet been studied), these results are certainly suggestive of the possibility that trivalent doping gives rise to a very different defect complex than does divalent doping.

SUGGESTIONS FOR FUTURE WORK

The present paper has aimed to determine what we know about defect interactions in fluorite-type oxides. It has therefore served to focus attention on questions which are, as yet, unanswered but which deserve special attention at this time. Among them are the following:

1) It appears that there may be a real difference between the elementary defect cluster in oxides doped with divalent and with trivalent cations. The most useful information on this question comes from relaxation experiments (dielectric and anelastic) and from diffuse scattering of neutrons. Thus far, however, virtually all relaxation experiments have been carried out on Ca^{2+}-doped oxides while diffuse scattering was mainly on Y^{3+}-doped samples. Accordingly, it is most desirable to close the gap in our knowledge by doing relaxation on M^{3+}-doped samples and diffuse scattering on M^{2+}-doped samples.

2) Careful conductivity measurements at high temperatures (and as a function of frequency) are needed, for various oxides, to establish where the break between stages II and III of the conductivity curve falls, and thereby, to obtain values of the association energy E_A.

3) It would be most helpful to have a more realistic model for vacancy motion in an electric field at high concentrations, which statistically takes into account those jumps involving higher activation enthalpies. In this way one could predict how the apparent enthalpy obtained from conductivity measurements increases with concentration, for a given dopant.

4) Comparison of precise diffusion measurements with conductivity offers the opportunity to obtain the so-called "Haven ratio", which gives insight into the diffusion mechanism [35]. While a few such measurements have been made [36], they would be of greatest value if carried out

in one system over the range from low to high dopant concentration. In this way, one can determine whether a
change in diffusion mechanism occurs when strong interactions and defect clustering become important.

Acknowledgement: This work was supported by the
National Science Foundation under grant number NSF-DMR74-
23877.

REFERENCES

1. F. Hund, Z. Phys. Chem. 199, 142 (1952).
2. A.M. Diness and R. Roy, Solid St. Comm. 3, 123 (1965).
3. T.Y. Tien and E.C. Subbarao, J. Chem. Phys. 39, 1041
    (1963).
4. E.C. Subbarao, P.H. Sutter and J. Hrizo, J. Am. Ceram.
    Soc. 48, 443 (1965).
5. A.B. Lidiard, in The Physics of Fluorite Compounds,
    ed. W. Hayes, (Oxford University Press, 1975) Chap. 3.
6. B.G. Hyde, D.J.M. Bevan and L. Eyring, Phil. Trans.
    Roy. Soc. (London) A259, 583 (1966).
7. L. Lynds, W.A. Young, J.S. Mohl and G.G. Libowitz,
    Adv. Chem. Series No. 39, (1963), p. 58.
8. J.W. Patterson, J. Electrochem. Soc. 118, 1033 (1971);
    J.W. Patterson, in Physics of Electronic Ceramics,
    Part A, ed. L.L. Hensh and D.B. Dove, (Marcel Dekker,
    New York, 1971).
9. H.L. Tuller and A.S. Nowick, J. Electrochem. Soc. 122,
    255 (1975).
10. J.B. Hardaway, J.W. Patterson, D.R. Wilder and J.D.
    Schieltz, J. Am. Ceram. Soc. 54, 94 (1971).
11. T. Takahasi, in Physics of Electrolytes, Vol. 2, ed.
    J. Hladik, (Academic Press, New York, 1972).
12. T.H. Etsell and S.N. Flengas, Chem. Reviews 70, 339
    (1970).
13. P. Kofsted, Nonstoichiometry, Diffusion and Electrical
    Conductivity in Binary Oxides, (Wiley, New York, 1972).
14. A.B. Lidiard, Handbuch der Physik 20, ed. S. Flügge,
    (Springer, Berlin, 1957), p. 246.
15. P. Süptitz and J. Teltow, Phys. Stat. Solidi 23, 9
    (1967).
16. J.M. Wimmer, L.R. Bidwell and N.M. Tallan, J. Am.
    Ceram. Soc. 50, 198 (1967).
17. R.E.W. Casselton, Phys. Stat. Solidi(a) 2, 571 (1970).

18. R.E. Carter and W.L. Roth, in Electromotive Force Measurements in High Temperature Systems, ed. C.B. Alcock, (American Elsevier, New York, 1968), p. 141.

19. M.J. Rice and W.L. Roth, J. Sol. St. Chem. 4, 294 (1972).

20. A.S. Nowick, in Point Defects in Solids, Vol. 1, ed. J.H. Crawford, Jr. and L.M. Slifkin, (Plenum, New York, 1972), p. 151.

21. A.S. Nowick and B.S. Berry, Anelastic Relaxation in Crystalline Solids, (Academic Press, New York, 1972).

22. J.B. Wachtman, Jr., Phys. Rev. 131, 517 (1963).

23. K.W. Lay and D.H. Whitmore, Phys. Stat. Solidi(b) 43, 175 (1971).

24. C. Bucci and R. Fieschi, Phys. Rev. 148, 816 (1966).

25. A.S. Nowick and W.R. Heller, Adv. Phys. 14, 101 (1965).

26. J.M. Dixon, L.D. LaGrange, L.D. Merten, C.F. Miller and J.T. Porter, J. Electrochem. Soc. 110, 276 (1963).

27. M.F. Lasker and R.A. Rapp, Z. Physik. Chem. 49, 198 (1966).

28. D.S. Park and A.S. Nowick, to be published.

29. W.W. Barker, Mat. Sci. Eng. 2, 208 (1967); W.W. Barker and O. Knop, Proc. Brit. Ceram. Soc. No. 19 (1971), p. 15.

30. M. O'Keeffe, in Chemistry of Extended Defects in Non-Metallic Solids, ed. L. Eyring and M. O'Keeffe, (North Holland, Amsterdam, 1970), p. 609.

31. S.P. Ray and V.S. Stubican, Private Communication.

32. J.G. Allpress, H.J. Rossell and H.G. Scott, Mat. Res. Bull. 9, 455 (1974).

33. S.F. Bartram, Inorg. Chem. 5, 749 (1966); M.R. Thornber, D.J.M. Bevan and J. Graham, Acta Cryst. B24, 1183 (1968); S.P. Ray and D.E. Cox, J. Solid St. Chem. 15, 333 (1975).

34. See, B.E.F. Fender in Chemical Applications of Thermal Neutron Scattering, ed. B.T.M. Willis, (Oxford University Press, 1973), Chapter 10.

35. A.D. LeClaire in Physical Chemistry X. Solid State, ed. W. Jost, (Academic Press, New York, 1970).

36. L.A. Simpson and R.E. Carter, J. Am. Ceram. Soc. 49, 139 (1966); W.D. Kingery, J. Pappis, M.E. Doty and D.C. Hill, J. Am. Ceram. Soc. 42, 393 (1959).

Ion Dynamics and Defect Structure of Cubic PbF_2.

J. B. BOYCE AND J. C. MIKKELSEN, Xerox Palo Alto Research Center.--The NMR relaxation rates of the mobile fluorine nuclei have been measured in the superionic conductor β-PbF_2 as a function of temperature. From room temperature to about 300°C these rates agree with the simple BPP theory and yield an attempt frequency of about 10^{15} sec^{-1} and an activation energy for fluorine motion of 0.74 eV, in good agreement with the $1/T_2$ measurements of Hwang et al.[1] (0.73 eV). Above about 300°C deviations from BPP theory are observed. This is the same temperature region in which the ionic conductivity begins to saturate[2] and a broad λ-type specific heat anomaly is observed.[3] This anomalous behavior is discussed in terms of the defect structure of PbF_2 at high temperatures.

[1]T. Y. Hwang, M. Engelsberg and I. J. Lowe, Chem. Phys. Lett. 30, 303 (1975).
[2]C. E. Derrington and M. O'Keefe, Science 246, 44 (1973).
[3]C. E. Derrington, A. Navrotsky and M. O'Keefe, Solid State Comm. 18, 47 (1976).

NMR Studies of F Ion Motion in Doped β-PbF_2.

T. Y. HWANG AND I. J. LOWE, Dept. of Physics, Univ. of Pittsburgh AND K. F. LAU AND R. W. VAUGHAN, Div. of Chem. Engr., California Inst. of Tech.--The results of a pulsed nuclear magnetic resonance study of the fluoride ion motion in two single crystals of β-PbF_2, doped with 0.02% and 0.12% by weight of NaF, are reported. For F^{19}, we measured the transverse relaxation time T_2, the spin lattice relaxation time T_1, and the spin lattice relaxation time in the rotating frame T_{1r} as a function of temperature from -50°C to 160°C. T_1 and T_2 were also measured as a function of pressure to 4 kilobars at 19°C, and T_1 was measured at several different frequencies between 10 and 60 MHz. T_1 is described by an activation energy of 0.205±0.01 ev/ion, and is frequency independent. T_{1r} and T_2 are described by activation energies of 0.29±0.02 and 0.27±0.01 ev/ion respectively. A T_{1r} minimum at 56°C yields a correlation time of 0.74 μsec for the F motion in the 0.02% crystal, and an estimated diffusion rate of 2×10^{-10} cm^2/sec.

Pressure dependence of T_1 and T_2 yield activation volumes of $\leq$0.2 cm^3/g-mole and 1.76±0.05 cm^3/g-mole respectively. These data, along with the measured magnetic field independence of T_1, and along with the measured T_1 being much shorter than predicted by F motion, suggests that T_1 is dominated by thermally excited electrons or holes.

E.S.R. of Mn^{2+} in Zirconia-Calcia Solid Solutions.
G. BACQUET, J. DUGAS AND C. ESCRIBE, Laboratoire de Physique des Solides, 31077 Toulouse-cedex France.--
E.S.R. experiments were carried out on single crystal or powdered samples of Zirconia-Calcia solid solutions doped with Mn^{2+}. With single crystal Calcia Stabilized Zirconia (CSZ) different spectra were recorded when the amount of CaO was 15 or 20 mole%. With CSZ-15 the Mn^{2+} spectrum is characteristic of a weak axial symmetry along each of the 3[100] directions. There is no O^{2-} vacancy associated to Mn^{2+} in nearest neighbor position and the hyperfine coupling constant value is A=89.6 G. The axial symmetry can be explained by the O^{2-} arrangement proposed by Carter and Roth for the "ordered" structure. With CSZ-20 we recorded an isotropic spectrum with A=82 G.

For powdered samples in the range 8-16 mole% CaO the A value is 89.6 G. Between 16 and 19 mole% CaO there are 2 different spectra, one with A=89.6G, the second for which A=82G. From 19 to 50 mole% CaO we obtained a unique spectrum with A = 82 G.

We think that in the cubic phase of CSZ there exist some microdomains of $CaZrO_3$, the number and the size of which are increasing with the quantity of Calcia.

As Mn^{2+} ions seem to go preferentially in such micro domains, the E.S.R. technique permit us to detect them from 16 mole% CaO.

Both kinds of samples used in our experiments were obtained at very high temperatures by melting of the constituent oxides.

Composition and Electrical Properties of R.F. Sputtered Stabilized Zirconia Thin Films. M. CROSET AND G. VELASCO, Thomson-CSF, Domaine de Corbeville, 91401 Orsay, France, AND J. SIEJKA, Groupe de Physique des Solides de l'E.N.S., Tour 23, 2 Place Jussieu, 75005 Paris, France.
--Thin calcia stabilized zirconia films were obtained by R.F. sputtering (from 300 to 10,000 Å). The composition

and relative distribution of the constituents of these films were analyzed by nuclear microanalysis and MeV ion back-scattering techniques. The stoichiometric composites are obtained in pure oxygen or in mixture of oxygen and krypton atmosphere ($PO_2 \geq 10^{-5}$ Torr). The stoichiometric deposits are very stable and do not change during annealing at 525°C in vacuum or in air. The electrical properties of the deposits are measured by d.c. and a.c. techniques (amplitude of a.c. signal 10 mV) using following structures: (i) Al/CSZr/Si, (ii) Pt/CSZr/Pt/PtO$_x$/Si or (iii) Pt/CSZr/ Ni/NiO/Si. The electrical properties i.e., a.c. conductivities are the same independently of the electrodes used. A conductance value of $10\Omega^{-1}$ cm^{-2} at 350°C for films 1000 Å thick was deduced from the a.c. measurements. The activation energy of a.c. conductivity was measured between 200 and 350°C and was found to be equal to 1.18 eV, which is in good agreement with the literature data reporting the activation energy of oxygen movement in the thick, sintered CSZr electrolytes of the same composition. The structures (ii) and (iii) were tested as an oxygen monitoring solid cells, where Pt/PtO$_x$ and Ni/NiO burried electrodes were used as a reference electrode for the potentiometric measurements of oxygen concentration. The Nernst law is found to be followed in the oxygen pressure range studied (10^{-6} - 10^{-2} atm. of oxygen) at temperatures as low as 320°C. Stability of the electrochemical measurements as a function of time, temperature and oxygen pressure, was found to be strongly related to the nature of reference, burried electrode; the physical meaning of this effect will be discussed. Some preliminary results concerning the improvement of speed of response of these thin solid oxygen sensors will be also reported.

Influence of Dopant and Dopant Concentration on Ionic Conduction and Stoichiometry in Solid Ceria Electrolytes. H. L. TULLER* AND D. S. TANNHAUSER, Physics Dept., Technion.--In a recent study[1] the region of temperatures and P_{O_2}'s for which $(CeO_2)_{0.95} (Y_2O_3)_{0.05}$ behaves primarily as an ionic conductor (i.e., $t_i \geq 0.99$) was established. This "electrolytic domain" was found to extend over a significant range of P_{O_2}'s for temperatures below 600-700°C suggesting possible use of doped ceria electrolytes in oxygen meters or fuel cells operating within these ranges. The present study was initiated to establish the dependence of the important parameters for electrolyte operation on the type and concentration of aliovalent impurities present in the ceria solid solutions. Results of electrical conductivity measurements performed over an extended range of temperature and P_{O_2} are reported for ceria doped with 2, 5 and $7^{m}/o$ Nd_2O_3 and 5, 7.5 and $10^{m}/o$ Y_2O_3. These results and results reported earlier[2] for CeO_2 doped with 1, 5 and $12^{m}/o$ CaO were used to establish the electrolytic domain boundaries, the ionic conductivity parameters and the defect formation energies for the various solid solutions. The dependence of these parameters on concentration and type of dopant will be discussed.

*Present address: Dept. of Materials Science, M.I.T.
[1]H. L. Tuller and A. S. Nowick, J. Electrochem. Soc. <u>122</u>, 256 (1975).
[2]R. N. Blumenthal, F. S. Brugner and J. E. Garnier, J. Electrochem. Soc. <u>120</u>, 1230 (1973).

<u>New Fluor Ion Conductors</u>. J. M. REAU, C. LUCAT, G. CAMPET, J. CLAVERIE, J. PORTIER AND P. HAGENMULLER, <u>Laboratoire de Chimie du Solide du C.N.R.S., Université de Bordeaux I, 33405 Talence, France</u>.--The graphical representation of the complex impedance of a solid electrolyte, measured with an a-c technique, allows for an accurate determination of the resistivity and the activation energy. This method is applied to a $Ca_{2-x}Y_xF_{2+x}$ solid solution ($0 \leq x \leq 0.38$). Electrical and structural data are correlated.

From these results and from considerations on the ionic polarizability, the study of ionic conduction in fluorides with fluorite-type structure leads to new anionic superconductors of formulating $Pb_{1-x}Bi_xF_{2+x}$ ($0 \leq x \leq 0.50$). The strong polarizability of the Pb^{2+} and Bi^{3+} cations give raise to a highly conducting material. In particular $Pb_{0.75} Bi_{0.25}F_{2.25}$ has a conductivity close to that of β-alumina.

New Lithium Ion Conductors.* B. A. BOUKAMP,
I. D. RAISTRICK, C. HO, Y-W. HU, AND R. A. HUGGINS,
Stanford University.
--Conductivity data for several new lithium conductors are
presented. The materials discussed fall into three groups.
The first class are solid solutions of lithium orthosilicate,
Li_4SiO_4, and lithium phosphate, Li_3PO_4. The unit cell of
Li_4SiO_4 contains two SiO_4^{-4} tetrahedra linked by 8 lithium
ions, which are distributed over 18 possible sites.
Although the conductivity of pure Li_4SiO_4 is rather low,
an improvement of 3 to 4 orders of magnitude has been obtained at the maximum dopant level studied, which corresponds to 30 mole percent Li_3PO_4.

The second materials group are based upon the anti-
fluorite structure. One family of compounds studied
includes Li_5AlO_4, Li_5GaO_4, and Li_6ZnO_4, which have a partially
occupied cation sublattice. Like many known anionic conductors having the fluorite structure, these materials
exhibit a transition to a more highly conducting state--at
about 380°C. D.T.A. experiments indicate a large endothermic
effect associated with the onset of this transition.

The third, Li_3N, has a unique structure with large
intersecting tunnels in 2 dimensions. High ionic conductivity
and negligible electronic conductivity are found at relatively low temperatures. Data on both pure and impure polycrystalline samples indicate that the conductivity parameters
can be improved by suitable doping.

Ionic conductivities of some of these materials at
several temperatures are tabulated below.

Conductivity Values at Various Temperatures; σ in $ohm^{-1} cm^{-1}$.

	25°C	200°C	400°C	450°C
Li_5AlO_4	--	2.3×10^{-7}	3.2×10^{-3}	0.17
$Li_4SiO_4 + 0.43\ Li_3PO_4$	4.5×10^{-7}	1.6×10^{-3}	0.92	0.18
Li_3N (96%)	2.3×10^{-7}	1.3×10^{-3}	0.086	0.17

*This work was supported by ARPA through ONR.

Ionic Conductivity in Lithium Aluminosilicate Solid
Electrolyte Materials.* R. M. BIEFELD AND R. T. JOHNSON, JR.,
Sandia Labs.--Studies of ionic conductivity in lithium
aluminosilicate, $Li_2O \cdot Al_2O_3 \cdot nSiO_2$, glass-ceramic materials
based on β-eucryptite (n=2) a one-dimensional tunnel
structure, are being pursued. These materials are of
interest as Li^+-ion conducting solid electrolytes for
thermal batteries because of their high ionic conductivity,
σ, at elevated temperatures (~600°C), low thermal expansion,
ease of fabrication, and variable composition. The effects
of composition and structure on σ in glass and glass-ceramic
materials are being examined. The highest conductivities
[$4 \times 10^{-2}(\Omega\text{-cm})^{-1}$ at 600°C] have been obtained with enhanced
Li_2O concentrations and with B_2O_3 substitutions for Al_2O_3.
Substitutions of Ga_2O_3 and GeO_2 for Al_2O_3 and SiO_2, respectively, have no significant effect on σ. Ionic conductivities
have been determined for five glass-ceramics with various
degrees of crystallinity. Results show no significant change
in σ below 60% crystallinity. However, as the percentage
crystallinity increases from 60 to 88%, the activation
energy, E, and the pre-exponential factor, σ_0, increase from
~0.7 to ~1.2 eV and from ~10^3 to ~10^5 $(\Omega\text{-cm})^{-1}$, respectively
[determined from $\sigma = \sigma_0 \exp(-E/kT)$]. These results suggest
that σ in β-eucryptite glass-ceramics may be modeled by
percolation theory. Unlike the other materials examined,
σ for certain B_2O_3 substituted glass-ceramics is greater
than σ of the corresponding glass. Present studies are
oriented toward further enhancing the conductivity as well
as understanding the mechanisms and relative importance of
lithium ion transport in the glass and crystalline phases
in these glass-ceramic materials.
*Prepared for the U.S. Energy Research and Development
 Administration under Contract AT(29-1)-789.

Li^+ Conduction in the Beta Alumina Structure.
G. C. FARRINGTON AND W. L. ROTH, General Electric Corporate
Research and Development.--Li^+ is one of several monovalent
cations known to completely replace Na^+ in Na^+ beta alumina.
The conditions under which ion exchange is brought about
determine whether the product is pure Li^+ beta alumina or
beta alumina containing both Na^+ and Li^+. Many investigators
have believed that this process of ion exchange leads to
cracking of polycrystalline Na^+ beta alumina and, therefore,

is not a satisfactory route for Li^+ beta alumina preparation. It also has been reported that Li^+ beta alumina has a single crystal ionic conductivity of only 1.6×10^{-4} (ohm cm)$^{-1}$ at 25°C, 90 times less than that of Na^+ beta alumina.[1,2] The work discussed in this paper indicates that neither of these points is true.

Polycrystalline Na^+ beta alumina can be successfully converted to its Li^+ counterpart by ion exchange in $LiNO_3$ at 400°C-600°C. The product is a physically stable Li^+ solid electrolyte having a single crystal conductivity of 2.8×10^{-3} (ohm cm)$^{-1}$ at 25°C, only five times less than Na^+ beta alumina at the same temperature. Li^+ beta alumina, therefore, is a highly conductive solid ionic conductor for Li^+. It is however unstable in contact with molten Li at 450°C.

Our work has also described a conductivity phenomenon in beta alumina which we have called "co-ionic conductivity." It is possible to prepare beta alumina containing both Na^+ and Li^+. Ionic conductivity in beta alumina having a Li^+/Na^+ ratio greater than one appears to involve primarily Li^+ motion, that is, Li^+ ions migrate through the lattice without altering its Na^+ content. This apparent preferential order- ing and stability of co-ionic Li^+-Na^+ beta alumina is in- fluenced by the ionic environment in contact with the solid electrolyte.

[1] R. H. Radzilowski, Y. F. Yao, and J. T. Kummer, J. Appl. Phys., 40, 4716 (1969).
[2] M. S. Whittingham and R. A. Huggins, N.B.S. Spec. Publ. 364, Solid State Chemistry (1972), 139.

Mechanism of Formation of the Sodium Gallate Super- ionic Conductors.   B. M. FOXMAN*, S. J. LaPLACA AND L. M. FOSTER, Thomas J. Watson Research Center, IBM Corporation. --Superionic conductors isomorphous with the β-aluminas result from the reaction of monoclinic gallium oxide with sodium carbonate at elevated temperatures (β" $\simeq$ 1100°C; β $\simeq$ 1390°C). Two series of experiments were conducted in order to explore the mechanism of the reaction. In the first, X-ray powder diffraction techniques were used to follow the development of discrete intermediate phases that appear as a function of temperature when Na_2Co_3 is decomposed in situ and reacted with intimately mixed Ga_2O_3 powders. In the second, mono-crystals of Ga_2O_3 were

exposed to sodium oxide vapors at 1100°C and the phases
present in the reaction zones were identified by X-ray.
The results of the two types of experiments are in agreement and show that two sodium gallate phases of nominal
composition $3Na_2O \cdot 5Ga_2O_3$ and $Na_2O \cdot 3Ga_2O_3$ appear as pre-
cursors in the formation of the rhombohedral $\beta"-Na_2O \cdot 6Ga_2O_3$
phase. $Na_2O \cdot 3Ga_2O_3$ is a previously unreported phase in the
$Na_2O-Ga_2O_3$ system. β and $\gamma Na_2OGa_2O_3$, which can be prepared
in separate experiments by reaction of 1:1 mole ratios of
the separate components, do not appear in the series leading
to the $\beta"$ phase. $3Na_3O \cdot 5Ga_2O_3$ is the first phase produced,
together with excess unreacted Ga_2O_3. $Na_2O \cdot 3Ga_2O_3$ then
forms and converts completely to $\beta"-Na_2O \cdot 6Ga_2O_3$ at about
1200°. X-ray diffraction experiments and examination of
the volume relationships between the various phases suggest
a topotactic pathway for the precursor stages of the
reaction. The detailed structural relationships between
the various phases will be discussed.
*Brandeis University

Reaction of $\beta"$-Sodium Gallate With Water. L. M.
FOSTER AND G. V. ARBACH, Thomas J. Watson Research Center,
IBM Corporation.--The interaction of moisture with the
rhombohedral $\beta"$ form of sodium gallate ($\beta"-Na_2O \cdot 5.7Ga_2O_3$)
has been studied. This material is completely analogous to
$\beta"-Al_2O_3$. Moisture pickup at room temperature from humid
atmospheres is extremely rapid in fine powders of this
material until it reaches about 1 molecule of water per
sodium atom. Further takeup is very slow.

The sorption is a function of particle size, and the
data are interpreted as an absorption along the c planes
to a depth of about 22 microns, after which it essentially
ceases. In large crystals this makes the crystal very sus-
ceptible to cleavage, even though the percentage weight
change is negligible. An explanation of this behavior is
advanced based on a dislocation structure.

The water takeup results in a 11-12% expansion of
the c axis parameter as a distinct change without intermediate values of the parameter. The water is completely
evolved again and the highly crystalline $\beta"$ structure is
restored by heating at 200-300°C.

It is calculated that there are sufficient open
sites in the $\beta"$-gallate to accommodate up to 2.8 water

molecules per Na atom. The fact that the only prominent
feature in the sorption curve is a marked decrease in water
takeup after a 1:1 ratio to the sodium is reached suggests
that the principal action of water is to form a coordination
complex with each sodium atom.

Electron Paramagnetic Dection of the Liquid-Like
Motion of Divalent Ions in Solid β-Sodium Gallate. R. S.
TITLE AND G. V. CHANDRASHEKHAR, IBM Research Center.--
We have, for the gallium analog of β-alumina, β-sodium
gallate, chosen diffusion conditions that favor the incorporation of paramagnetic impurities into the sodium bearing
mirror planes. The EPR spectra of Cu^{2+}, Mn^{2+} and Eu^{2+} that
have been so incorporated have been studied. For all three
impurities, the EPR spectra at or near room temperature
show pronounced relaxation effects caused by the motion of
the ions. From the relaxation effects the correlation time
of the motion is found to be of the order of 10^{-11} sec or
less. Correlation times of this magnitude are characteristic
of ionic motion in liquids. In fact, the EPR spectra of
these ions in solid β-sodium gallate are identical to those
seen in liquid solutions containing these ions. In the case
of Cu^{2+} we see an increase in the correlation time if motion
impeding impurities are added to the material. The motional
correlation times we find for the unimpeded motion agree
with the theoretical prediction of Rice and Roth.

New Solid Electrolytes.* H. Y-P HONG, J. A. KAFALAS,
K. DWIGHT AND J. B. GOODENOUGH, Massachusetts Institute of
Technology.--A solid electrolyte should consist of a rigid,
three-dimensional network, stabilized by electrons donated
by mobile ions which partially occupy a two or three-dimen-
sionally linked interstitial space. Such networks may be
classified into three groups, depending on whether the
three-dimensional, rigid skeleton is built from tetrahedra,
octahedra, or a combination of the two. Each group is then
divided into subgroups, depending on whether the anions are
bonded to two, three, or four network cations. The system
$K_{1+2x}Mg_{1-x}Si_{1+x}O_4$ represents the first group, where MgO_4
and SiO_4 tetrahedra are shared by corners in such a way that
each oxygen is bonded by two skeleton cations to form a

three-dimensional network. Systems $Na_{1+2x}Ta_2O_5FO_x$ and
$NaSbO_3 \cdot 1/6$ NaF represent the second group, where the three-
dimensional networks are formed by sharing corners and/or
edges of TaO_5F and SbO_6 octahedra. The system $Na_{1+x}Zr_2Si_x$
$P_{3-x}O_{12}$ represents the third group, where the three-
dimensional skeleton is made from corner-shared SiO_4 (or PO_4)
tetrahedra and ZrO_6 octahedra. Relations between activation
energy, bottleneck size and chemical bonding will be discussed. The ionic conductivity of $Na_3Zr_2Si_2PO_{12}$ is 0.3
Ω^{-1} cm^{-1} at 300°C, which is competitive with the best β''-
alumina. The chemical stability, ceramic processing and
physical properties of $Na_3Zr_2Si_2PO_{12}$ will also be reported.
*This work was sponsored by the Defense Advanced Research
Projects Agency and NSF/RANN.

Anomalous Properties of Cu_3VS_4. N. LeNAGARD,
G. COLLIN, O. GOROCHOV, Laboratoire de Chimie Minérale
(Associé au C.N.R.S. N° 200) 4 Avenue de l'Observatoire,
75270 Paris Cédex 06, France AND H. ARRIBART, A. WILLIG,
B. SAPOVAL, Laboratoire de Physique de la Matière Condensée,
Ecole Polytechnique (Equipe de Recherche du C.N.R.S.),
91120 Palaiseau, France.--Sulvanite (Cu_3VS_4, space group
P43m) presents several unusual properties: electrical
transport, X-ray structure and N.M.R., N.Q.R. relaxation.
Single crystals have been grown by vapor transport reaction
using Cl_2. Susceptibility measurements show that Cu_3VS_4 is
diamagnetic. Conductivity at low current density ($j \lesssim 1A.cm^{-2}$)
shows that Cu_3VS_4 is a semiconductor ($\sigma \sim 1$ ohm^{-1} cm^{-1} at
300°K and $\sigma \sim 10^{-3}$ ohm^{-1} cm^{-1} at 4.2°K). At room temperature and for higher current density ($j \gtrsim 1A.cm^{-2}$) anomalous
dc transport occurs: after a given time, the voltage across
the sample increases (at constant current) by one order of
magnitude. After this "polarization time", which depends
on the current density, the system is nonlinear and exhibits
hysteresis. Such "polarization" phenomena do not occur at
low temperature.
        X-ray structure shows that the average positions of
Cu atoms are not the center of the faces but the slightly
shifted (0.14 Å) 12h positions of the P43m space group.
        Anomalous nuclear relaxation of $^{51}V(I=7/2)$,
$^{63}Cu(I=3/2)$ and $^{65}Cu(I=3/2$ is observed in powders as well
as in single crystals. In single crystal $T_1(^{51}V) = 10$ sec
at 1.5°K and $T_1(^{51}V) = 0.25$ sec at 300°K. The low

temperature T_1 is anomalously short. This is also true of
the T_1 of [63]Cu and [65]Cu which are of the order of 10 sec
measured by pulsed pure N.Q.R. at 1.5°K.
    Anomalous dc conductivity could be due to partially
ionic conductivity. Anomalous rates of nuclear relaxation
at low temperature demonstrate that charge motion still
exists at low temperature (possibly tunnelling of Cu^+ ions
between the four nearby 12h positions).

On the Transport Mechanism in Sulfate-Based Solid
Electrolytes. A. LUNDEN, E. BOWLING, B. HEED, B. E. MELANDER,
L. NILSSON, K. SCHROEDER AND C. A. SJOBLOM, Chalmers Univ.
of Technology, Gothenburg, Sweden.--The sulfate-based solid
electrolytes differ in at least one aspect from the groups
of electrolytes that have obtained main attention. While
in other classes the high mobility is limited to only a few
ionic species, all mono- and divalent cations have a high
mobility in the sulfate-based electrolytes. This fact is
of technical interest, since it increases the number of
possible anode materials in electrochemical cells; e.g.,
magnesium and zinc can be used. It should also be considered
when the transport mechanism is discussed. Another interesting feature is that high-temperature X-ray studies of
fcc Li_2SO_4 and the bcc $LiNaSO_4$ and $LiAgSO_4$ show very few
lines, which, however, are sharp.
    This discussion of transport mechanism will be
restricted to the high-temperature phase which is formed at
573°C for pure Li_2SO_4. It is remarkable that both the transformation energy and the volume increase are much larger
than the changes occurring when the salt melts at 860°C.
    Forland and Krogh-Moe[1] reported in 1957 that the
sulfate ions form a face centered cubic, or pseudo-cubic,
lattice, in which the size of the possible cation positions
depends on the spatial orientation of the sulfate ions.
In all cases the two tetrahedral positions (1/4, 1/4, 1/4;
3/4, 3/4, 3/4) are of course small compared with the "octahedral" (1/2, 1/2, 1/2) position. Oye has suggested that
every second sulfate ion layer should be rotated 90°,
which would make the tetrahedral positions equally sized.
If the transport mechanism is assumed to be jumps between
adjacent positions: tetrahedral-octahedral-tetrahedral,
it is difficult to understand why large ions such as
potassium are very mobile, since the K^+ ion is much too

large to fit in a tetrahedral position.

Andersson and Bovin[2] have recently suggested a "crystobalite" structure for fcc Li_2SO_4 (i.e., the Si ions of SiO_2 are replaced alternatively by S and Li atoms, which thus are in tetrahedral positions). Large three-dimensional tunnels are then obtained, and it is assumed that half of the cations are in the tunnels. These ions are thus very mobile, while the Li ions in tetrahedral positions are less mobile.

Work is in progress to compare the different models with the physical properties of the fcc phase, such as the large solubility of Na^+, Ag^+ and Zn^{2+} ions, low solubility of K^+ and larger ions, influence of the composition on the mechanical properties, etc.

We have reported recently on tests of electrochemical cells with sulfate-based electrolytes.[3,4]

[1]T. Forland and J. Krogh-Moe, Acta Chem. Scand. 11, 565 (1957).
[2]S. Andersson and J. O. Bovin, private communication.
[3]B. Heed and A. Lundén, Power Sources 5, 573. Ed. D. H. Collins, Academic Press, London.
[4]B. Heed, A. Lundén and K. Schroeder, 10th Indersociety Energ Conversion Engineering Conference, Aug. 1975, p.613.

Structural Aspects and High Partial Cu^+-ionic Conductivity in Compounds of CuTeX (X=Cl,Br,I). U. v.ALPEN, J. FENNER, J. MARCOLL* AND A. RABENAU, Max Planck-Institut, W. Germany.--Preparation, phase relationships, and crystallographic data for the compounds CuTeX (X=Cl,Br,I) have been reported previously.[1] These compounds in the isotypic series CuTeX (X=Cl,Br,I) crystallize with tetragonal symmetry in space-group $I4_1/amd$. The recent results of the complete X-ray structure analysis for the CuTeX compounds show helicoidal arrangements of the copper and tellurium atoms, symmetry related by a fourfold screw axis (4_1) and a statistical population of Cu-atomic sites. The latter is a hint for a copper ionic-conductivity. Therefore transference experiments, a-c conductivity measurements, and d-c polarization measurements have been performed for the CuTeX compounds. The investigations showed CuTeI to be a mixed conductor with a high partial ionic Cu^+-conductivity while CuTeCl and CuTeBr are pure Cu^+-ionic conductors. Between room temperature and 250°C the total (a-c) conductivity was

determined to be 1-2 orders of magnitude above the values
reported for the binary copper-halides. The electronic
contribution to the conductivity as derived from d-c polarization experiments was found to be at least more than one
order of magnitude below the value of the total conductivity
in the case of CuTeI while the partial electronic conductivity of CuTeBr and CuTeCl are shown to be negligible compared to their ionic conductivities.
*Now at Universität Dortmund.
[1]A. Rabenau, H. Rau and G. Rosenstein, Z. anorg. allg.
Chem. $\underline{374}$, 43 (1970).

The State of Order of the Conducting Ions in Hollandites.

H. U. BEYELER, T. HIBMA AND C. SCHULER,
Brown Boveri Research Center, Baden, Switzerland.--
In strictly one-dimensional channel structures ionic motion
over any macroscopic length is very likely to be impeded by
crystal faults and impurities. We have found that even in
optically perfect single crystals of hollandite ($K_{2x}Mg_xTi_{8-x}$
O_{16}) the true DC conductivity is unmeasurably low.
       On the other hand the low dimensionality of the con-
ducting path makes hollandites the ideal model system for a
detailed study of local cationic ordering by diffuse X-ray
studies: in 1 dimension we don't have to introduce ad hoc
model assumptions as is necessary for the interpretation of
highly complex 2 and 3 dimensional diffuse X-ray data.
       In hollandite $K_{1.6}Mg_{0.8}Ti_{7.2}O_{16}$ prominent diffuse
scattering was found which could be shown to originate from
a partial order of the cations in the substoechiometrically
filled channels. The interpretation of the diffuse intensity
pattern leads unambiguously to the following conclusions:
a) the order is primarily one-dimensional, i.e., there is
only a weak correlation between the order in different
channels, b) there is a strong tendency towards formation
of a supercell containing five regular channel sites,
c) the order in the supercell is both occupational and displacive: the supercell consists of a vacancy and four
occupied sites (average occupancy of 80% in accordance with
the stoechiometry), the ions adjacent to the vacancy are
displaced towards the latter by about 20% of the distance
between sites and the next nearest neighbors of vacancies
are displaced accordingly; d) the correlation length of
this supercell ordering is about 30 Å, i.e., about 2

supercells. The outstanding feature of these findings is the extremely large "relaxation" of the ions towards the vacancy. The 20% corresponds in fact to more than halfway towards a completely equidistant distribution of the ions in the channels. It may safely be assumed that ion-ion electrostatic repulsion is the prime cause for this moving apart of the ions as the radius of the K^+ ions (1.33 Å) is appreciably smaller than half the distance between two crystallographic channel sites (1.45 Å).

In the hollandites $Rb_{1.5}Mg_{.75}Ti_{7.25}O_{16}$ and $Cs_{1.33}Mg_{.67}Ti_{7.33}O_{16}$ the situation is very similar except for the length of the supercell which contains 3 resp. 2 occupied plus one empty site in accordance with the lower average occupancy. The displacements of the ions adjacent to the vacancy decreases with increasing cationic radius: the larger the ions the more they "feel" the periodic structure of the channel framework.

We have thus found evidence that in hollandites and possibly in other superionic conductors as well, ion-ion interactions lead to a considerable shift of the equilibrium positions of the mobile ions and to the establishment of a very distinct short range order. In view of these findings models not including both occupational and displacement short range order appear inadequate for a microscopic under-standing of superionic conductors.

Diffusion in the Intercalation Compounds of the Layered Disulfides. R. R. CHIANELLI, B. G. SILBERNAGEL, AND M. S. WHITTINGHAM, Exxon Research and Engineering Co. --In intercalation compounds of the layered transition metal disulfides, atomic and molecular guest species are incorporated between the layers of the host sulfide lattice by a topochemical reaction with only minor structural changes. Unlike the beta aluminas, there are only minor restraints to separation of the layers to incorporate ions of differing sizes and, in addition, the composition of the mobile specie can be readily varied. Using X-ray, optical microscopy, electrochemical and nmr techniques, we have examined the effects of structural changes and physical properties on the diffusion of both ionic and molecular guests. These will be illustrated by discussion of the compounds Li_xTiS_2 for $0 \leq x < 1$ and $NH_3 \cdot TiS_2$. Both possess well-defined structures with the guest atoms in Li_xTiS_2 occupying octahedral sites between the layers, whereas the ammonia molecules reside in trigonal prismatic sites. For Na_xTiS_2, the sites filled depend on composition markedly effecting the diffusion of Na^+.

Intercalation of alkali atoms involves essentially complete electron donation to the layers, as indicated by a very small Knight shift and a large free energy of reaction, ~50 kcal/mole. On the other hand, ΔH_f for $NH_3 \cdot TaS_2$ is not more than 10 kcal/mole, and both guest-host and guest-guest interactions make significant contributions to the bonding process. Optical and electrochemical studies give high values for the chemical diffusion coefficient, e.g., $\tilde{D} \sim 10^{-7}$ cm^2/sec at 25°C for Li_xTiS_2. Self-diffusion studies using pulsed nmr techniques indicate hopping times of $\sim 10^{-10}$ sec for $NH_3 \cdot TiS_2$ and $\sim 10^{-7}$ for $LiTiS_2$, significantly larger than those observed in the presence of a chemical concentration gradient. These differences between $\tilde{D}$ and D^* are consistent with the predictions of the Darken Equation. The mechanisms of fast diffusion in these materials will be discussed and compared with other superionic conductors.

The high values of diffusion coefficient for lithium ions permits the use of TiS_2 as the cathode of high energy density reversible batteries. Current densities in excess of 10 ma/cm^2 can be maintained at 25°C with high utilization.[1,2]
[1]M.S. Whettingham, J. Electrochem. Soc. 126, 315 (1976).
[2]M. S. Whettingham, Science (June 1976).

Single Crystal Studies of Cubic Phases of VD_x by Neutron Diffraction.* F. REIDINGER, J. J. REILLY, R. W. STOENNER, Brookhaven National Laboratory.--The high diffusion rates of hydrogen in certain metals justify the classification of these metal hydrides as class I superionic conductors.[1] The favorable relation between the scattering amplitudes of vanadium and deuterium ($-b_D = 13b_V$) and the existence of cubic phases with low and high deuterium content at ambient temperatures make vanadium deuteride a promising compound for a neutron diffraction study of ions in the conducting state.

Refinements of data from single crystals of $VD_{.038}$, $VD_{.047}$ and $VD_{.79}$ confirm the tetrahedral site (0, 0.5, 0.25) as the preferred site of deuterium with moderately aniso-tropic thermal parameters: $U_{11} = 2U_{22}$; $U_{33} = U_{22}$.

Difference Fourier maps show additional density at the octahedral site(0,0.5, 0.5) and unexpectedly, at a site (0,y,y) with orthorhombic symmetry. The latter site can be viewed as the saddle point of the diffusion process dividing the shortest path between two tetrahedral sites.

The additional density can be described either by partial atoms or by an expanded probability density function including third and fourth order Hermite polynomials[2,3], which are indicative of anharmonic thermal vibrations.

We plan further studies, systematically varying composition and temperature, to investigate a possible relation between the occupancy of the octahedral site and the transition from the α (or α') to the β phase.

*Research performed under the auspices of the U.S. Energy Research and Development Administration.

[1] M. J. Rice and W. L. Roth, J. Solid State Chem. 4, 294 (1972).
[2] Least-Squares Program ORJFLS by C. K. Johnson, Oak Ridge National Laboratory, P.O. Box X, Oak Ridge, Tennessee 37830.
[3] C. K. Johnson and H. A. Levy, International Tables for X-ray Crystallography, vol. 4, Birmingham: Kynoch Press (1974).

AUTHOR INDEX

FORMULA INDEX

SUBJECT INDEX